Cereals and Legumes in the Food Supply

Cereals and Legumes in the Food Supply

EDITED BY
JACQUELINE DUPONT and ELIZABETH M. OSMAN

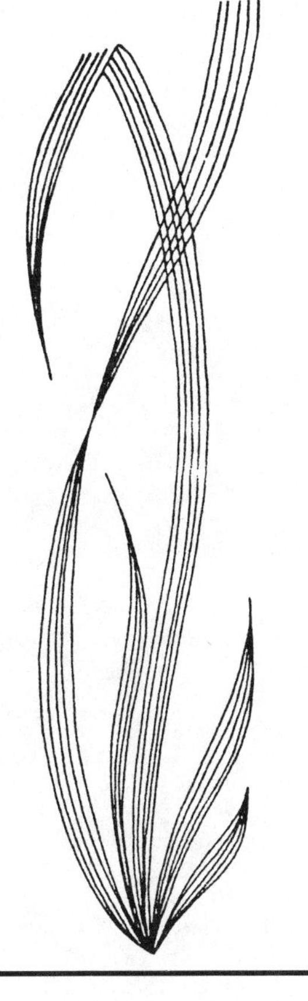

Iowa State University Press / Ames

Jacqueline Dupont is chair, Food and Nutrition Department, Iowa State University.
Elizabeth M. Osman is emeritus professor, Food and Nutrition Department, Iowa State University.

Printed in the United States of America from camera-ready copy provided by the editors

First edition, 1987

Library of Congress Cataloging-in-Publication Data

Cereals and legumes in the food supply.

Includes index.
1. Cereals as food. 2. Legumes as food. I. Dupont, Jacqueline, 1934– . II. Osman, Elizabeth M.
TX557.C39 1987 641.3′31 87–16796
ISBN 0–8138–1578–9

Contents

Preface

The production and use of cereals and legumes for direct human consumption has burgeoned in the last 25 years. Most of the world's countries can produce enough of these foods for a healthy diet for their people. The breadth of the technology--from genetics through marketing--involved in this endeavor was discussed at a symposium held at Iowa State University in June, 1983. In the time since, interest in the biotechnology of these food products has continued to grow. The presenters at the symposium reviewed their presentations in 1986 and prepared them for this book.

Authors with expertise in all areas of this broad field have considered their specialty in light of the whole. The book can orient the newcomer or the specialist to the broad scope of knowledge needed to provide food for nourishment and culture.

Acknowledgments

The 1983 symposium and the proceedings publication were supported by the following contributors:

- Nutritional Sciences Council of Iowa State University

The Helen and Wise Burroughs Fund

Dr. Wise Burroughs provided an endowment in support of the Nutritional Sciences Council which contributed generously to both the symposium and the publication. Dr. Burroughs died December 16, 1986 and this book is a tribute to his memory. His research in animal science resulted in significant developments and application of technology in modern production of beef cattle. Dr. Burroughs' scientific work provided the basis for utilization of by-products as feed. His most widely acclaimed research concerned the anabolic effects of estrogens in beef cattle.

- The Department of Food and Nutrition of Iowa State University

The Bernice Kunerth Watt Fund

Dr. Bernice Watt gave a gift for the support of food science to the department. Her gift made final work on the publication possible. Dr. Watt died March 8, 1984 and this book is a tribute to her memory. She joined the Consumer and Food Economics Research Division of the U.S. Department of Agriculture in 1941 as a research nutrition analyst. She became the leader of the Nutrient Data Research Center where she helped with the development of food composition tables (Handbook 8) that have been used worldwide. The tables list the quantities of 50 different nutrients in 3,000 different types of food.

- The Home Economics Development Fund of Iowa State University

- The Helen LeBaron Hilton Fund of Iowa State University

- The Campbell Soup Company

- Pioneer Hi-Bred International, Inc.

·Best Foods Division of CPC International

·The Quaker Oats Company

Sponsors of speakers were:

Molecular Genetics, Inc.
United States Department of Agriculture
U.S. Food and Drug Administration
General Foods
A.E. Staley Manufacturing Co.

The preparation of the manuscript was completed through the devoted efforts of Bettye Danofsky and Karen Hethcote.

Contributors

Dr. Isaac Akinyele
Department of Human Nutrition
University of Ibadan
Ibadan, Nigeria

Dr. Thomas J. Aurand
A. E. Staley Manufacturing Company
220 Eldorado Street
Decatur, IL 62525

Dr. C. E. Bodwell
Protein Nutrition Lab
Beltsville Human Nutrition Center
Building 308, Room 214 BARC-East
USDA
Beltsville, MD 20705

Dr. Adolph S. Clausi
General Foods Corporation
250 North Street
White Plains, NY 10625

Dr. Bert L. D'Appolonia
Department of Cereal Chemistry and Technology
North Dakota State University
Fargo, ND 58105

Dr. Joel W. Dick
Department of Cereal Chemistry and Technology
North Dakota State University
Fargo, ND 58105

Dr. Donald N. Duvick
Pioneer Hi-Bred International, Inc.
(730 NW 62nd Ave.)
Box 85
Johnston, IA 50131

Dr. David R. Erickson
American Soybean Association
(777 Craig Road)
Box 27300
St. Louis, MO 63141

Dr. David Fellers
USDA Western Regional Research Center
800 Buchanan Street
Berkeley, CA 94710

Dr. Kenneth Gilles
Marketing and Inspection Services
U.S. Department of Agriculture
228-W Administration Bldg.
Washington, D.C. 20250

Dr. Louis Grivetti
Department of Geography
University of California-Davis
Davis, CA 95616

Dr. Ellis L. Gunderson
Division of Chemical Technology
Bureau of Foods
Food and Drug Administration
Washington, D.C. 20204

Dr. Theodore Hymowitz
Department of Agronomy
University of Illinois
Turner Hall
1102 S. Goodwin Street
Urbana, IL 61801

Dr. Jean Kinsey
Department of Agricultural and Applied Economics
University of Minnesota
St. Paul, MN 55455

Dr. Roger Kleese
Plant Products
Molecular Genetics, Inc.
10320 Bren Road East
Minnetonka, MN 55343

Dr. Charles Kolar
Ralston Purina Company
Checkerboard Square
St. Louis, MO 63164

Mr. Richard Leviton
8 Skinner Lane
S. Hadley, MA 01075

Dr. Edmund W. Lusas
Food Protein R & D Center
Box 183
Texas A & M University
College Station, TX 77844

Dr. Joseph Rakosky
J. Rakosky Services, Inc.
5836 Crain Street
Morton Grove, IL 60053

Dr. Khee Choon Rhee
Food Protein R & D Center
Box 183
Texas A & M University
College Station, TX 77844

Dr. Stanley H. Richert
Ralston Purina Company
Checkerboard Square
St. Louis, MO 63164

Dr. F. Edward Scarbrough
Bureau of Foods
Food and Drug Administration
Washington, D.C. 20204

Dr. Clyde E. Stauffer
631 Christophal Drive
Cincinnati, OH 45231

Mr. Aubrey J. Strickler
Hubinger Company
1211 West 22nd Street, Suite 527
Oakbrook, IL 60521

Dr. Hwa L. Wang
Fermentation Laboratory
USDA--Northern Regional Research Center
1815 N. University Street
Peoria, IL 61604

Dr. Bill D. Webb
USDA-SEA FR SR-Rice Quality Lab
Rt. 7 Box 999
Beaumont, TX 77066

Dr. William T. Yamazaki
U.S. Grain Marketing Research Laboratory
1515 College Street
Manhattan, KS 66502

Dr. Vernon L. Youngs
Spring & Durum Research Service
USDA, Agriculture Research Service
Cereal Chemistry & Technology
North Dakota State University
Fargo, ND 58105

The Gift of Osiris

LOUIS GRIVETTI

"--Man Doth Not Live By Bread Alone--"(1).
"--Give Us This Day Our Daily Bread--"(2).
"--A Jug of Wine, A Loaf of Bread, and Thou--"(3).
"--Here Is Bread Which Strengthens Man's Heart, And Therefore Called The Staff Of Life--"(4).
"-- A Crust of Bread and Liberty--"(5).
"--Hey Man, You Got Any Bread--"(6)?

Throughout history, cereals have played political, religious, symbolic roles far beyond their basic importance as food for animals and humans. Sharp political and geographical divisions exist between the wheat north and rice south of China (7). The shape, form, and composition of pastas reflect regional, geographical differences within Italy (8).

In Judaism, the leavened *challah* is symbolic of life; the unleavened *matzoh* symbolic of speed in leaving ancient Egypt during the Exodus (9). In Christianity, bread is the symbolic body of Christ, whether as the Roman Catholic mass wafer, or the Greek Orthodox, *prosferon* (10). Greek Easter breads, called *tsourekia* or *lambropsomo*, are baked with hard boiled eggs: red is symbolic of the blood of Christ, while the egg represents life contained within the tomb. On Easter morning the eggs are cracked to release the spirit of Christ (11). Iron cookware of basques has an equilateral chi cross in the lid; when bread is baked in such containers, this cross of Christ is embossed on top of the finished product. Sometimes, bread, symbolism, and religion are mixed as in instances where dough is used to prepare breads in the outline form of a mosque, the Islamic house of worship.

It is with birth, death, and resurrection, however, that cereals reach their most interesting, complicated symbolism. Wheat and barley have been associated with magic and ritual since pre-history. As a young college student at Berkeley, I was attracted to the exciting writings of James George Frazer and his *Golden Bough*, especially his essays on ancient religious harvest rites (12); behavior and activities still practiced in modern Europe, the Mediterranean, and the Americas through manufacture and display of protective charms made of wheat

Cereals and Legumes in the Food Supply, edited by Jacqueline Dupont and Elizabeth M. Osman Iowa State University Press, Ames, Iowa 50010 straw, the so-called celtic corn dollies, or festive creatures such as the Jul Bock Christmas animal of Sweden.

In ancient Eqypt barley and wheat were used to predict the gender of unborn human infants. Two clay bowls were filled with Nile mud, one planted with barley, the other wheat. The woman wishing to know the sex of her fetus urinated over both bowls. If barley grew first, she would bear a son; if wheat first, a daughter; if neither grew, she was not pregnant (13).

This ancient obstetrical technique was tested scientifically some years ago by my friend and colleague, Paul Ghalioungui. He concluded: urine from non-pregnant women always prevented growth of barley or wheat; urine from pregnant women caused growth in 40% of test cases; but no relationship was found between infant gender and whether barley or wheat grew first (14).

While cereals have been used to predict life, they also have been associated with death. Lifeless in appearance, dry seeds are symbolic of the dead. When buried in earth, however, new life reemerges, perpetuating belief in resurrection and the continuity of life after death. In ancient Egypt trays were crafted from wooden strips, bent to form the outline shape of the agricultural God, Osiris. Completed trays were filled with Nile mud and planted with quick-growing wheat or barley. Such Osiris beds were placed atop the sarcophagus and the tomb sealed. The seeds germinated in the moist mud, struggled upward searching for light, attained a height of a few centimeters, then withered (15).

These Osiris beds are the basis for the recent 19th-20th century myth that ancient Egyptian cereals are capable of germination after being in tombs for thousands of years. I could think of nothing more delightful or exciting to report than to provide details on successful germination experiments with ancient Egyptian grain. All such scientific trials, however, have failed despite occasional reports, to the contrary, in the popular press (16).

Given the Osiris bed custom in ancient Egypt, is it not curious that 20th century Egyptian Greek Orthodox Christians also plant barley or wheat in small pots filled with Nile mud to commemorate a dead God? In this instance, however, it is Christ rather than Osiris. Planting occurs on December 12, Saint Spyridon's Day; the sprouts emerge on or about December 25 (Nativity), as a symbolic reminder of the birth of Christ who rose from the dead (17).

The title of this paper, The Gift of Osiris, while taken from our book (13) on ancient Egyptian diet and nutrition, also reflects scientific attempts to understand the processes or mechanisms for plant domestication, especially the domestication of cereals. The rippling grain fields of the American heartland are familiar to all. But there is a second world region that should also attract our interest, where winds, also, produce amber waves of grain. This region is the flat to rolling slopes of southwestern Turkey and northwestern Syria in the Eastern Mediterranean. There, extensive stands of grain are wild, undomesticated. Some researchers suggest that the earliest management of barley and wheat production occurred there, a process that ultimately contributed to the stability of the human food supply.

These stands of wild grain, plentiful even today, present an enigma, a paradox that baffles archaeologist, geographer, historian, and scientist alike. Most would agree that domestication could not have taken place during periods of environmental or social stress, such as drought or human conflict, because there would not have been time to domesticate. Most also would agree that domestication could only occur during periods of leisure and social ease since domestication is a long, biological process. But why would domestication occur during periods of social and environmental ease in southwestern Turkey where wild grain was so plentiful? Why would there be a need to domesticate in the midst of food plenty?

To evaluate this enigma we start with a given and work backward. Barley and wheat were domesticated. Who was responsible? Probably inhabitants of the eastern Mediterranean, although recent archaeological data from southern Egypt suggest a possible second site for consideration (18,19). When did domestication occur? Certainly by 7-5,000 B.C., perhaps even 17,000 B.C., but the answers for how and why remain elusive, controversial.

How curious this gift of Osiris, the domestication of grain. Why does it continue to tantalize and puzzle researchers? The myths of two societies, ancient Greece and ancient Egypt, provide interesting insights for consideration. Both myths suggest that domestication was given to humans as a gift by super-human, cosmic dieties, because humans were basically kind and good. No one now--especially myself--suggests super-human, cosmic dieties taught humans to cultivate grain. Nevertheless, this theme remains popular in America today in the "fringe" press and television science fiction.

These myths, however, deserve retelling because the process is associated with kindness, and ultimately the development of civilization. Whatever our cultural-ethnic-religious origins, whether we are scientist or interested layman, we are part of a great chain of knowledge, linking myth and oral tradition with writing, and ultimately 20th century logic and precision. These myths, therefore, are part of our heritage; they are part of us, even today.

West of Athens, along the northern shore of the Gulf of Salamis, lies the archaeological site, Eleusis. At an undefined misty time before the dawn of history, three brothers playing there were interrupted by an old woman seeking information of her lost daughter. Unknown to the brothers the old woman was the great earth diety, Demeter, searching for Core (sometimes called Persephone). One of the brothers, Triptolemus (his name means three-times daring), mustered a polite response and informed Demeter he had seen a beautiful girl being abducted, carried into a nearby cave.

Armed with this information, Demeter ultimately retrieved Core and struck a bargain with the kidnapper, Hades: Core/Persephone would remain underground for a portion of the year (a period thereafter to be called winter), but she would be permitted to emerge from the underworld to bless the earth again with her beauty (a period thereafter to be called Spring). For his role in these events--essentially for being polite to a distraught old woman--Triptolemus received from Demeter the gift of grain, whereupon, he civilized his contemporaries and provided them with a solid, reliable food base of barley and wheat (20,21). This, and other novel Greek myths build upon ancient themes linking ancient Greece with ancient Egypt. Central to the Egyptian myth is Osiris, the gentile, respected, earth-dwelling God of Agriculture represented by the lush, fertile, black soil of the Nile valley. Osiris, married to Isis, his sister, was slain by his evil brother, Seth.

Seth, the personification of evil, was represented by the arid, sterile, sandy desert wastes along each side of the Nile valley. Osiris, slain by Seth, was resurrected to mate one last time with Isis; he sired Horus the avenging falcon. Horus defeated his evil uncle in a terrible cosmological battle (the eclipses of sun and moon are reminders of how Horus and Seth maimed each other in their last great conflict). Osiris, before his final earthly departure, took pity on humans for their frailty and provided them with the gift of grain, the gift of a sustained, stable food supply (22).

Between 1964 and 1967, and for a portion of 1969, I worked as a field anthropologist with the Vanderbilt University Nutrition Program in Egypt. One aspect of our work was to document the cultural and dietary roles of food, comparing and contrasting practices of both ancient and contemporary Egypt. Our objectives were ambitious; our data sources extraordinarily abundant. Egypt, unique among all ancient societies, has 5200 years of sustained, recorded history that reflects changing patterns of food procurement and diet. Numerous successive civilizations along the Nile valley have contributed to this vast food history, among them, ancient Egyptian, Libyan, Ethiopian, Persian, Greek, Roman, Byzantine, Arab, Turkish, even French and British armies of occupation.

Egypt is unique, too, because of its location at the junction between Africa, Asia, and Europe; unique because of the aridity that permits preservation of fragile food-related artifacts usually destroyed through decomposition at most other world localities. In our study of diet and change we utilized data of eight different types and validity: 1) foods identified from intestinal contents of preserved mummies, 2) remains of actual foods preserved by desiccation and found in either sealed or pilfered tombs, 3) models of food or food-related activities, 4) depictions of food procurement, production, or preparation, evidenced by tomb art, 5) papyrus documents with data on rations, menus, and food-habits of either nobility or poor, 6) identification of banquet foods served to the deceased, 7) religious texts concerning human food use, and 8) ancient, medieval, and contemporary travel accounts that mentioned Egyptian food and dietary patterns (23).

Specific examples of the wealth of data available for analysis include:

Cereals: harvesting wheat; measurement of grain fields and production for taxation; transportation of grain to threshing floors; threshing grain using oxen; winnowing; stamping grain to make bread flour; stonegrinding grain to make bread flour; actual bread loaves dating to nearly 3500 B.C.; silo storage of grain; 20th century hand threshing, winnowing, preparation of wheat pastries, and manufacture of wheatbased snack foods sold in modern Cairo markets.

Fowling: capture of birds for food using throw-sticks or nets; use of trained cats to retrieve downed water fowl.

Wine production: Picking grapes; treading grapes to extract juice for fermentation; actual grapes--now raisins; even the result of over-indulging.

Having examined myth and archaeological data on ancient food production, let us turn to a brief exploration of the engima of domestication, then conclude with evaluation of why domestication occurred.

Some have argued the case for "sudden insight" as the impetus for domestication, as if one ancient human told another: "--What a fine day Og, let's go out and cultivate grain so our people can have enough to eat next year--". Others have written that desiccation drove humans and animals into river valleys, where, during the pleistocene ice retreat, mutual proximity led humans to domesticate plants and animals (24). Still others have reported that humans responsible for domestication already were settled near rivers or lakes, where they fished for food. Given such a setting, and probable prior knowledge and use of local plants for dye, fiber, narcotics, or fish poisons, such "settled" fisher-folk would have taken easily to cultivation, and thus, would have first domesticated plants (25). Another view is that domestication resulted from necessity, based on the overriding need to feed an expanding human population (26).

In my view these considerations do not address the central issue: why domesticate during periods of food affluence? Harlan first raised this issue in two classic papers (27,28). His conclusions strike at the core of the domestication engima: starving people die, they do not cultivate or domesticate. Further, well-fed people, living in zones of vast wild food wealth, have no reason to domesticate for food. Harlan was curious: he left his desk and laboratory, and went to southwestern Turkey. There, he strode in stands of wild wheat armed with a simple flint sickle. Using his flint sickle and collecting basket, he began to harvest wild grain. In two hours work he collected enough to feed 20 to 30 people for one month! With such a minor effort in a zone abundant in wild food wealth, Harlan concluded that the domestication of wheat and barley must have occurred elsewhere, in a zone where grain and food supply was limited.

As a graduate student in 1970, I was introduced to the writings of the botanist, Edgar Anderson. In his delightful book, Plants, Life, and Man, he challenged readers to be keen observers, to carefully describe how and why traditional, non-industrial societies use wild plants (29). Whereas Harlan believed food was the key to domestication, and was forced to turn away from southwestern Turkey as the domestication center because of the enigma of wild wheat abundance, Anderson did not believe food was at the core of the domestication process.

Thus, human use of edible wild plants became one of the central themes in my doctoral fieldwork during 1973-1975 in the Kalahari desert of Botswana, southern Africa (30-32). There, I worked among the Batlokwa Ba Moshaweng, an agro-pastoral-hunter-gatherer society living east of Gaborone, the capital of Botswana. On numerous plant gathering expeditions with my hosts, I collected notes on more than 130 species of edible wild plants used regularly by the Tlokwa. But of these, only one was dug up from the surrounding bushlands, removed, and transplanted at Tlokwa homesteads.

While only one food plant was dug up, however, nearly 50 other species regularly were removed from the bush, transplanted at homesteads, and tended. All these had non-food purposes. The prinicipal reason for transplanting and tending these wild plants--the first steps in the domestication process--was for decoration: to decorate household compounds. Other non-food uses for transplanted species included: dye, fiber, magical charms, and medicine. Food was not an important consideration in the Tlokwa transplanting decision: color and beauty were the keys, as predicted by Anderson in 1952 (29).

Initially, I had difficulty accepting these findings; that color and beauty could be the central keys for transplanting, ultimately domestication. In my deterministic logic, the cause had to be food. The Tlokwa knew how to carefully transplant wild species; why not transplant edible varieties? Since the Tlokwa ate and relished more than 130 wild plants, why not bring these tasty, useful dietary items under cultivation in household gardens? If not transplant, why not at least mark localities for favorite edible species? In two years of fieldwork we always wandered about in a seemingly hit-or-miss manner when collecting. The Tlokwa, in my opinion, did not economize their time or maximize their search efforts by minimizing the time spent out in the bushlands. Why, indeed?

I asked my elderly male and female instructors in Tlokwa ethnobotany. Their replies can be condensed and paraphrased something like this:

> "Louis, you have worked among us now for two years. Still you are as ignorant as when you first came to us, slow to learn like one of our youngest children. Louis, really you are probably less smart. At least Tlokwa children eventually learn our ways and you seem to always ask silly questions. The joy, Louis, is in the search. There is much pleasure in walking long distances all day, talking and sharing news of the past week with friends. Surely, you have worked among us long enough to know this; surely you do not suggest that we dig up plants just so we can have food nearby; surely you cannot be serious? Why would we ever do such a thing?"

The Tlokwa had a great sense of patience with me. In their eyes I was a very slow learner. They asked me to wait for the rains, wait to see the results of their transplanting efforts. When the rains came, the results were magnificant! The transplanted desert species bloomed with color along Tlokwa courtyards, walkways, and fences. And viewing, this array of color I could understand Tlokwa attraction to these species. The Tlokwa, indeed, were logical; they were, indeed systematic. Their viewpoint, however, was quite different from my western, food-oriented training.

Those days of 1973-1975 were a joy for me and my family, days remembered fondly. Many years have now passed. Most of my colleagues, and almost all of the students I teach at the University of California, Davis, are interested in the technical aspects of food production, whether improving productivity, increasing nutritional quality of human diet, nutrient bio-availability, or other food-related approaches to resource maximization. My colleagues and students tell me, of course early humans domesticated plants for food--what other reason could there be? But the Anderson thesis and the Tlokwa data offer a different insight. Thus, we are challenged: was domestication for food, or was it the result of non-food purposes, related to beauty, color, and decoration? I am inclined to believe the latter.

In Montana in 1982, near the town of Three Forks at the headwaters of the Missouri River, I was on an outing collecting agate with my father, daughter, and niece. The agate-strewn road took us through the center of a wheat field. Without prompting, each little girl stopped, walked to the edge of the road, picked a single stem of summer wheat, ran her fingers along the soft, fuzzy crown, and admired the beautiful complexity and delightful geometry of that grain head.

At such moments we are provided brief insights to the past, glimpses of an essential human characteristic first described by Anderson: human attraction to the form, color, and design of plants. While Anderson suggested decoration and beauty were the keys to plant domestication, what, too, if the selected species also were edible? Might not multi-purpose species be favored in a selection process, favored over those offering only a single function? Have our domesticated plants come to us from antiquity primarily because of beauty, and secondarily because of edibility?

As a scientist, I am no longer bothered by Anderson's thesis. Where my colleagues seek rational, food-related behavior as the determinant for domestication, I see rationality to Anderson's suggestion, once we accept his basic premise: domestication was not originally associated with food production. If Anderson is correct--and the Tlokwa would certainly agree--domestication easily could have occurred during periods of leisure and food wealth, and the stands of abundant wild wheat in southwestern Turkey no longer would be enigmatic.

In conclusion, as a student of Middle East culture, history, and nutrition, and as an observer of human nature, I visualize and sense the attraction to the geometric beauty of grain. It was the gift of Osiris that ultimately fed humans, and permitted technological development into the 20th century. But the gift of Osiris was more basic, it was one of plant beauty; a gift to be enjoyed by all humans, from antiquity to the 20th century, one for each of us to treasure, forever.

REFERENCES

1. Deuteronomy 8:3.
2. Matthew 6:9-13.
3. Omar Khayyam, Rubaiyat of Omar Khayyam, The Astronomer-Poet of Persia, rendered into English verse by Edward Fitzgerald. MacMillan, London (1898).
4. Matthew Henry, The Family Bible, or, Complete Commentary and Exposition on the Sacred Texts of the Old and New Testament by the Reverend Matthew Henry. Thomas Kelley, London. [Psalm 104] (1938).
5. Pope, Alexander, Pope's Satires and Epistles. Edited by Mark Pattison. Clarendon Press, Oxford [Satire VI:Book II:Line 220] (1872).
6. Anonymous. Carnaby Street, London (1967).
7. Chang, K-C., Food in Chinese Culture. Anthropological and Historical Perspectives. Yale University Press, New Haven (1977).
8. Root, W., The Food of Italy. Atheneum, New York (1971).
9. Exodus 12:15.
10. Hastings, J., Encyclopedia of Religion and Ethnics. 13 Vols. Charles Scribner's Sons, New York. [Vol. 6. pp. 425-435] (1951).
11. Freedman, M., Food habits and dietary change. A study of first, second, and third generation Greek-Americans of Sacramento, California. M.S. Thesis. Department of Nutrition, University of California, Davis (1980).

12. Frazer, J. G., The Golden Bough. 12 Vols. MacMillan and Co., London (1935).
13. Darby, W. J., P. Ghalioungui, and L. Grivetti, Food: The Gift of Osiris. 2 Vols. Academic Press, London. Vol. 2, pp. 482 (1977).
14. Ghalioungui, P., S. Khalil, and A. R. Ammar, Medical History 7:241 (1963).
15. Darby, W. J., P. Ghalioungui, and L. Grivetti, Vol. 2, p. 483, (1977).
16. Darby, W. J., P. Ghalioungui, and L. Grivetti, Vol. 2, p. 483, (1977).
17. Darby, W. J., P. Ghalioungui, and L. Grivetti, Vol. 2, p. 483, (1977).
18. Wendorf, F., R. Said, R. Schild, Science 169:1161 (1970).
19. Wendorf, F., R. Schild, N. Elhadidi, A. E. Close, M. Kobusiewicz, H. Wieckowska, B. Issawi, and H. Haas, Science 205:1341 (1979).
20. Pausanias, Guide to Greece. Vol. 1. Central Greece. Translated by P. Levi. Harmondsworth:Penguin (1971).
21. Graves, R., The Greek Myths. Vol. 1, Penguin, Baltimore (1955).
22. Plutarch, Isis and Osiris, in Moralia. Vol. 5, Translated by F.C. Babbitt, Harvard University Press, Cambridge (1936).
23. Darby, W. J., P. Ghalioungui, and L. Grivetti, Vol. 1, pp. 21-47, (1977).
24. Child, V. G., The Neolithic Revolution. pp. 23-25, 67-72, in Man Makes Himself. C.A. Watts, London (1951).
25. Sauer, C. O., Agricultural Origins and Dispersals. New York: American Geographical Society (1952).
26. Flannery, K. V., Annual Review of Anthropology 2:271 (1973).
27. Harlan, J., and D. Zohary, Science 153:1074 (1966).
28. Harlan, J., Archaeology 20:197 (1967).
29. Anderson, E., Plants, Life, and Man. Little, Brown, and Co., Boston (1952).
30. Grivetti, L. E., Dietary Resources and Social Aspects of Food Use in a Tswana Tribe. Ph.D. Dissertation. Department of Geography, University of California, Davis, (1976).
31. Grivetti, L. E., American Journal of Clinical Nutrition 31:1204 (1978).
32. Grivetti, L. E., Ecology of Food and Nutrition 7:235 (1979).

I

Increasing Yields and Food and Nutritional Quality through Breeding

1

Increasing Yield and Quality of Cereals through Breeding

DONALD N. DUVICK

INTRODUCTION

Yields of the cereal grains have increased many-fold over the past 50 years. We would like to know if yields will continue to increase and what factors such as weather, fertilizer or improved varieties have been responsible for the increased yields. We also would like to know whether new varieties are as reliable as the old ones and what changes in grain quality, if any, are associated with the new varieties. I shall present data dealing with the above questions about yield and reliability of yield for wheat, maize and grain sorghum. I also will give a brief discussion of possible future trends for grain quality of wheat and maize.

YIELD GAINS, 1930-1982

50-year Trends

In the approximately 50-year period since 1930, grain yields of wheat, maize and sorghum have increased at an average rate of between two and three percent per year, calculated as a percentage of the 50-year mean (Table 1.1). Soybeans, in contrast to the cereal grains, have increased in yield at an average rate of about one percent per year.

Critical Events Affecting the Trends

Synthetic nitrogen fertilizer. Certain critical events can be identified as affecting production and breeding of the three cereal grains (Table 1.2). The most important event for all three crops was the increased availability of large quantities of cheap synthetic nitrogen fertilizer starting in about the mid-1950s (1). Heavy use of this fertilizer on maize, wheat and sorghum stimulated a large number of cultural and breeding changes which eventually resulted in higher rates of gain.

Changes in wheat culture. Semi-dwarf wheat cultivars with good standability were introduced in the late 1950s. They allowed farmers to add large amounts of nitrogen fertilizer without suffering the

Cereals and Legumes in the Food Supply, edited by Jacqueline Dupont and Elizabeth M. Osman

TABLE 1.1. Rates of Gain in Yield of U.S. Wheat, Maize, Grain Sorghum and Soybean, 1930-1982

Crop	Mean[a] Yield bu/A	Rate of gain	
		bu/A/yr[b]	% of mean
Wheat	22.3	0.45	2.0
Maize	55.3	1.74	3.1
Sorghum	32.3	1.07	3.3
Soybeans	22.3	0.30	1.3

[a]For years 1930-1982.

[b]Regression coefficients (b) from the linear regression equation $Y = a + bX$. Coefficient of determination values (r^2) are 0.91, 0.93, 0.86, and 0.89 for wheat, maize, sorghum and soybeans.

TABLE 1.2. Critical Events in Production and Breeding of Wheat, Maize and Grain Sorghum

Crop	Approximate year	Event
Wheat	1958	Semi-dwarf cultivars, abundant nitrogen fertilizer
Maize	1960	Single cross hybrids, abundant nitrogen fertilizer
Sorghum	1956	Hybrids, abundant nitrogen fertilizer, wide scale irrigation
Sorghum	1968	Greenbug, mildew, midge, maize dwarf mozaic virus

consequence of severe lodging that typically happened when the old varieties were heavily fertilized. But the semi-dwarfs probably would have had little advantage over the old varieties if cheap nitrogen fertilizer had not become available at about the same time the semi-dwarfs were introduced (2).

Changes in maize culture. Availability and use of increased amounts of nitrogen fertilizer encouraged maize farmers to increase planting rates. This in turn brought about more stalk and root lodging and also made hybrids more susceptible to barrenness (ear abortion) when hot dry weather coincided with flowering time. Breeders thus were stimulated to develop hybrids with better resistance to root and stalk lodging as well as better resistance to barrenness induced by crowding and heat stress. Breeders found that progress in selecting for high yield and stress resistance could be made more quickly with single cross hybrids than with double cross hybrids. Single crosses therefore replaced double crosses in the 1960s and so, as with wheat, interactions among new farming practices and new varieties eventually resulted in higher annual rates of gain in yield (3).

Changes in culture of grain sorghum. Much the same story can be told for annual yield gains in grain sorghum except that for sorghum two additional factors turned up at about the same time as abundant supplies of synthetic nitrogen fertilizers. Hybrid sorghum was introduced in the mid-1950s; this immediately put achievable yields on a much higher plane. Relatively inexpensive and simple irrigation technology also appeared in the 1950s. The three new factors: easy irrigation, hybrid vigor and inexpensive nitrogen fertilizer made it possible for farmers in the sorghum belt, stretching from Texas to South Dakota, to treat sorghum as a high input, high yield crop rather than as a crop for marginal dryland farming conditions.

But unfortunately for grain sorghum, after 10 or 15 years of intensive cultivation and extremely rapid gains in per acre yields a series of problems appeared. They included a shift in greenbug (*Toxoptera graminum* (Rond.)) genotype to a race that could damage sorghum as well as wheat, introduction of sorghum downy mildew (*Peronosclerospora sorgi* (Weston and Uppal) C.G. Shaw) from the tropics and the spread of midge (an insect, *Contarinia sorghicola* (Cop.)) and maize dwarf mosaic virus (MDMV). Greenbug hit the northern two-thirds of the sorghum belt, mildew and midge affected the deep southern parts and maize dwarf mosaic virus generally was most troublesome in the middle portions. Within a few years breeders were able to counter greenbug, mildew and MDMV with resistant hybrids but by the time this holding action had been completed, high energy prices had made irrigation of sorghum less desirable in the view of many farmers, so sorghum to a large extent again was planted on marginal acres rather than on highly productive irrigated land (4).

Soybeans. The legume, soybeans, has had no event in its culture during the past 50 years comparable in importance to introduction of cheap nitrogen fertilizer for the cereal grains. Continual improvements in breeding and farming practices have brought about uniform but relatively slow improvements in yield of soybeans over the past half century.

Rates of Gain Before and After Critical Events

When annual yield gains are calculated for each of the cereal grains for the periods before and after the landmark events described above, it can be seen that absolute rates of gain (bushels per acre per year; bu/A/yr) changed greatly in the period after each landmark event (Table 1.3). Rates of gain in bushels per acre per year were about 50% greater for wheat and about 100% greater for maize in the post-1960 periods. The absolute rate of gain for sorghum was about six times as great during the period 1956-1968 as in the 25-year period before 1956. Since 1968, sorghum yields have been highly erratic and gains have been negligible. In contrast to the cereal grains, the soybean rate of gain after 1960 was not greatly different from that before 1960.

Although annual yield gains in bushels per acre were increased for each of the cereal grains in the period following introduction of low-priced nitrogen fertilizer, yield gains calculated as percent of the mean yield for the period were reduced for wheat and maize (Table 1.3). This is because mean yield in successive years is a cumulative

TABLE 1.3. A Comparison of Rates of Gain in Yield, Before and After Critical Events in Breeding and Production of Wheat, Maize and Grain Sorghum, and Before and After 1960 for Soybeans

Crop	Years	Mean Yield bu/A	Rate of Gain bu/A/yr[a]	Rate of Gain % of mean
Wheat	1930-1958	16.3	0.31	1.9
	1958-1982	29.2	0.44	1.5
Maize	1930-1958	34.5	1.07	3.1
	1960-1982	83.4	2.16	2.6
Sorghum	1930-1956	16.2	0.36	2.2
	1956-1968	42.0	2.23	5.3
	1968-1982	53.6	0.10	0.2
Soybeans	1930-1960	19.0	0.29	1.5
	1960-1982	26.9	0.32	1.2

[a]Regression coefficients (_b_) from the linear regression equation $Y = a + bX$.

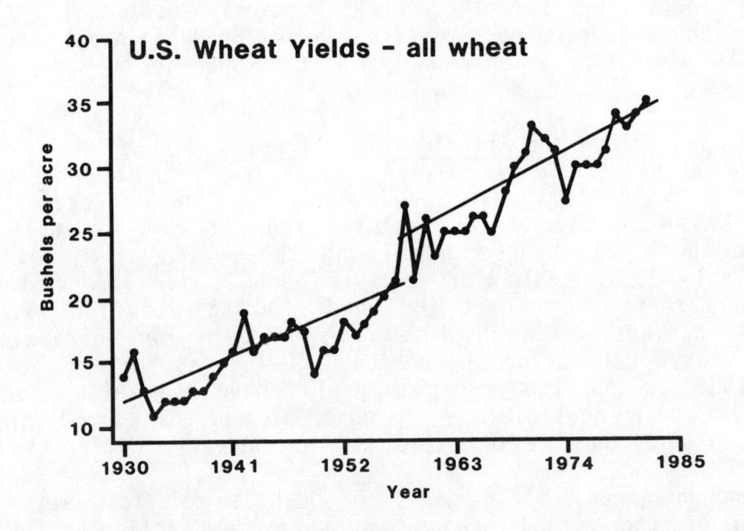

FIG. 1.1. Grain yields of U.S. wheat from 1930 to 1982. Linear regressions plotted for 1930-1958 (r^2=0.62, _b_=0.31 bu/A/yr) and 1958-1982 (r^2=0.74, _b_=0.44 bu/A/yr) (5).

figure if rate of gain is positive. Mean yield will therefore grow faster than the rate of gain which generates it, unless rates of gain are very high or increasing rapidly, and/or the time for accumulation of higher mean yield is very short. Some of these exceptional factors occurred with sorghum in the period 1956-1968; rate of gain was very high and the time period was relatively short. Therefore the gain in percent of period mean was twice as high in 1956-1968 as in the preceeding period. Soybeans resembled the cereal grains in that soybean yield increases calculated as percent of period mean are smaller for 1960-1980 than for 1930-1960, even though gains in bu/A/yr are larger in 1960-1980 than in 1930-1960.

Figures 1-4 detail the data summarized in Table 1.3. They show clearly that (1) wheat yields have moved to a higher plane and are rising at a slightly steeper rate since 1958, (2) the rate of increase in yield of maize has been much greater since 1960, (3) grain sorghum yields have been extremely variable since 1965, and (4) soybean yields through the years have increased at a reasonably steady rate with no sudden shift in level or rate, post-1960.

ESTIMATES OF THE GENETIC PORTION OF TOTAL YIELD GAINS

Methods for Estimation of Genetic Gain, and Variation in Size of the Estimates

Comparisons of yields of a series of varieties or hybrids introduced over several years allow one to make estimates of genetic gain, the gain in yield due to breeding alone. (Varieties from different eras may be compared to a standard check variety in yield trials grown through the years, or they can all be grown together in one yield trial.) Estimates of genetic gain can be compared with total yield gain to give estimates of the percentage of total yield gain which is due to breeding, i.e., to genotype. Table 1.4 lists examples of such calculations.

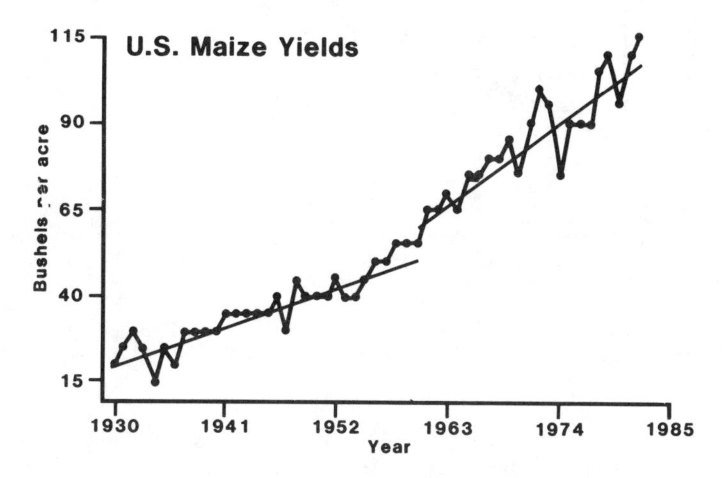

FIG. 1.2. Grain yields of U.S. maize from 1930 to 1982. Linear regressions are plotted for 1930-1960 (r^2=0.87, b=1.07 bu/A/yr) and 1960-1982 (r^2=0.81, b=2.16 bu/A/yr) (5).

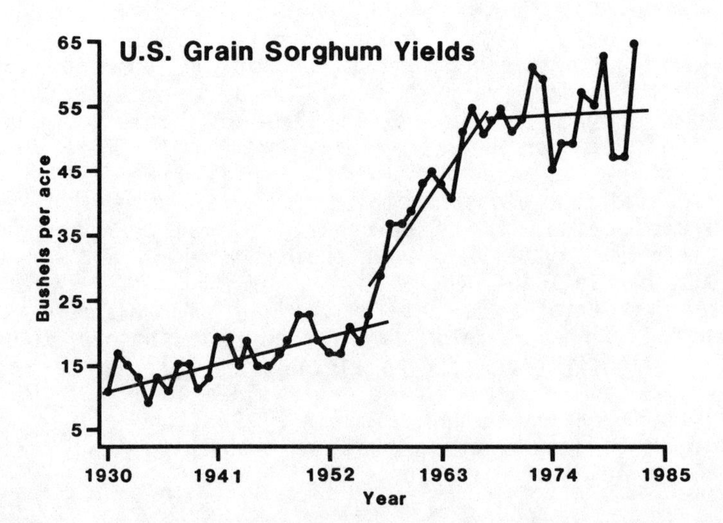

FIG. 1.3. Grain yields of U.S. grain sorghum from 1930 to 1982. Linear regressions plotted for 1930-1952 (r^2=0.50, b=0.39 bu/A/yr), 1952-1965 (r^2=0.92, b=2.68 bu/A/yr) and 1965-1982 (r^2=0.02, b=0.13 bu/A/yr) (5).

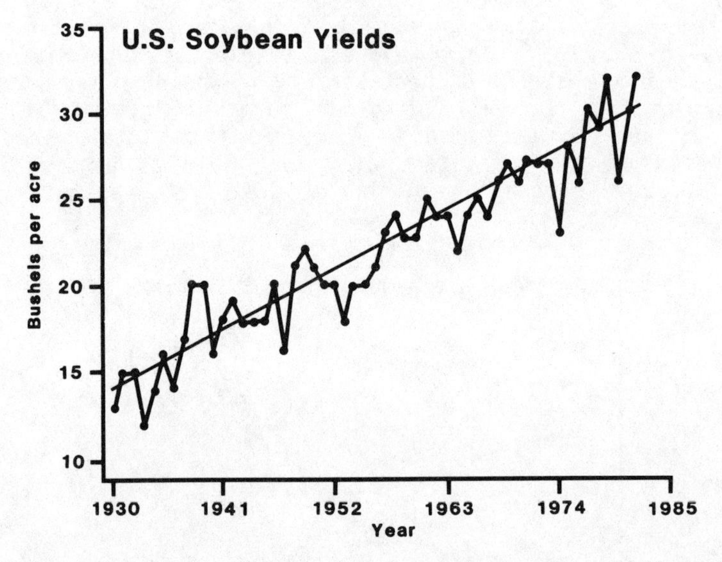

FIG. 1.4. Grain yields of U.S. soybeans from 1930 to 1982. Linear regressions plotted for 1930-1982 (r^2=0.89, b=0.30 bu/A/yr) (5).

Such methods of estimation are far from being an exact science, however. Estimates of the genetic portion of total yield gain for wheat, made by six different sets of investigators, have ranged from 20% up to 55% (Ref. 1, and Table 1.5). Seven different estimates of the genetic portion of total yield gain for corn have ranged from 33%

TABLE 1.4. Comparison of Total and Genetic Rates of Gain in Yield of Wheat, Maize and Grain Sorghum

Crop	Time span	Total rate of gain bu/A/yr[b]	Total rate of gain %/yr	Genetic rate of gain[a] bu/A/yr[b]	Genetic rate of gain[a] %/yr	Genetic % of total
Wheat	1958-1980	0.44	1.5	0.22[d]	0.7	50
Maize	1955-1980[c]	2.23	2.6	1.47	1.7	66
Sorghum	1956-1980	0.91	1.9	1.14	1.6	84

[a]Data from Refs. 2, 3, 6.
[b]Regression coefficients (b) from the linear regression equation Y = a + bX.
[c]Iowa data.
[d]Est. from %/yr genetic gain.

TABLE 1.5. Estimates of Genetic Portion of Total Yield Gains for Wheat[a]

Author	% Genetic
Jensen (7)	50
Sim and Araji (8)	55
Salmon, et al (9)	40
Reitz and Solmon (10)	20
Hueg (11)	53
Schmidt (2)	50[b]

[a]From Ref. 2.
[b]Calculated from Ref. 2.

TABLE 1.6. Estimates of Genetic Portion of Total Gains for Maize[a]

Author	% Genetic
Darrah (3)[b]	33
Russell (13)	79
Russell (13)	63
Duvick (14)	57
Duvick (14)	60
Duvick (2)	89
Duvick (2)	71

[a]From Ref. 3.
[b]Ref. 12.

up to 89% (Duvick, 1983 and Table 1.6). Three different estimates for sorghum have ranged from 28 up to 84% (Ref. 4 and Table 1.7).

Reasons for Variation in Size of Estimates for Wheat

Reasons for the variations in size of estimate of genetic gain are many but the most important reasons probably have to do with the category or class of material in the species that is being studied

TABLE 1.7. Estimates of Genetic Portion of Total Yield Gains for Grain Sorghum[a]

Author	% Genetic
Wellman and Hassler (15)	28
Maunder (16)	34[b]
Miller (4)	84[c]

[a]From Ref. 4.
[b]Prediction for future gains.
[c]Calculated from Ref. 4.

and with the growing conditions that prevailed or that were imposed when comparisons were made between old and new varieties or hybrids. For example, genetic yield gains per year for the winter wheats have on average been larger and more consistent than those for the spring wheats over the past 20 years (2).

But gains within these broad classes of wheats are not necessarily uniform if different categories are compared. For example the winter wheats in the southern region of the hard red winter wheat belt have shown more impressive yield gains in recent years than have those in the northern region. This may be because unfavorable environments for winter wheat in the northern region reduce the effectiveness of selection for high yield potential.

To further complicate the analysis, in both regions gains of new varieties compared to long time checks appeared to be especially great in the early years of the comparisons. This was largely because the newer wheats had superior stem rust resistance in the presence of severe stem rust epidemics. When this disease was no longer an important factor due to widespread planting of the resistant new wheats, apparent advantage of the newer varieties dropped off markedly.

The soft red winter wheats of the eastern states may be compared to the hard red winter wheats of the mid-west. The eastern wheats, under a different disease spectrum, made sharp genetic yield increases compared to long time checks in the early 1960s but genetic yield gains seem to have been erratic in the past few years, perhaps due to appearance of new disease complexes.

Reasons for Variation in Size of Estimates for Maize

Growing conditions of an experiment also can affect the proportion of genetic gain that is attributed to breeding, as is shown by a recently concluded experiment designed to examine changes in maize hybrids introduced during the past 50 years (3). The experiment showed that on average the newer the hybrid the higher the yield. However, the greatest advantage of the newer hybrids was at high planting rates typical of recent years. The old hybrids made their highest yields at low planting rates. The low planting rate of the experiment was typical of planting rates of the 1930s, and the high planting rates were typical of those used in the 1970s. Trials conducted only at the low planting rate would give too low an estimate of gains due to breeding, because the new hybrids would not express their full potential. Trials conducted only at the high planting rates would give too high an estimate, because the old hybrids would not express their full yield potential. To get the most accurate estimate of gains due to breeding, one should use yields of the old hybrids at the old (low) planting rate and yields of the new hybrids at the new (high) planting rates.

Investigation of other factors in the maize experiment showed additional reasons for higher yields of the new hybrids. The newer hybrids had stronger roots than the older hybrids, especially at higher rates of planting, which promote root lodging. The newer hybrids were more resistant to stalk rot and consequently had fewer broken stalks. The newer hybrids were more resistant to premature death, a phenomenon with largely unknown cause which usually occurs in mid-September, about one month before harvest time. The newer hybrids were more tolerant to second brood infestation of the European corn borer. (Second brood infestation usually results in stalk tunneling by the borer larvae, followed by deterioration of the stalk, stalk breakage and yield loss.) The newer hybrids tended to have more ears per plant than the old hybrids at low planting rates and they also had less tendency to have barren stalks at medium or high planting rates. The new hybrids yielded more than the old hybrids when yield levels were severely reduced due to hot dry weather. But the new hybrids also yielded more than the old hybrids in fields with high yield potential due to favorable weather. And finally, the newer hybrids tolerated low levels of soil nitrogen better than the old hybrids even though they also were able to outyield the old hybrids in the presence of high levels of soil nitrogen.

The results of this study show therefore that newer maize hybrids outyield older ones in part because the new hybrids are better able to take advantage of favorable weather, high soil fertility and high planting rates but also, and probably most importantly, because the new hybrids are tougher than the old ones; they are better able to withstand the stresses of hot, dry weather, low soil fertility and insect or disease attack. Similar advantages can be demonstrated for the newer varieties and hybrids of wheat and sorghum (4, 6).

Thus, the presence or absence of stress conditions, either through experiment design or through chance effect of the season, can affect the estimate of gains due to breeding. Gains will tend to appear greatest if stresses are present, least if stresses are absent.

CHANGES IN QUALITY OF CEREAL GRAINS

Introductory Remarks

I will not attempt to document changes in quality of the cereal grains over the past 50 years. That would require a second full length paper. I will make a few general remarks, however, based on information given to me in the course of day-to-day contacts with breeders of the cereal crops and with those who are knowledgeable about the uses of these crops.

Possible Changes in Milling and Baking Quality of Wheat

I believe most wheat breeders would agree with me in saying that the greatest advances in yield of wheat in recent years have been in those classes of wheat or in those parts of the country where the pressure for high milling and baking quality is the lowest. I believe many breeders also would agree with the statement that there is a tendency for newer varieties of wheat to be released primarily because they have high dependable yield and pest resistance, and in spite of average or below average milling and baking quality. It will be interesting to see how this trend develops. No doubt a complex series of interactions between millers, bakers and farmers, with each group trying to maximize profits, will determine the eventual outcome.

Changes or Lack Thereof in Maize Grain Nutrient Composition and Breakage Potentials

Ten or 15 years ago most of us in plant breeding expected that maize hybrids with increased percentages of the limiting amino acids lysine and tryptophane (*opaque-2* hybrids) soon would dominate the market. These expectations were not realized, although good *opaque-2* hybrids were introduced about 10 years ago. The hybrids were not accepted by the farmer nor were they demanded by feed manufacturers. Reasons for lack of success of these hybrids are numerous and complex but important factors surely were that the high lysine hybrids yielded 10-15% less than their normal counterparts, the grain was soft textured and hard to handle in mills, it required separate storage facilities in grain elevators and terminals and probably most important of all, soybean meal has been plentiful and relatively low in price.

Total grain protein of maize has stayed very close to 9% over the years, according to experts in the milling industry. Some concern was expressed years ago, that as yields per acre went up protein percent would go down. But those predictions were made on the basis of experiments comparing grain from trials giving different yield levels at constant rates of nitrogen fertilizer application. An increase in nitrogen fertilizer usually gives an increase in grain nitrogen percentage. Apparently, farmers have increased nitrogen fertilizer applications in proportion to grain yield increases, such that protein percentage of maize grain has stayed the same over the years.

An entirely different quality factor for maize is the tendency for its grain to break up when handled repeatedly, as for example when it moves from farm to elevator to barge to overseas terminal. When farmers switched from slow drying of ear corn in a cool corn crib to rapid drying of high moisture shelled grain in a hot drying bin, grain buyers noticed a sharp drop in the quality of the grain that reached overseas terminals. It was much more likely to be broken and powdery. But in the past few years it is likely that grain quality has been improving. Because of high energy costs, farmers now allow grain to reach lower moisture percentages before harvest and they also seem to be drying the harvested grain at lower temperatures. Both changes will give sounder grain.

SUMMARY AND CONCLUSIONS

Per acre yields of wheat, maize and sorghum have increased at average rates of two to three percent per year during the past 50 years. The per acre yield gains for wheat and maize show no signs of leveling off whereas sorghum yields have plateaued in recent years. For all three crops, introduction of plentiful amounts of low priced synthetic nitrogen fertilizer stimulated introduction of a host of associated cultural and breeding changes that eventually resulted in markedly increased absolute rates of gain in grain yield. For each crop, improvements due to breeding have accounted for about 50% of the total yield gain. A large portion of the gains from breeding is due to introduction of varieties with superior adaptation to modern agronomic practices, in particular adaptation to high rates of nitrogen fertilizer. But the new varieties and hybrids also are superior to the old ones in absence of favorable growing conditions. They have better disease and insect resistance and they are better able to withstand the stresses of unfavorable weather and low soil fertility. Grain quality of wheat is likely to be slightly lower in the future, in regard to its milling and baking traits. Grain quality of corn probably will stay constant in regard to percentage of essential amino acids and total protein but it probably will improve in regard to resistance to kernel breakage during handling. The changes in wheat quality will be due to breeding, changes in corn breakage will be due primarily to changes in grain harvesting and drying practices.

REFERENCES

1. United States Department of Agriculture, "Agricultural Statistics 1977". U.S. Government Printing Office, Washington, D.C. (1977).
2. Schmidt, John W., in "Genetic Contributions to Yield Gains of Five Major Crop Plants" (W. R. Fehr, ed). CSSA Special Publication No. 7, Madison, Wisconsin, (1983).
3. Duvick, D. N., in "Genetic Contributions to Yield Gains of Five Major Crop Plants" (W. R. Fehr, ed.), CSSA Special Publication No. 7, Madison, Wisconsin, (1983).
4. Miller, F. R. and Kebede, Y., in "Genetic Contributions to Yield Gains of Five Major Crop Plants" (W. R. Fehr, ed). CSSA Special Publication No. 7, Madison, Wisconsin, (1983).
5. United States Department of Agriculture, "Agricultural Statistics 1982" and previous volumes. U.S. Government Printing Office, Washington, D.C., (1982).
6. Austin, R. B., Bingham, J., Blackwell, R. D., Evans, L. T., Ford, M. A., Morgan, C. L., Taylor, M., Jour. Agric. Sci., U.K. 94, 675-689 (1980).
7. Jensen, N. F., Science 201, 317-320 (1978).
8. Sim, R. J. R. and Araji, A. A., The economic impact of public investment in wheat research in the Western Region. Idaho Agric. Expt. Stn. Res. Bull. 116, p. 27 (1981).
9. Salmon, S. C., Mathews, O. R. and Leukel, R. W., Adv. Agron. 5, 1-151 (1953).
10. Reitz, L. P., and Salmon, S. C., Hard red winter wheat improvement in the Plains: A 20-year summary. USDA Tech. Bull. No. 1192. 117 p. (1959).
11. Hueg, W. F., Jr., in "Agronomists and Food: Contribution and Challenges", Amer. Soc. Agron. Special Publication No. 30, pp. 73-85, Madison, Wisconsin, (1977).
12. Hallauer, A. R., Egypt. Jour. Genetics and Cytology 2, 84-101 (1973).
13. Russell, W. A. Comparative performance of maize hybrids representing different eras of maize breeding. Proc. 29th Ann. Corn and Sorghum Res. Conf. 29, 81-101 (1974).

14. Duvick, D. N., Maydica XXII, 187-196 (1977).
15. Wellman, A. C. and J. B. Hassler, Quarterly Serving Farm, Ranch and Home. (Univ. of Nebraska, College of Agric. and Home Econ.) 16, 23-25 (1969).
16. Maunder, A. B. Meeting the challenge of sorghum improvement. Proc. 24th Ann. Corn and Sorghum Res. Conf., 24, 135-151 (1969).

2

Increasing Yields and Food and Nutritional Quality through Breeding: Grain Legumes

THEODORE HYMOWITZ

INTRODUCTION

With approximately 650 genera and 18,000 species, the Leguminosae is the third largest family of flowering plants, after the Compositae and Orchidaceae, and second only to the Gramineae in economic importance. Leguminous plants are found in virtually every ecological zone and man has exploited this natural resource for meeting his economic and social needs. Legumes are used for timber, pasture, green manures, forage crops, cover crops, chemicals, esthetic value and as food.

Of the thousands of known legume species only about 12 are used extensively as grain legumes (Table 2.1). The grain legumes are those plants of the Leguminosae used as food in the form of unripe pods, immature seeds, or mature dry seeds, directly or indirectly.

Today, dietitians and food scientists evaluate the grain legumes in terms of their nutritional contents. However, when the grain legumes were first domesticated by man, the main considerations were adaptability to a particular region (Table 2.1), reliability of yield, taste acceptability and whether the seeds could be transported and/or stored easily.

Grain legumes are rich sources of protein averaging about 20-25% protein, with soybeans having about 40% protein. Soybeans and peanuts are also excellent sources of vegetable oil averaging about 20 and 40%, respectively. In general, the grain legumes are rich in lysine but poor in methionine content, thereby complementing the reverse amino acid pattern found in the cereals. Additionally, all of the major grain legumes fix their own nitrogen, thereby reducing the cost of nitrogen fertilizer inputs by farmers.

Worldwide grain legume production is small (ca. 10%) relative to that of cereals; however, the role of grain legumes in human nutrition is more important than their relative acreage. Of the 22 major crops in the world, the cereal grains wheat, maize, rice, barley, sorghum, oats, millets and rye are number 1, 2, 3, 5, 10, 13, 19, and 21, respectively, in total production (1) (Table 2.1). Soybeans and peanuts are number 8 and 22 on the production list. However, for total protein production, soybeans, peanuts, common beans and peas are number 3, 10, 13 and 15 among the major crops of the world (1) (Table 2.3). Hence, the grain legumes provide supplementary protein and other nutrients to diets based on cereal grains and/or starchy root and tuber

Cereals and Legumes in the Food Supply, edited by Jacqueline Dupont and Elizabeth M. Osman

TABLE 2.1. Scientific Name, Common Name and Region of Diversity of the Major Grain Legumes (1)

Specific Name	Common Name	Region of Diversity
Arachis hypogaea	Peanuts	South Africa
Cajanus cajan	Pigeon peas	E. Africa, Indian subcont.
Cicer arietinum	Chickpeas	SW Asia, Ethiopia, India
Glycine max	Soybeans	China
Lens culinaris	Lentils	SW Asia, Mediterranean
Phaseolus lunatus	Lima beans	Peru
Phaseolus vulgaris	Common beans	Mexico, Guatemala
Pisum sativum	Peas	SW Asia
Vicia faba	Broad beans	Asia, Mediterranean
Vigna angularis	Adzuki beans	China, Japan
Vigna radiata	Mung beans	India, SE Asia
Vigna unguiculata	Cow peas	W Africa, India

TABLE 2.2. World Production of Major Crops (1)

Crop	Million Metric Tons	Crop	Million Metric Tons
Wheat	458	Cottonseed	46
Maize	451	Oats	44
Rice	413	Banana	40
Potato	257	Oranges	38
Barley	158	Yam	32
Sweet Potato	146	Apples	31
Cassava	127	Coconuts (nuts)	29
Soybean	88	Millets	29
Cane sugar	78	Beet sugar	28
Sorghum	71	Rye	24
Tomato	50	Peanut	19

TABLE 2.3. Protein Production of Major Crops (1,2)

Crop	Million Metric Tons	Crop	Million Metric Tons
Wheat	49	Rye	2.2
Maize	41	Beans	2.2
Soybean	32	Millets	2.9
Rice	23	Peas	1.3
Barley	10	Sunflower	1.3
Sorghum	6.4	Cassava	1.0
Potato	5.1	Yams	0.6
Cottonseed	5.0	Coconut	0.5
Oats	4.8	Banana	0.3
Peanut	4.0	Cane sugar	0.0
Sweet potato	3.0	Beet sugar	0.0

crops. In addition, soybeans are by far the largest source of fat or vegetable oil in the world (3) (Table 2.4).

The major complaint about the grain legumes is their relatively low yield per unit area when compared to the cereals. The low yields reflect the evolutionary background of the *Leguminosae*. There are three biochemical processes that take place in all grain legumes which operate against high grain yields. These processes are photorespiration, nitrogen fixation and photosynthetic energy relationships. Photorespiration occurs in the light and consumes about 30% of the products of photosynthesis. Maize functions without photorespiration. At present, there is no known benefit of photorespiration.

The adage that there is no free lunch is quite apt when applied to the nitrogen fixation process. The symbiotic relationships between the legume plant and the *Rhizobium* bacterial organism take place in all grain legumes. The legume-inhabiting bacteria form nodules on the roots which convert atmospheric nitrogen into a form that the plant can assimilate for its use. The plant in turn furnishes the necessary carbohydrate or energy to the bacterial organism. This diversion of

TABLE 2.4. World Fats and Oils Production, 1980-1981 (3)

Commodity	Thousand Metric Tons
Soybeans	12,223
Tallow grease	6,047
Palm	4,990
Butter	4,927
Sunflower	4,781
Rapeseed	3,841
Lard	3,827
Coconut	3,261
Cottonseed	3,206
Peanuts	2,996

carbohydrates by the plant for use by the bacterial organism reduces potential grain production by about 10%.

The maize plant produces a seed which is mostly starch. It takes about 1.2 kg of glucose to produce 1 kg of carbohydrate which has about 4.0 million calories of stored energy. The soybean plant produces a seed which on the average contains about 40% protein and 20% oil. It takes about 2.2 kg of glucose to produce 1 kg of protein which has about 4.8 million calories of stored energy and 2.8 kg of glucose to produce 1 kg of oil which has 9.5 million calories of stored energy. Therefore, a major reason why soybean yields are lower than maize is because of the energy intensive process of producing a high protein and oil grain.

The soybean is a good example for demonstrating the above situation. Table 2.4 summarizes soybean production statistics in the U.S. from 1930 to 1980. The table reveals that soybean yields per hectare in the U.S. doubled in 50 years; however, the major increase in soybean production came about by an increase in the amount of land planted to the crop. Within the same time period maize production per hectare increased by almost fivefold. Obviously, if soybeans and other grain legumes are to compete successfully with cereals, yields will have to increase per unit area relative to the cereals or, alternatively, the nutritional quality of the grain legumes will have to be improved so as to increase the value of the crops.

TABLE 2.5. U.S. Soybean Production Figure, 1930 to 1980 (3)

Year	Planted (1,000 hectares)	Yield (kg/ha)	Production (million metric tons)
1930	430	874	376
1940	1,923	1,089	2,094
1950	5,523	1,458	8,052
1960	9,462	1,579	14,940
1970	16,900	1,794	30,319
1980	28,035	1,774	49,734

The rest of this paper will be devoted to providing examples of research devoted to improving the yield and quality of soybeans. The soybean is being used as the model grain legume because a great deal of research has been conducted on the crop and, secondly, soybeans are by far the most important grain legume grown in the U.S.

SOYBEAN RESEARCH PRIORITIES

The primary objective of soybean breeders in the U.S. is the development of high yielding cultivars having good agronomic characteristics and resistant to economically important pathogens and pests. Soybean breeding research is dictated by three factors germplasm resources, analytical capabilities and incentives.

A rational starting point in any breeding program is the accumulation of germplasm to provide access to the genetic variation of the breeding material. The U.S. *Glycine* collection contains about 10,250 accessions (Table 2.6). The collection is maintained by the U.S. Department of Agriculture, in Urbana, Illinois (MG 000 to IV); at Stoneville, Mississippi (MG V to X); and at Ames, Iowa (cytogenetic collection). The collection is divided into several parts:

TABLE 2.6. Composition of the U.S. *Glycine* Collection

Category	Number of Accessions
Foreign accessions (*G. max*)	8,850
Domestic cultivares (*G.max*)	350
Wild soybean (*G. soja*)	560
Perennial *Glycine* species	275
T-lines	100
Cytogenetic	100
Total	10,235

1. Foreign Accessions (*Glycine max*). This collection is commonly known as the Forage Crop (F.C.) and Plant Introduction (P.I.) collection. It includes seed of "land races" or improved strains from Japan, Korea, China and other countries.

2. Domestic Cultivars (*G. max*). This collection contains representative samples of publicly released cultivars in the U.S. and Canada.

3. Wild Soybeans (*Glycine soja*). This is a collection of *G. soja* obtained from Japan, Korea, China, the U.S.S.R. and Taiwan.

4. Perennial *Glycine* Species. This is a collection of perennial species closely related to the soybean. Most of the accessions are from Australia.

5. T-lines. The type collection comprises soybean lines with specific traits with known inheritance.

6. Cytogenetic Collection. This collection includes lines with cytological abnormalities such as translocations, inversions, deficiencies, and trisomics.

A prerequisite for the desired progress in a soybean breeding program devoted to nutritional quality changes is the availability of simple, rapid and efficient analytical techniques. NMR, GLC, HPLC, nearinfrared reflectant, electrophoretic, single and double beam spectrophometric and fiber-optic colorimetric instruments commonly found in chemistry departments are now cluttering up soybean breeding laboratories. Power problems are now commonplace in buildings designed 20 years ago for traditional plant breeding research. In the future, research facilities for breeders will include built-in cold rooms, culture rooms with transfer hoods and telephone lines for linking laboratory micro- and mini-computers to mainframe computers. To paraphrase a popular cigarette advertisement: "We've come a long way baby."

In addition to simple, rapid and efficient analytical methods, the breeder must also establish breeding methods that can be used to solve the yield and nutritional quality problems. Of course, the breeding methods used will depend upon (a) the diversity of the components in the germplasm collection, and/or (b) the nature of the inheritance of the components under study. For example, soybean breeders have observed consistent associations between yield and the two major products of the soybean, the oil and the protein-rich meal. High yield is associated with high oil content and low protein content of the seed. High oil content is correlated with low protein content of the seed. Hence, a soybean breeder, depending on the genotypes being utilized, might find it difficult to select for both high protein and yield in a single line. Another example concerns the selection for high oil content in seeds. Investigations by workers at Illinois and North Carolina have shown that oil synthesis in soybean seed is determined principally by the genotype of the maternal parent producing the seed rather than by the genotype of the seed. Hence selection for high oil content among individual seeds on a single plant is likely to be ineffective (4, 5).

Lastly, farmers are paid for their soybeans on a weight basis. At present they are more concerned with seed quality such as diseasefree seed, than with nutritional quality. Only if incentives, i.e., bonuses, were paid to farmers for growing specific types of soybeans, and market outlets were assured over a long period of time, might soybean breeders markedly change their priorities in soybean breeding programs. Therefore, in the absence of marketing alternatives, any nutritionally improved soybean cultivar will have to compete in the market place with the cultivars being grown now.

Breeding for changes in yield and nutritional quality in the soybean is a complex problem. Nevertheless, investigators in the area have been successful and the results of a few of those investigations are summarized in the next section.

YIELD AND QUALITY CHANGES

Yield And Protein

Two decades ago Brim and co-workers at North Carolina State University began a study to determine if the recurrent selection method could be utilized to increase the protein content in seed (6). In 1961, a population of 247 F_4 lines and parents was tested for yield and chemical composition. From the 247 lines, 102 were chosen for high percentage seed protein regardless of yield performance and were designated as cycle zero (C_0). The 102 lines were randomly intermated and the resulting crosses were advanced one generation to form cycle one (C_1). Selection of lines for high protein content, intermating the lines and then selfing them has succeeded through 7 cycles of selection. Eight lines from cycle 7 were advanced without selection from the F_2 to F_5 generation and then tested against Ransom and Bragg, two commercial cultivars. In 1982, 21 years after the start of the experiment, soybean composite population NC-1 was released by North Carolina State University (Table 2.7). NC-1 should be useful to soybean breeders as a source of high protein breeding lines that have good yield potential (7).

Kunitz Trypsin Inhibitor

More than 60 years ago, Osborne and Mendel (8) demonstrated that unheated soybean meal is inferior in nutritional quality to properly heated meal. The trypsin inhibitor proteins in raw mature soybean seed are one of the major factors responsible for poor nutritional value of unheated meal (9). Several different trypsin inhibitors are present in soybeans. However, much of the trypsin inhibitor activity is thought to be due to the Soybean Trypsin Inhibitor A_2 (SBTI-$_2$) which was first crystallized by Kunitz and is commonly known as the Kunitz trypsin inhibitor (10).

At Illinois, seed from the U.S. Department of Agriculture soybean germplasm collection was screened by polyacrylamide gel electrophoresis for the absence of electrophoretic forms of the Kunitz trypsin inhibitor. Two plant introductions, P.I. 157440 and P.I. 196168 do not have the Kunitz trypsin inhibitor (11). The lack of the protein is inherited as a recessive allele (12).

Bernard and Hymowitz, utilizing the backcross breeding technique, developed near isolines of the cultivars Williams and Clark lacking the Kunitz trypsin inhibitor. In 1982, the near isolines were tested in Urbana (Table 2.8). The soybean lines lacking the Kunitz trypsin inhibitor germinate normally and the subsequent plants grow, flower and set seed just like the soybean seed that has the inhibitor. Feeding experiments will be conducted in Urbana to determine if the near isolines of Williams and Clark lacking the Kunitz trypsin inhibitor are nutritionally superior when compared to the cultivars containing the inhibitor.

Linolenic Acid

Linolenic acid is believed to be partly responsible for the undesirable flavor in refined soybean oil. Soybean oil processors believe that the cost of processing out the high linolenic acid in soybean oil is a deterrent to soy oil competitiveness in U.S. and foreign markets. A soy oil with 2 to 3% linolenic acid would improve the

TABLE 2.7. Comparison of NC-1 With Two Cultivars for Yield and Protein Content, North Carolina (7)

Genotype	Yield (ka/ha)	Protein (%)
NC-1 (8 lines)	2704-2018	50.2-47.9
Bragg	2771	42.1
Ransom	2778	43.6

TABLE 2.8. Yield of Kunitz Tryspin Inhibitor Free Near Isolines of Cultivars Williams and Clark Tested at Urbana, Illinois, 1982.

Genotype	Yield (kg/ha)	Genotype	Yield (kg/ha)
Williams	3132	Clark	2587
Williams 79	3172	Clark 63	2601
Williams 82	3219	L81-4870	2466
L81-4583	3172	L81-4871	2695
L81-4584A	3178	L81-4873	2480
L81-4584B	3192		
L81-4585	3199		
L81-4590	3219		
L81-4593	3273		
L81-4594	3320		

competitiveness of soy oil and probably improve consumer acceptability. The soybean germplasm collection has been screened for genotypes lacking linolenic acid but without success.

Hammond and Fehr (13) initiated a project to improve the stability of soybean oil by lowering the percentage of linolenic acid in seeds. They treated seeds of line FA9525, a line low in linolenic acid, with ethyl methanesulfonate. The ancestry of FA9525 traces back to P.I. 80476 and P.I. 85671. A M_4 plant selection from FA9525 was designated as A5 and released in 1982. This line has the lowest percentage of linolenic acid in its seed oil of any plant introduction, experimental line or variety tested for the character (Table 2.9). The line should be useful as parental material in soybean oil quality breeding programs (14).

Lipoxygenase-1

Lipoxygenase has been implicated as the principal cause of the undesirable flavors of soybean products, especially soy milk. Heat treatment of soybeans prior to oil extraction increases the stability of the oil, presumably due to the inactivation of lipid oxidizing enzymes such as lipoxygenase, but has the undesirable consequence of reducing the solubility of the protein. Soybean seeds contain at least three lipoxygenase isozymes commonly designated lipoxygenase-1, -2, and -3 (15, 16). Lipoxygenase-1 is the most reactive with free linoleic acid, and is at least 36 times more stable than lipoxygenase-2 at 69°C.

TABLE 2.9. Oleic, Linoleic, and Linolenic Acids in A5, Its Parent FA9525, and Two Cultivars at Ames, Iowa, 1981 (14)

Genotype	Oleic	Linoleic	Linolenic
	----------	%	----------
A5	39.8	42.9	4.1
FA 9525	39.1	42.2	6.2
Corsoy	23.5	54.6	8.3
Weber	19.7	55.0	10.6

Hildebrand and Hymowitz (17) screened the U.S. Department of Agriculture soybean germplasm collection for genotypes lacking lipoxygenase-1. The spectrophotometric assay based on conjugated diene formation at 234 nM was used in the screening procedure. Two introductions, P.I. 133226 and P.I. 408251, lack lipoxygenase-1 activity. The lack of activity of lipoxygenase-1 is inherited as a simple recessive allele (18). Lipoxygenase-1 does not appear vital to the soybean since seed lacking the enzyme germinates normally and the plants grow, flower, and set seed like those with the enzyme. Recently Nielsen et al. (19) reported soybean lines lacking lipoxygenase-2 and lipoxygenase-3. The incorporation of the null-alleles for lipoxygenase into one genotype should be useful for studying the role, if any, of lipoxygenase in causing undesirable flavors in soybean products.

SUMMARY

In summary, recent advances in yield increase per unit area of wheat, rice and maize has raised hopes that similar results may be possible with grain legumes. Funds, which were lacking in the past for intensive research and production investigations on the grain legumes, are becoming increasingly available. Modification of components of legume seed through plant breeding is on the horizon. The availability of soybean lines with increased protein content, low linolenic acid, lacking the Kunitz trypsin inhibitor and lipoxygenase should be of particular interest to the scientific community for developing cultivars with improved nutritional quality, improved flavor and perhaps increased yield.

REFERENCES

1. FAO Production Yearbook. Volume 35. Food and Agricultural Organization of the United Nations, Rome, Italy, (1981).
2. Morrison, F. B. Feeds and Feeding. The Morrison Publishing Co., Ithaca, New York, (1948).
3. Soya Bluebook. American Soybean Association, St. Louis, Missouri, (1982).
4. Brim, C. A., Schutz, W. M., and Collins, F. I., Crop Sci. 7, 220-224 (1967).
5. Singh, B. B., and Hadley, H. H., Crop Sci. 8, 622-625 (1968).
6. Brim, C. A., and Burton, J. W., Crop Sci. 19, 494-498 (1979).
7. Burton, J. W., and Brim, C. A., Crop Sci. 23, 191 (1983).
8. Osborne, T. B., and Mendel, L. B., J. Biol. Chem. 32, 369-377 (1917).
9. Borchers, R., Anderson, C. W., Mussehl, F. E., and Moehl, A., Arch. Biochem. 19, 317-322 (1948).
10. Rackis, J. J., Sesame, H. A., Mann, R. K. Anderson, R. L., and Smith, A. K., Arch. Biochem. Biophys. 98, 471-478 (1962).
11. Hymowitz, T., Orf, J. H., Kaizuma, N., and Skorupska, H., Soybean Genet. Newsl. 5, 19-22 (1978).
12. Orf, J. H., and Hymowitz, T., Crop Sci. 19, 107-109 (1979).
13. Hammond, E. G. and Fehr, W. R., Merr. Fette Seifen Anstrichm. 77, 97-101 (1975).

14. Hammond, E. G. and Fehr, W. R., Crop Sci. 23, 192 (1983).
15. Christopher, J., Pistorius, E. K., and Axelrod, B., Biochem. Biophys. Acta 198, 12-19 (1970).
16. Christopher, J., Pistorius, E. K., and Axelrod, B., Biochem. Biophys. Acta 284, 54-62 (1972).
17. Hildebrand, D. F., and Hymowitz, T., J. Am. Oil Chem. Soc. 58, 583-586 (1981).
18. Hildebrand, D. F., and Hymowitz, T., Crop Sci. 22, 851-853 (1982).
19. Nielsen, N. C., Wilcox, R. J., and Davies, C. S., J. Am. Oil Chem. Soc. 60, 707 (1983).

3

Genetic Engineering: Prospects for Improving Crop Species

ROGER A. KLEESE

INTRODUCTION

In the relatively recent span of a half dozen years, much has been written about the emerging technologies of genetic engineering and what these will do for the improvement of crop agriculture. We have been promised plants that will not require fertilizer, resist a variety of stress conditions, have increased yields, have improved nutritional quality, have pest resistance and which can grow under a wide range of environmental conditions. Genetic engineering has been suggested as a solution for almost every kind of plant breeding problem.

The success story in manipulating genes in microorganisms grown under carefully controlled conditions has been taken as evidence that we will be able to do the same thing to crop plants. However, this position does not fully consider the complexity of the organisms and of the environments in which they grow. Most agronomically important traits are controlled by a number of genes which must function at the proper time and place in the plant's life cycle. We have almost no information about the number of genes, their location, their interaction, or the kinds of gene products which control these traits.

I would like to give my perception of the prospects for using some of these new technologies to improve crop species. I will attempt to describe two technologies--recombinant DNA and plant tissue culture --where we stand today in being able to use these, and then speak to some of the applications being made. I will use corn as an example for quite a bit of the presentation, since that is what I am familiar with and I believe it represents a good example of the kinds of problems and issues which we would face in working with any crop species. I would also remind you of my background and possible bias. I am a plant breeder who has had one foot in the lab and who is now attempting to bring together some of these laboratory technologies with conventional plant breeding to produce improved crop varieties.

There are two general areas of laboratory research which we normally consider in the realm of genetic engineering. These are recombinant DNA and plant tissue culture. As you will see, each of these has the potential to contribute directly to improving crop plants. In addition,

Cereals and Legumes in the Food Supply, edited by Jacqueline Dupont and Elizabeth M. Osman

there is a third area which may come about by the coupling of the two technologies.

RECOMBINANT DNA TECHNOLOGY

The recombinant DNA technology is actually a fusion of ideas from molecular biology, genetics, biochemistry and chemistry. It involves the restructuring and editing of genetic information. No doubt the factor which has contributed most to the development of the recombinant DNA technology has been the discovery of a class of enzymes known as restriction enzymes. These enzymes, found in microorganisms, cleave DNA at specific sites determined by the base sequence. There are a number of different enzymes, each with its own site specificity. This allows the molecular biologist to choose an enzyme that will cleave in the desired region and to leave the cut end of the DNA complementary to the cut end of a piece of DNA to be inserted. Using these enzymes together with a knowledge of gene structure, function, and regulation, it has been possible to remove genes from one organism, insert them into the DNA of another and to obtain expression of the transferred gene.

Basic Ingredients for Gene Transfer

If the objective is to transfer a gene from one organism to another, there are several ingredients which must be at hand. These are:

1. A DNA vector that can replicate in living cells,
2. A DNA fragment or gene to be inserted into the vector,
3. A method of joining the gene to the vector,
4. A method of introducing the joined molecule into the host,
5. A method of detecting those cells that carry the desired recombinant molecule.

These ingredients are fairly well in place in certain microorganisms, especially *E. coli*. *E. coli* contains small circular stretches of DNA known as plasmids which serve quite well as vectors. Today certain of these plasmids can be manipulated so they contain drug resistance markers used for their detection, as well as the necessary regulatory genetic information responsible for the expression of a foreign gene. A number of microbial genes, including both bacterial and viral genes, have been isolated and transferred into these plasmids and subsequently expressed in the bacterium. In addition, several human genes have been inserted into the bacterial plasmids and expressed in the bacterium. In most of these examples, and this is especially true of the human genes, it has been possible to identify and isolate the gene because of our knowledge of the gene product. For example, the cloning of human insulin and human growth hormone was accomplished because it was first possible to isolate insulin and human growth hormone and then to biochemically work our way back toward the gene.

I have already made reference above to the restriction enzymes used to join the gene in question to the vector. By using the appropriate enzymes, the cut ends of the vector and the ends of the gene to be inserted can be tailored and fitted together.

Moving the vector into *E. coli* is a fairly straightforward procedure. Although not well understood, it is possible to incubate bacterial cells in the presence of the vector and to cause the bacterial cells to take up vector molecules. Then, by the use of drug resistance markers on the vectors, one can plate out the population of bacteria exposed to the vector and screen for those colonies which possess resistance. This leads to the identification of the transformed cells which carry the vector and the gene being transferred.

I would caution you that the above is not quite as straightforward as I have laid it out. The molecular biologist may spend considerable time and effort in tailoring the gene so that it will be properly expressed in the host and this may involve manipulating regulatory elements which effect both transcription and translation of the gene. In certain applications of gene transfer, such as the production of human insulin, the objective is to produce as much gene product as possible. This will require rather novel approaches of "cutting and pasting" various stretches of DNA involving the regulatory elements for that gene. A certain amount of this is by trial and error, but it is becoming a more exact process.

Opportunities and Problems for Gene Transfer in Crop Plants

What are the prospects for single gene transfer in plants and why might we be interested? There are several single gene traits in crop species which have significant practical benefit. In corn, resistance to certain diseases is controlled by a single gene, as well as some aspects of chemical composition such as the amount and kind of starch, or protein quality in the endosperm. It's possible for the plant breeder to use a back crossing procedure to transfer these genes into lines of choice. However, to do this requires a number of generations and a matter of several years. In some cases, it is desirable to transfer a gene from a wild or related species. The success here will greatly depend upon the cross compatibility of the two species. The cross compatibility varies in degree and, as a result, may require a great deal of effort before a successful cross can be made. Today the genetic base for any crop species is limited to that species and those relatives which are cross compatible with it. It would seem a worthy goal to ask whether we might broaden the gene base by including species which with present techniques are cross incompatible.

What are the constraints to gene transfer in crop species using recombinant DNA technology? Today we do not have good methods of gene isolation. The plant genes which have been isolated biochemically produce a gene product with which we are familiar and which we can use to help isolate the gene. This approach has value in manipulating genes which control chemical composition since we can study appropriate biochemical pathways and identify genes coding for enzymes or study the gene product itself in the case of a storage protein. And, in fact, corn genes which have been isolated are those genes which control carbohydrate composition or protein composition in the endosperm. However, there is a great deal about biosynthetic pathways in plants which we do not understand and which has to be researched before we develop a systematic approach to altering chemical composition in crop plants.

In general, we do not have a good understanding of plant gene structure and function. However, there is some very good work in this area with zein, the corn storage protein (1, 2). This has been possible because of the ability to isolate zein, a gene product, and by the fact that relatively pure messenger RNA for zein could be isolated from the developing corn endosperm. Zein genes have been cloned and their DNA sequence analyzed. Also, several zein messenger RNAs have been used to obtain DNA clones. This work has provided information about the structure of certain zein genes indicating regions of variability as well as a conserved region which is repeated several times. It appears that zein is made up of several proteins coded by a multigene family.

The molecular biology of genes controlling the important agronomic traits such as standability, maximum stable yield and maturity, is untouched. Even for some of the single gene disease resistance we have no clues as to what the gene product is or how to isolate the gene. If a suitable gene transfer system were in place today, we simply would not have very many genes of practical value to transfer.

Perhaps the area of plant molecular biology receiving the most attention today is the design of suitable vectors. Much of this work is with plasmids of Agrobacterium tumefaciens. This bacterium is the pathogen which causes crown gall in susceptible species. The bacterium possesses a plasmid which naturally invades the host cells and a portion of which is stably integrated into the host DNA. Because of this feature, a number of researchers have been interested in the use of the plasmid, Ti, as a vehicle for genetic engineering plant cells. An example of its use is the successful transfer of the Phaseolus vulgaris storage protein, phaseolin, into sunflower cells (3).

Unfortunately, the natural host range of A. tumefaciens is limited to broadleaf species. Some effort is being made to broaden the host range so the grasses including cereals might be targets for the Ti plasmids. A somewhat different approach to developing a vector for corn may involve the use of controlling elements. These interesting stretches of DNA have the ability to move from gene to gene and to alter the genes which they reside near or within. Because of this transposable nature, it may be possible to use these elements to lead us to genes of importance or alternatively to carry genetic information with them to alter agronomically important genes.

One of the major questions in devising a scheme for transferring genes into plant cells is that of which cells to choose as recipient cells. If we attempt to mimic transformation in bacteria, it seems that we need protoplasts, that is plant cells which have had the cell walls removed, as the recipient cells. The fundamental problem with this approach is that plants of most crop species cannot be regenerated from protoplasts. Several horticultural crop species including potato, tomato and carrot can be regenerated from protoplasts. However, among the field crop species, neither soybeans nor any of the cereals except pearl millet can be regenerated from protoplasts. I think we will be able to solve this problem for those species such as corn where there is a considerable amount of effort. But we cannot assume that solving it for one species will directly lead us to the solution for others since it has been possible for a number of years to regenerate plants from protoplasts of several species, yet these regeneration procedures are not successful with most crop species.

A special problem which we face in the successful transfer of a gene into a crop species involves the regulation of that gene so that it functions as we would like over the growth and development of the plant. The objective for many of the practical applications of gene transfer into bacteria has been to maximize production, that is, to cause that bacterium to produce as much of the gene product from the foreign gene as is possible. In most of the examples that we might imagine in crop species, this would not be the objective. The gene would have to function in the proper tissues at the proper time and in concert with a host of other genes which together control a genetically and physiologically complex process. We really do not know how easy or difficult this will be until we begin to make these kinds of manipulations.

PLANT TISSUE CULTURE

Tissue culture is simply the growing of plant cells using microbial techniques. The cells are grown aseptically and may be grown either on solid media, generally agar, or in suspension cultures. One provides a complete medium for those cells which are not photosynthesizing and attempts to regulate differentiation and development by the addition of auxins, hormones or related compounds. Cultures growing in solid media are generally referred to as callus cultures and may be either organogenic or embryogenic. Organogenic cultures maintain a fair degree of organization, with clumps of cells seeming to develop directly into leaf-like structures or root-like structures. In contrast, embryogenic cultures seem to have reverted to an earlier stage of differentiation and possess clumps of cells which have an embryonic morphology. In general, the plant scientists would like to be able to manipulate the culture so as to maintain cells in a completely undifferentiated state and then to bring about differentiation at any time by controlling the media and/or the environment. Conceivably this would allow the maintenance of cultures over an extended period of time omitting the frequent reestablishment of cultures.

Suspension cultures are, as the name suggests, clumps of cells growing in liquid media, generally in some kind of a shaker flask so they are aerated. Protoplast cultures may come from whole plant, callus or suspension cultures in which the cell walls have been removed enzymatically.

One of the primary applications of tissue culture is for the selection of useful traits. Potentially one has the advantages of the microbial geneticist in that large populations of cells can be grown under rigidly controlled environmental conditions. This should enable one to introduce a selective agent into the media at the appropriate concentration and to select from among thousands or millions of cells those rare mutants which possess the desired trait. This can be done either in solid or liquid media. Because of the many cells in a clump of callus tissue, it is possible to miss mutant cells which might be surrounded by thousands of other cells in that callus. For this reason, a suspension culture is more attractive since smaller groups of cells or, ideally perhaps, even single cells may exist. In addition, a suspension culture provides a somewhat continuous culture system in which a single desired mutant gives rise to additional mutant cells, and in time the population in the flask begins to shift in the direction of the mutant types. Until recently all selection in corn has been with callus cultures since whole plants could not be regenerated from cells grown in suspension. Green et al. (4) have now reported on the regeneration of plants from suspension cultures, thus making it possible to do selection work in these cultures.

Perhaps the most ideal situation would be protoplasts existing as single cells, each with the capability of regenerating into a new plant. In this system every cell would be autonomous and represent a separate unique individual.

Difficulties of Growing Crop Plants in Culture

There are a number of problems in growing crop plant cells in tissue culture. There is a significant variation in culturability among species and among genotypes within species. For example, it is easy to grow tobacco in culture. One can initiate cultures from almost any part of the plant, grow them either on solid media or in suspension, and regenerate whole plants from protoplasts. With corn, we can produce both organogenic and embryogenic callus cultures of some lines, but not others. We have regenerated plants from a number of inbred lines, but there is a significant difference in our ability to regenerate plants of one line versus another.

It is extremely important to be able to manipulate a number of different genotypes in culture for the crop species being studied. For the greatest commercial application, it is necessary to work with the agronomically important lines which cover the market area of interest. This would be true for corn as well as any other crop species.

A major problem we face in working with cells in culture is not having a good understanding of plant development. If we understood the role of various chemicals and environmental conditions on plant development, then we might better control the growth of cells in culture and manipulate them developmentally in a way which we choose.

Another issue which needs to be resolved is the question of how much and what kind of genetic variability arises while cells are in culture. We know that cells grown in culture over an extended period of time develop gross chromosomal aberrations which lead to genetic instability. If the objective is to select for a trait in culture using cells of an elite inbred line, then one does need some genetic variation so that the desired mutant can be recovered. On the other hand, it is important that the essence of the elite inbred line stay intact. We have much to learn about controlling the genetic variability while cells are in culture.

Selecting in Tissue Culture for Improved Crop Varieties

Examples of the kinds of traits which might be selected are biochemical mutants, disease resistance and herbicide resistance. Selection for biochemical mutants is done, in principle, in the same way that microbial geneticists have selected for mutants. It requires some understanding of the biochemical pathways and the enzymes and their regulation which are involved in those pathways. Selection for these kinds of mutants seems especially well-suited to tissue culture as opposed to conventional plant breeding since these selective agents generally are materials which one would not be able to apply under field conditions. I shall defer further consideration of selection for biochemical mutants to the next section, "Using New Technologies to Improve Protein Quality."

In those diseases which produce toxin, it is possible to incorporate toxin into the tissue culture medium as a screen for cells resistant to the toxin. Gengenbach et al. (5) used the toxin from *Helminthosporium maydis* race T as a selective agent to screen for corn cells resistant to this toxin. They successfully isolated resistant cells from culture and regenerated plants which, as whole plants, possessed the disease resistant reaction. This resistance was transmitted maternally to the progeny.

The farmer's selection of the appropriate herbicide is based on a number of factors. Two of these are the weed spectrum to be controlled and the compatibility with the crop species. In some situations, the best herbicide for controlling the weeds is not compatible with the crop and thus forces the farmer to choose an herbicide or method of weed control which is less satisfactory. Some herbicides persist in the soil beyond the season in which they are applied. In the next growing season, the farmer will be restricted to growing a crop which is compatible with the herbicide he used the preceeding year. This has always been a problem for the farmer that wanted to use Atrazine to control weeds in corn, while at the same time wanting to rotate his crops from corn to soybeans, since soybeans are sensitive to Atrazine.

Several examples of selecting for herbicide resistance in tissue culture have been reported (6). As new, more potent herbicides are developed, we may see the need to develop additional crop tolerance. This could have the advantages of allowing use of smaller quantities of chemical while controlling a broader range of weed species.

I believe that selection for important traits in tissue culture represents the greatest opportunity today for applying some aspect of these new technologies to improving crop species. The above examples demonstrate the technical feasibility of this approach. It is important to point out that a number of agronomically important traits are not amenable to selection in tissue culture. Certain traits simply have to be selected under environmental conditions found in the field and expressed only in whole plants. There may be opportunity to work in the laboratory on some aspect of a more complex whole plant trait, but it is naive to assume that all agronomic traits can ultimately be disassembled into their component parts and subjected to selection and improvement under laboratory conditions.

USING NEW TECHNOLOGIES TO IMPROVE PROTEIN QUALITY

Because of the orientation of this symposium to problems of food and nutrition, I would like to consider how some of these new technologies might be used to attack a specific problem, that being protein quality in corn. Corn has, on the average, about 9% protein in the kernel. The protein is not of good quality for monogastric animals since it is deficient in lysine and tryptophan. Monogastric animals, such as swine and poultry as well as humans, must have their diet supplemented with other sources of protein which provide the necessary quality for a balanced diet. The major storage protein of corn is zein. It is this specific protein, or group of proteins, which is greatly deficient with respect to lysine and tryptophan. The other proteins of the kernel have a fairly good balance of the essential amino acids.

In 1964, a gene was discovered, opaque-2, which significantly improved the protein quality. Both lysine and tryptophan were elevated, with lysine approximately 50% above that of normal corn. This was brought about by a decrease in the amount of zein and a compensatory increase in the amounts of the other kernel proteins. A number of major seed corn companies, as well as several public institutions, initiated research programs to understand the mode of action of opaque-2 and to begin to develop commercial hybrids carrying this gene. However, there is a deleterious side effect from opaque-2. Kernels possessing the gene were very soft textured, prone to disease susceptibility and insect damage, dried slowly and had yields reduced by 10-15%. Breeding efforts to improve the kernel texture, generally, have not been successful. As a result, today there is very little research in the United States on the use of opaque-2 for improving protein quality in corn.

One of the new approaches which has been used to alter amino acid composition in plants is to select in tissue culture for amino acid over-producers. Efforts to do this are dependent upon an understanding of the biosynthetic pathway leading to the amino acids in question.

Both lysine and threonine are synthesized in the same pathway. It is known that lysine plus threonine will inhibit cell growth in corn tissue cultures. By culturing corn cells in the presence of lysine plus threonine, it seemed reasonable to expect that a mutant which could tolerate the high levels of these amino acids might be selected. In 1982, Hibberd and Green (7) reported their successful isolation of a mutant from corn which produced extremely high levels of free threonine. Plants were regenerated from these cultures, seed was produced and the mutant was found to breed true. The free threonine in kernels was 75-100 times that found in normal kernels. Although the more desirable mutant would have been a lysine overproducer, it should be noted that threonine is probably the third most limiting amino acid in corn. More importantly, this successful isolation of an amino acid overproducing corn clearly demonstrates the potential for selecting amino acid overproducers for either lysine or tryptophan.

As mentioned earlier, there is a significant amount of research currently underway aimed at understanding the molecular biology of the zein genes. One approach for using this information would be to make base substitutions in the DNA which would lead to lysine or tryptophan residues in the zein and to ultimately transfer this altered zein gene back into the corn plant. There are several problems which must be addressed. One of these is the number of zein genes, which may be as high as 100. To successfully improve lysine or tryptophan in the kernel, it would seem necessary, if possible, to choose a zein gene which is more functional than the average of all zein genes. Alternatively, it may be possible to understand zein gene regulation so that one could enhance the activity of the zein gene which is modified. Another problem which must be considered is the question of where in the zein molecule a lysine or tryptophan substitution might be made without disturbing the three-dimensional conformation of the zein molecule (8).

There is obviously still a great deal that we do not understand about the structure, function and regulation of zein genes. Nonetheless, there has been much progress in a relatively short period of time and it lays the groundwork for making precise changes in zein proteins.

NEW TECHNOLOGIES AND COMPLEX AGRONOMIC TRAITS

What is the possibility of using these new technologies to manipulate some of the most important agronomic traits which are complex genetically and biochemically? The example I will use is that of maturity in corn. Corn hybrids tend to be localized on the basis of maturity in the central and northern corn belt. This is an obvious requirement, especially in the northern corn belt where the growing season is limited and the farmer must select hybrids which will mature within the growing season. The comparison I want to make is between two corn hybrids, 3732 and 3780, which have the same maturity, that is they reach harvest moisture at the same time in the fall (9). Both 3732 and 3780 flower at approximately the same time so a superficial inspection of these hybrids would suggest they grow and develop at the same rate. A closer look indicates this is not the case.

Dry matter accumulation in the grain ceases well before harvest time. This point in the life cycle of the plant is called physiological maturity. The time from flowering to physiological maturity is the period for grain fill.

If we compare physiological maturity of the two hybrids there is a marked contrast betwen 3732 and 3780. Hybrid 3732 reaches physiological maturity much earlier than Hybrid 3780. One might expect that with its extended length of grain fill 3780 would probably yield more than 3732. However, in southern Minnesota both hybrids yield approximately the same. This suggests that the rate of grain fill in 3732 is much more rapid than that in 3780.

It is interesting to consider the possibility of combining the genetic systems which control grain fill in both of these hybrids such that one would have the rapid rate of grain fill from 3732 combined with the extended period of grain fill in 3780. Another interesting feature in comparing these hybrids is the rate of dry down or the loss of kernel moisture from physiological maturity until harvest. Since both hybrids reach harvest moisture at about the same date, it is apparent that 3780 must dry much more rapidly than 3732. This rate of dry down has important economic considerations for the farmer who must pay ever increasing fuel costs for artificially drying corn that is harvested by combine.

Much as we would like to precisely manipulate the genetic control for rate of grain fill, rate of dry down and other aspects of maturity, we do not have our hands on specific genes in corn which will control these particular traits. If we are to consider using some of the molecular biology approaches mentioned above, it is essential that we begin to locate genes which control the various aspects of maturity, find out how many genes there are and begin to understand their functions. To do this we will also have to begin to understand the biochemistry and physiology of the various aspects of maturity.

I believe we can answer some of these questions about complex agronomic traits. However, this will require a very close working relationship between some combination of plant breeders, geneticists, physiologists, biochemists, or molecular biologists. Each scientist should develop an understanding and appreciation of the research techniques and strategy used by scientists in the other fields. This is essential if they want to ask the right questions using the right genetic materials and techniques.

What has been said for maturity in corn can be said for corn in general. We need to understand just how the corn genome is put together and organized and how it functions to control growth and development. The use of restriction enzymes may help us to cut apart the genome to develop physical maps of the chromosomes which will allow us to begin to associate a certain region of a particular chromosome with a trait of interest. It may be possible to develop DNA probes specific for a particular trait which can be used as spot tests to analyze which plants in the segregating population have the gene or trait of interest. Plant breeding materials that contrast in traits of value will be especially useful in tying the molecular biology to these traits.

A unique approach for finding genes of interest may involve controlling elements. As mentioned earlier, these elements can move about the corn genome and alter the genes they reside near or within. By observing variation in the traits of interest in plants possessing controlling elements, it may be possible to identify genes which control these traits. Then, with DNA probes specific for the controlling elements one might fish out the gene of interest.

Instead of concentrating most of our effort on single genes, we may find more interest in working with blocks of genes. Corn has been under intense selection for the past 75 years and under some form of selection for well over 1000 years. This may have resulted in blocks of genes which stay together to function as a unit in controlling some of the complex traits of interest.

Special genetic stocks which have a known extra chromosome or which lack a chromosome, plus stocks with duplicated or deficient chromosome segments, may be useful in these molecular analyses. Isolation of specific whole chromosomes also may provide source material for the molecular analysis of corn chromosomes. Conceivably it could lead to the ability to transfer whole or partial chromosome segments.

The use of molecular biology to analyze the corn genome may lead us in very different directions from those found with microbes. We should be open to this possibility.

SUMMARY

In summary, I want to comment about the relationship between some of these new biotechnologies and plant breeding as we know it today. The plant breeder will continue to provide the main thrust in improving crop species by identifying important market needs, by developing a source of elite materials both for his/her own program and for laboratory manipulations, by selecting for those traits which are much better evaluated under field conditions and which may never be amenable to laboratory manipulations, and by providing the necessary evaluation of improved materials, which ultimately must be a field activity.

The new technologies could impact crop improvement by enabling us to select the traits which can not be manipulated under field conditions, by widening the germ plasm base for a crop to include not only the relatives of that species but perhaps also genes which are not even in the plant kingdom, by teaching us a great deal more about gene structure and function and genetic control of traits of interest and by providing techniques which assist the plant breeder in identifying those rare recombinants in the segregating populations under investigation.

Finally, to move the new technologies along most effectively, we need a greater understanding and communication between field and laboratory scientists, and we need scientists who can move between the two approaches. This is beginning to happen, but it needs to be fostered as much as possible.

REFERENCES

1. Marks, M. D., and Larkins, B. A., J. Biol. Chem. 257, 9976-9983 (1982).
2. Pedersen, K., Devereux, J., Wilson, D. R., Sheldon, E., and Larkins, B. A., Cell 29, 1015-1026 (1982).
3. Kemp, J. D., Merlo, D. J., Sutton, D. W., Barker, R. F., Slightom, J. L., and Hall, T. C., J. Cell. Biochem., Suppl. 7B, 245 (1983).
4. Green, C. E., Armstrong, C. L., and Anderson, P. C., in "Advances in Gene Technology: Molecular Genetics of Plants and Animals" Miami Winter Symposium Series Vol. 20. (A. Fazelahmad, K. Downey, J. Schultz, and R. W. Voellmy, eds.), pp. 147-157, Academic Press, New York, (1983).
5. Gengenbach, B. G., Green, C. E., and Donovan, C. M., Proc. Natl. Acad. Sci. U.S.A. 74, 5113-5117 (1977).
6. Meredith, C. P., and Carlson, P. S., in "Herbicide Resistance in Plants" (H. M. LeBaron and J. Gressel, eds.), pp. 275-291. John Wiley & Sons, New York (1982).
7. Hibberd, K. A. and Green, C. E., Proc. Natl. Acad. Sci. U.S.A. 79, 559-563 (1982).
8. Argos, P., Pedersen, K., Marks, M. D., and Larkins, B. A., J. Biol. Chem. 257, 9984-9990 (1982).
9. Kleese, R. A., and Duvick, D. N., in "Genetic Improvement of Crops: Emergent Techniques" (I. Rubenstein, B. Gengenbach, R. L. Phillips, and C. E. Green, eds.), pp. 24-43. Univ. of Minnesota Press, Minneapolis, MN, (1980)

II

Legume Products: Factors Affecting Physical, Nutritional, and Flavor Characteristics

4

Extraction, Processing, and Properties of Oils

DAVID R. ERICKSON

INTRODUCTION[1]

The subject is extraction, processing and properties of oils in legume products but the only legume currently of commercial and practical interest as a source of oil is the soybean. This means the discussion can be devoted to this product while recognizing that the principles being discussed will be equally applicable to other legumes that might become important in the future, such as the winged bean, and additionally applicable to all other oilseeds or oil-bearing products.

The use of soybean oil goes back many centuries, but its major growth to its current position as the world's number one edible oil is of fairly recent origin. In fact, the growth has been largely in the past 40 years and this is shown in Table 4.1.

The approximate world wide distribution of soybean production as of 1980-81 is shown in Table 4.2. The most dramatic growth has been in North and South America and the five leading countries as of 1981 are shown in Table 4.3.

The United States has led the world in soybean production and in soybean oil utilization. The growth of soy oil utilization in the

TABLE 4.1. World Production of Soybean Oil (1940-1980)

Year	Million Pounds	Metric Tons (X100)
1935-39 (1)*	2,060	934
1945-49 (1)*	3,060	1,388
1950-54 (1)*	4,110	1,864
1960 (2)	7,265	3,295
1970 (2)	13,142	5,960
1979-80 (3)	31,620	14,340

*Averaged/year for time frame.

1. Adapted from Erickson, D.R., JAOCS 60:303A, 1983, with permission of American Oil Chemists' Society, Champaign, Illinois.

Cereals and Legumes in the Food Supply, edited by Jacqueline Dupont and Elizabeth M. Osman © 1987 Iowa State University Press, Ames, Iowa 50010

TABLE 4.2. 1980-81 World Distribution Soybean Production (4) (in 1,000 metric tons)

North America	49,770
South America	19,822
Asia	9,899
Europe	656
Soviet Union	525
Africa	192
Oceania	70

TABLE 4.3. World Soybean Production--1980-81 Five Leading Countries (4)

	Metric Tons (X1000)	Million Bushels
U.S.	48,772	1792.4
Brazil	15,500	569.6
Peoples Republic of China	7,880	289.6
Argentina	3,500	128.6
Canada	73	26.2

U.S. is shown in Table 4.4, and in Table 4.5 is shown the dominance of soybean oil in four classes of U.S. edible oil products.

The United States has led the world in soybean production and in soybean oil utilization. The growth of soy oil utilization in the U.S. is shown in Table 4.4, and in Table 4.5 is shown the dominance of soybean oil in four classes of U.S. edible oil products.

TABLE 4.4. Growth of SBO Utilization in U.S. (SBO as % Total Domestic Disappearance Edible Fats/Oils)

Year	Percentages
1950 (1)	27%
1960 (1)	36.7%
1970 (5)	56.5%
1980 (6)	66.5%

TABLE 4.5. Dominance of SBO in Four Classes of U.S. Edible Oil Products 1980/81 (7)

Shortening	63.3%
Margarine	82.3%
Salad/Cooking Oils	80.8%
Prepared Dressings	95+%

This growth has not been without problems. Soybean oil has probably been the subject of more research than any other oil. Recently the American Soybean Association, in conjunction with the USDA, published a soy oil bibliography covering mainly the period of 1966 through 1979 with some selected and limited references going back to 1936 (8). This provided nearly 2,400 entries for the bibliography. In other words, this means that for this period there was, on the average, a publication approximately every other day about soy oil.

The soybean is different from most other oilseeds in that the oil content is about 35-40% of the value and the remainder is in the value of the high protein meal. The approximate composition of the soybean is protein 40%, lipid 20%, cellulose and hemicellulose 17%, sugar 7%, crude fiber 5% and ash 6%.

Soybean oil also has a somewhat unique composition in comparison with most other common vegetable oils except for the oil from the newer varieties of rapeseed. This uniqueness is in its higher content of linolenic acid. A comparison of the fatty acid average composition and range of compositions of soybean oil are shown in Table 4.6. The average composition of soybean oil, as compared to other vegetable oils, is shown in Table 4.7.

EXTRACTION OF SOYBEAN OIL

The vast majority of soybean oil at present is produced by solvent extraction. A schematic diagram of the unit processes used in modern solvent extraction plants are shown in Figure 4.1 (11). The basic technical principles in this processing are not new and have remained rather static as to changes going back to work started in Europe in the early 1900s. The changes that have come about within the past

TABLE 4.6. Fatty Acid Composition of Soybean Oil (9)

Component acid	Fatty acid composition, wt. %	
	Range	Average
Saturated		
Lauric	---	0.1
Myristic	<0.5	0.2
Palmitic	7-12	10.7
Stearic	2-5.5	3.9
Arachidic	<1.0	0.2
Behenic	<0.5	---
Total	10-19	15.0
Unsaturated		
Palmitoleic	<0.5	0.3
Oleic	20-50	22.8
Linoleic	35-60	50.8
Linolenic	2-13	6.8
Eicosenoic	<1.0	---
Total	---	80.7

TABLE 4.7. Fatty Acid Composition of the Five Leading Vegetable Oils (10)

			Sunflower		Rapeseed		
	Soybean	Palm	North	South	hi. Er.	lo. Er.	Cottonseed
C12	--	0.1	--	--	--	--	--
C14	0.1	1.0	--	--	--	--	0.8
C16	10.3	43.5	5.9	5.4	2.6	4.8	22.7
C18	3.8	4.3	4.5	3.5	0.9	1.6	2.3
C16:1	0.2	0.3	--	0.2	0.3	0.5	0.8
C18:1	22.8	36.6	19.5	45.3	11.2	53.8	17.0
C20:1	0.2	0.1	--	--	7.5	1.0	--
C22:1	--	--	--	--	48.1	0.1	--
C18:2	51.0	9.1	65.7	39.8	12.8	22.1	51.5
C18:3	6.8	0.2	--	0.2	8.5	11.1	0.2

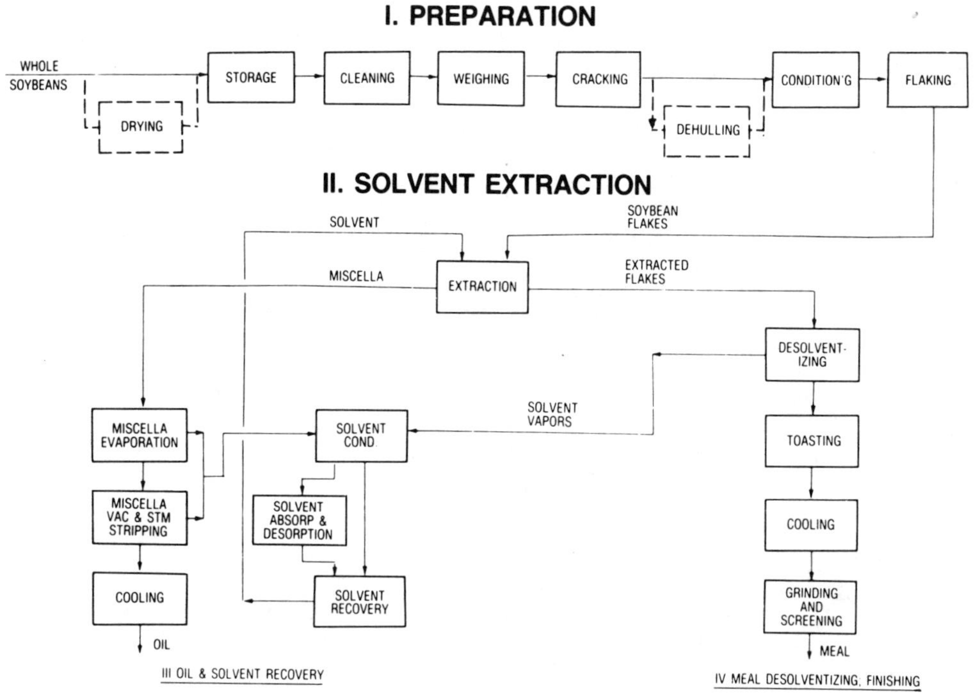

I. PREPARATION

WHOLE SOYBEANS
STORAGE
CLEANING
WEIGHING
CRACKING
CONDITION'G
FLAKING
DRYING
DEHULLING

II. SOLVENT EXTRACTION

SOLVENT
SOYBEAN FLAKES
MISCELLA
EXTRACTION
EXTRACTED FLAKES
DESOLVENT-IZING
SOLVENT VAPORS
MISCELLA EVAPORATION
SOLVENT COND.
TOASTING
MISCELLA VAC & STM STRIPPING
SOLVENT ABSORP & DESORPTION
COOLING
COOLING
SOLVENT RECOVERY
GRINDING AND SCREENING
OIL
MEAL
III OIL & SOLVENT RECOVERY
IV MEAL DESOLVENTIZING, FINISHING

FIG. 4.1. Typical soybean solvent extraction process.

two decades have largely been increases in size of extractors and extraction plants and with relative reduction in numbers of plants. This is particularly true in the U.S. and this change is seen in Figure 4.2.

Other important changes relate to the growing worldwide concern about energy costs and more attention being given to control of solvent losses for the same reasons.

The newer technical considerations since 1976 in soybean oil extraction are research and development directed toward basically two things, which are the effect of the extraction process on the refinability of soybean oil and alternate processes for extraction.

In the area of the effects of the extraction process on refinability, this has been largely driven by the current interest in physical or steam refining of soybean oil, which will be discussed later. Briefly the proposed changes are to destroy the possibility of enzymatic action prior to and during extraction. This leads to more efficient degumming of the oil in preparation for steam refining. The theoretical basis and research results of this have been published (12, 13).

The current published processes that could be considered for such enzyme inactivation are:

1. "Alcon process"--Cooking of flakes prior to extraction (13).
2. Continuous steam pressure cooking of whole soybeans prior to extraction (14).
3. Microwave vacuum heating[2].

This is not intended as a complete listing of possible alternatives as there are probably other conceivable routes to accomplish enzyme destruction.

The alternative extraction processes now under study and worthy of mention are:

1. Aqueous extraction (15, 16).
2. Alternative common solvents (17).
3. Super critical carbon dioxide extraction (18).
4. Flake expander prior to extraction[3].

The use of an aqueous extraction process for soybeans is attractive because it eliminates the problem of solvent safety and thus could require much less capital investment. It also might be more suitable for small plants. The major drawbacks currently are inefficient extraction, deemulsification problems, cleanliness requirements (dairy type process), and higher operating costs for water removal. Appropriate combinations of new and innovative techniques quite possibly will make this a viable process in the future.

The use of alcohols or aqueous alcohols for extraction has been demonstrated. The advantages are said to be safer operation due to higher flash points and miscibility of the solvents with water, lower gum content in the extracted oil, and reduction of enzyme activity and protein solubility during extraction requiring less toasting of

---

2. McKinney H. F., McDonnell Aircraft Co., St. Louis, MO. Private communication.

3. Hendrick, W. B., Hendrick Consultants, Ft. Worth, TX. Private Communication.

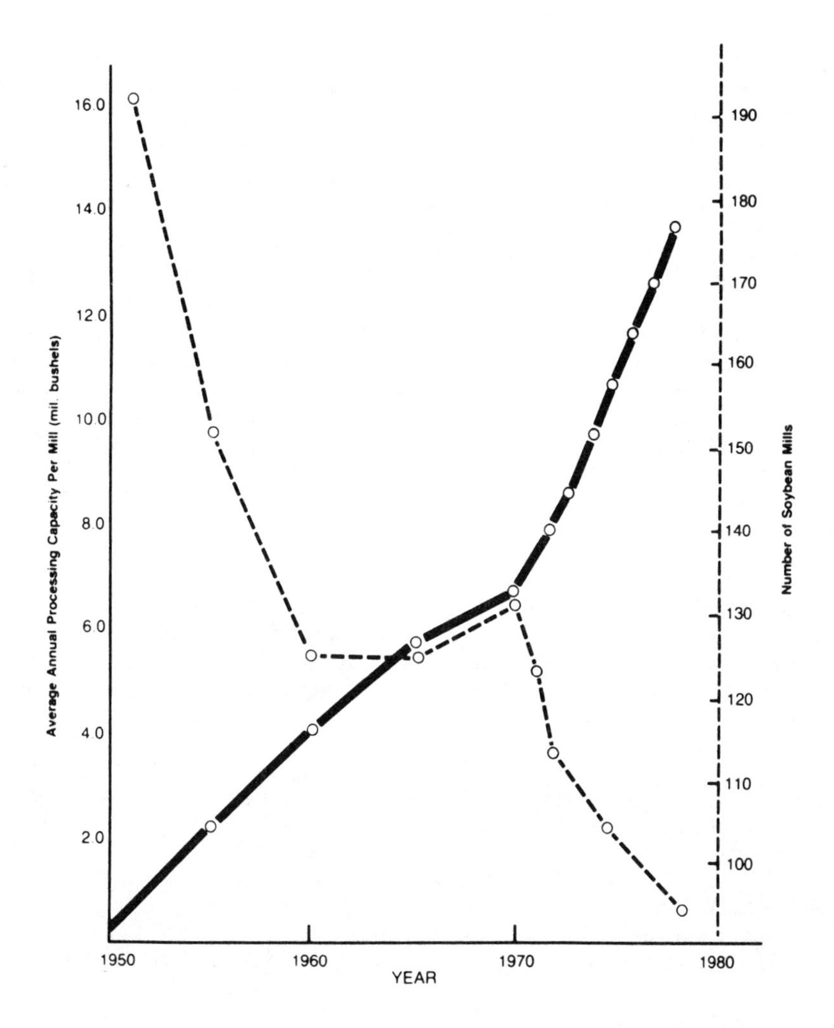

FIG. 4.2. Average annual capacity and number of U.S. soybean mills.

the meal. In the latter case the advantage is for production of meal for animal feed but a disadvantage in production of meals with a high protein solubility for use in soy protein production.

It is claimed these solvents can be used in retrofitting existing hexane plants with some modification, however, there may be a problem with increased corrosion and other unrecognized problems.

The basic principle of super critical carbon dioxide extraction has been demonstrated. It remains to be seen whether one can scale up from current, relatively small batch processes to the continuous multi-thousand ton per day requirements of the modern soybean industry.

The flake expander is a treatment of soybean flakes, which can be thicker than normal, through an extrusion process making a "full fat pellet" which is said to be quite porous. The "pellets" are said to be easier to extract than normal flakes allowing an increase of 30% in extractor throughput, a richer miscella, and better drainage, thus reducing the load on the desolventizer/toaster. The latter two advantages should reduce energy costs. The effect of the expander on enzyme destruction or protein solubility has not been published.

To summarize, there is really nothing new at this time in soybean extraction except for some manipulations within the current basic solvent process.

PROCESSING OF SOYBEAN OIL

Like soybean oil extraction, the processing of soybean oil is done largely with processes that are not particularly new. They are only improvements in application of well known and long practiced basic techniques.

The unit processes used in soybean oil processing are shown in outline form in Figure 4.3. The principle steps in the refining of soybean oil are degumming, neutralization, bleaching, hydrogenation and deodorization (steam refining).

REFINING OF SOYBEAN OIL

A comparison of the average composition of crude soybean oil and fully refined soybean oil are shown in Table 4.8 and depict what needs

TABLE 4.8. Average Compositions for Crude and Refined Soybean Oil (19)

	Crude Oil	Refined Oil
Triglycerides, %	95-97	>99
Phosphatides, %	1.5-2.5	0.003-0.045
Unsaponifiable matter, %	1.6	0.3
Plant sterols, %	0.33	0.13
Tocopherols, %	0.15-0.21	0.11-0.18
Hydrocarbons (Sequalene), %	0.014	0.01
Free fatty acids, %	0.3-0.7	<0.05
Trace metals		
Iron, ppm	1-3	0.1-0.3
Copper, ppm	0.03-0.05	0.02-0.06

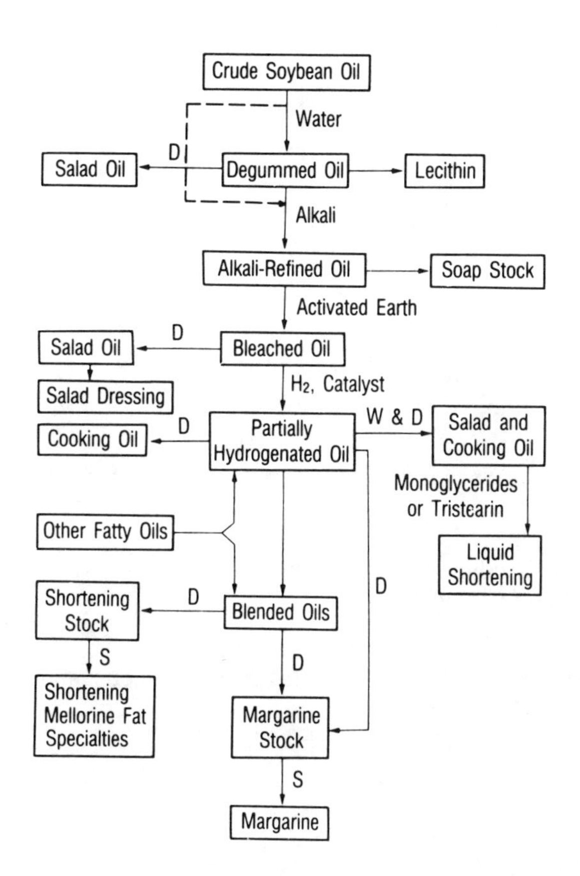

FIG. 4.3. Schematic diagram for manufacture of edible soybean oil products. D = deodorization, W = winterization, S = solidification, and H_2 = hydrogen gas.

to be removed in the refining process (19). Generally speaking, soybean oil is a "low-loss" oil in terms of refining and presents no particular problem in refining, but does require judicious use of adjusted treatment practices to maximize final refined soybean oil quality.

The initial steps of the degumming, neutralization and bleaching have been covered in an excellent paper by Wiedermann (20). Of interest at this point is Table 4.9 taken from the Wiedermann paper which shows the effects of soybean quality and the extraction process on food oil quality, which was alluded to in the section on soybean oil extraction.

Nominal working conditions for degumming of soybean oil are shown in Table 4.10, for caustic refining in Table 4.11, and waterwashing in Table 4.12 (29).

TABLE 4.9. Crude Oil Quality, Affect of Abuse Characteristics (20)

Abuse Characteristics	Increase In*
Weed Seed	4, 6
Immature Beans	6
Field Damaged Beans	1, 2, 3, 5
Splits (loading/transport/unloading)	1, 2, 3
Bean Storage (time/temp./humidity)	1, 2, 3
Conditioning Beans for Extraction	1, 2, 4, 5
Solvent Stripping Oil (overheating)	2, 4
Oil from Stripper (overheating)	2
Crude Oil Storage (time/temp.)	3, 4

*1 Total Gums/Phospatides
2 Non-Hydratable Phospatides
3 Free Fatty Acids
4 Oxidation Products
5 Iron/Metal Content
6 Pigments

TABLE 4.10. Working Conditions for Batch-Continuous and Continuous Degumming (20)

Batch-Continuous
- 2% water in mix tank
- 30 min-1 hr @ 60°-70°C (140°-160°F)
- Centrifuge

Continuous
- Preheat oil 60°-70°C (140°-160°F)
- Add 2% water
- 10-15 min mix-residence time
- Centrifuge

Acid Pretreatment
- 0.05-0.2%, 75% phosphoric acid
- Mix 4 hrs in Day tank @ 33°C (90°F)
- 1 min @ 70°-90°C (160°-195°F)/Super Mix

TABLE 4.11. Working Conditions for Continuous, Alkali, Wet Refining (20)

Proportion	16-18 Baume Caustic plus 0.12-0.15 for Crude Oil 0.10-0.12 for Degummed Oil
To Cold, 33°C (90°F) Oil	
5-10 Min Contact Time	
Heat to 75°C (165°F)	
Centrifuge	

TABLE 4.12. Water Washing Conditions (20)

Single
15% Hot (93°C/200°F), Soft Water
Mix
Centrifuge
Double
Twice with 10% Hot/Softened Water

BLEACHING OF SOYBEAN OIL

Proper bleaching is a critical step in soybean oil processing. Bleaching of vegetable oils is normally done for color reduction and such reduction is the usual method for determining dosage of bleaching earth. The final color of fully refined soybean oil normally presents no problem since adequate reduction is accomplished due to the combined effects of bleaching and deodorizing and hydrogenation when applied.

The modern U.S. practice of bleaching soybean oil is adjustment of bleaching earth dosage so as to achieve a zero peroxide value in the bleached oil using an acid-activated earth. Table 4.13 shows a comparison of such activated earth and their effects on peroxide value in a laboratory situation and Table 4.14 shows a comparison of a neutral

TABLE 4.13. Comparison of Some Activated Bleaching Clays Available in the World Market (20)

	Total Acidity			Bleach[b]	
	%	KOH[a]	pH	PV	Color
Filtrol 105	0.42	4.8	3.0	0.0	1.3R
Vega Plus	0.39	4.5	3.0	0.2	1.3R
Optimum X-FF	0.32	3.7	3.5	1.2	1.4R
Filtrol 54	0.15	1.7	3.5	1.2	1.4R
Optimum FF	0.04	0.4	3.0	1.0	2.0R
Refined Oil	-	-	-	4.0	5.2R

[a]mgKOH/g.
[b]Laboratory Bleach, 1% Earth; Lovibond Index.

TABLE 4.14. Plant Bleaching Test; Comparison of Two Activated Bleaching Earths[a] (20)

	Batch #	pV	Color
Refined Oil to Bleacher		2.2	7.9 R
Filtrol 54	1	1.7	3.5 R
(0.15% Acidity)	2	1.3	3.2 R
(3.5 pH)	4	1.5	3.3 R
	6	0.5	3.1 R
Filtrol 105	1	0.4	2.0 R
(0.42% Acidity)	2	0.0	2.0 R
(3.0 pH)	4	0.0	1.8 R

[a]0.5% earth added @ 180°F; temperature raised to 220°F; hold 20 minutes and pressed out; atmospheric conditions; 6-batch capacity press.

[b]Lovibond Red Index.

and acid activated earth on a plant scale (20). Exhaustive removal of bleaching clay from the oil is very important since any residual earth could act as a very active pro-oxidant.

The theoretical basis of this bleaching practice is to eliminate all oxygen-containing compounds, as shown conceptually in Figure 4.4 (20).

A key consideration after this type of bleaching is prevention of any new build-up of peroxides through either immediate deodorization or by protecting the bleached oil against further thermal or oxidative abuse.

HYDROGENATION

Soybean oil is an easy oil to hydrogenate when properly refined and bleached as previously described. The two principal catalyst poisons found in improperly refined and bleached soybean oil are soaps and phosphatides. The theory and practice of hydrogenation are well known and described (21) and the process for hydrogenation of soybean oil has been thoroughly covered by Hastert (22). Hydrogenation conditions for soybean oil to produce a series of base stocks for subsequent production of a wide variety of products has been well covered by Latondress (23).

DEODORIZATION

Deodorization is again not a new process, but effective deodorization is critical to obtaining the best quality of soybean oil. The optimum conditions for deodorization of soybean oil have been well described by Gavin (24).

Normal commercial deodorization conditions are shown in Table 4.15 (25). Most specifications for deodorized soybean oil show a 0.05% FFA maximum. Better quality is achieved at maximum levels of 0.03% and such soybean oil normally has a maximum color value of about 10

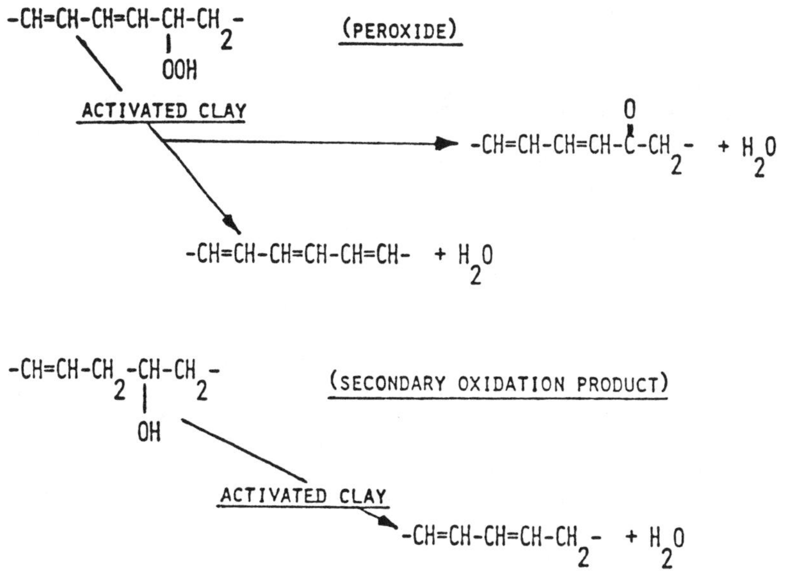

$-CH{=}CH{-}CH{=}CH{-}CH(OOH){-}CH_2-$ (PEROXIDE)

ACTIVATED CLAY

$\rightarrow -CH{=}CH{-}CH{=}CH{-}C({=}O){-}CH_2- + H_2O$

$\rightarrow -CH{=}CH{-}CH{=}CH{-}CH{=}CH- + H_2O$

$-CH{=}CH{-}CH_2{-}CH(OH){-}CH_2-$ (SECONDARY OXIDATION PRODUCT)

ACTIVATED CLAY

$\rightarrow -CH{=}CH{-}CH{=}CH{-}CH_2- + H_2O$

FIG. 4.4. Decomposition and dehydration or pseudo neutralization of peroxides and secondary oxidation products. (20)

TABLE 4.15. Commercial Deodorization Conditions (27)

Absolute pressure	1-6 mm Hg
Deodorization temperature	210°-274°C
Holding time at elevated temperataure:	
Batch type	3-8 hr
Continuous and semicontinuous types	15-120 min
Stripping steam: wt % of oil	
Batch type	5-15%
Continuous and semicontinuous types	1-5%
Product free fatty acid:	
Feed, including steam refining	0.05-6%
Deodorized oil	0.02-0.05%

yellow and less than 1 red. It is a normal practice that citric acid be added to the cooling section of the deodorizer at a level of not less than 50 PPM.

PHYSICAL REFINING

There is currently a great interest in physical or steam refining of soybean oil mostly to avoid potential pollution problems inherent in the more conventional wet alkali refining and also for other reasons. The key to production of high quality soybean oil by physical refining is in exhaustive degumming before introduction of the oil into the deodorizer.

This topic was covered very well in a conference (26) held in 1981. From this conference there seemed to be different opinions about whether the most desirable refining process for soybean oil was caustic or physical. At this point the best advice is to carefully analyze both processes in terms of initial investment, quality of available crude soybean oil, byproduct disposal, flexibility, energy efficiency, local pollution control situation, etc.

Only by laying out each alternative on a side-by-side comparison can a reasonable decision be made for any individual plant as to which is the better process.

UTILIZATION OF SOYBEAN OIL

Properly refined bleached and deodorized (RBD) soybean oil is an excellent salad oil requiring no winterization or dewaxing. As an RBD oil it can be, and is, used in salad dressings and mayonnaise, and as a liquid oil in margarine or shortening formulations.

Utilization of hydrogenated soybean oil in shortenings and margarines has steadily increased in the U.S., despite the fact that hydrogenated soybean oil crystallizes in the beta form. It has been found that this beta tendency can be somewhat overcome by the use in formulation of two and three component base stocks made of hydrogenated soybean oils. Such formulation information and guidelines for the hydrogenation of different base stocks have been published by Latondress (23).

SOYBEAN OIL FLAVOR

No discussion about soybean oil would be complete without discussing flavor. As pointed out earlier in this discussion, soybean oil is nearly unique in its content of linolenic acid. The only other common edible oils with a comparable or higher linolenic content are the newer varieties of rapeseed oil.

It is interesting to review the history of the development of soybean oil as an edible product. This has been done previously by Dutton (27), but a somewhat different perspective is deemed also to be of interest. As with Dutton, I would like to confine the discussion to U.S. history, however, this could be applicable in context to experience in other countries.

Soybean oil was first being used extensively in the U.S. in the late 1940s and early 1950s. It was introduced essentially into a market based on cottonseed oil. Early attempts to process soybean oil were with techniques then being applied to cottonseed oil. This gave a soybean oil of poor quality, which was not acceptable to the U.S. consumer used to cottonseed oil. There seemed no problem with hydrogenated soybean oil in products at that time.

There was originally thought to be something strange about fully refined soybean oil because it apparently developed an off flavor that was not recognized as oxidative in nature. This flavor was said to be like the flavor or odor of the original crude soybean oil and was thus called "reversion" flavor because the flavor of the refined reverted back to that of the crude oil. This nomenclature persists to the present time, despite later evidence showing that the flavor was indeed due to oxidation.

This evidence was found through use of better and improved analytical techniques. With this knowledge in hand, ordinary techniques for preventing oxidation could then be applied. One of the earlier techniques applied was the use of citric acid, which overcame the effects of copper and iron as oxidation catalysts.

Earlier research on the flavor of soybean oil led to the theory that the linolenate component might be responsible. This was deemed by some to be proved by the USDA experiment (28), where an interesterification of linolenic acid into cottonseed oil at a level equivalent to that of soybean, gave an oil which was identified by taste panels as a soybean oil.

Armed with this information, a liquid soybean oil with a reduced level of linolenate (3-4%) was re-introduced to the U.S. market in the early 1960s in the form of a lightly hydrogenated and winterized (LHW) soybean oil, which quickly became accepted. In addition, it was used as the liquid oil component of salad dressings, mayonnaises, margarines and shortenings.

Starting in the 1970s, improvements achieved in processing of soybean oil produced a refined, bleached and deodorized (RBD) oil that was acceptable for uses other than retail salad/cooking oils. As of this date even this latter use is being served in some instances in the U.S. by RBD oil. I hasten to add, however, that while pan frying is quite prevalent in the U.S. home, there is very little deep fat frying involving multiple repeated use.

Another issue needing discussion is the so-called "room-odor" problem with LHW soybean oil. This may have been a problem several years ago and may even then have been of interest or concern only to expert tasters. In 1978 the American Soybean Association compared LHW soybean oil, sunflower, and corn oil in an in-home use test where the oils were presented as unidentified oils to a national sampling representing the U.S. consumer. All of the oils were deemed highly acceptable and the only statistically significant difference as the preference for soy over corn, because soy was said to be "lighter." There were no differences among the oils as to comments about off odors upon heating. In other words, while some experts may consider LHW soybean oil to have a "room-odor" problem, it was not detected by the U.S. housewife (29).

The "linolenate theory" has not been universally accepted as being the principal cause of off flavor in soybean oil but it was certainly instrumental in the introduction of LHW soybean oil and the consequent improvement in the acceptance of soybean oil.

Other theories, such as residual phosphatides and lipoxygenase activity have also been put forth and may indeed be important in the development of off flavors. Certainly one would expect modern processing techniques, such as the bleaching mentioned before, would eliminate phosphatides and also the oxygenated products which could result from lipoxygenase activity.

In my opinion, the application of the refining practices mentioned above have gone a long way in overcoming development of objectionable off flavors in RBD soybean oil. This viewpoint is certainly reinforced by actual utilization data, at least in the U.S.

Even under the best conditions of optimal processing, an expert experienced in oil tasting could probably identify each individual refined common vegetable oil as presented. This may be of consequence, however, only to the expert tasters since the consuming public has no such expertise and may only object if the oil is less than reasonably bland. The fact that soybean oil may have a different or possibly objectionable low-level flavor to an expert taster thus may have no meaning unless the consumers also object. This latter can only be found out by actual consumer testing using unbiased methodology.

Pragmatically speaking, there now exist processing techniques to largely overcome the problem of "reversion flavor." I must also add that this does not overcome the inherent potential development of oxidation off flavors due to the high level of polyunsaturation in RBD soybean oil or the exclusion of any concern about "reversion flavor" where less than optimum refining methods are used.

All the highly polyunsaturated oils, such as safflower, sunflower, corn oil, rapeseed oil and soybean oil would not be recommended for heavy duty deep fat frying or for other applications demanding high stability toward oxidation. One should always consider hydrogenated counterparts of these oils for such use.

The American Soybean Association is currently sponsoring research by soybean plant breeders directed toward reduction of the linolenic acid content in the soybean itself. Progress has been slow because it has been nearly impossible to find a natural variety with a low linolenic content which greatly slows selective breeding. Varieties with linolenate contents of about 3-4% have been developed and grown for further testing.

We have discussed linolenic acid in a negative sense until now, but there is growing evidence that it probably has a unique essentiality in human nutrition and may be a required constituent in the human diet (30). Evidence is also growing that the class of Omega-3 polyunsaturated fatty acids, which includes linolenic acid, may be of nutritional importance and desirable in the diet (31). These latter two things may be interrelated. They certainly bear watching in the future.

REFERENCES

1. USDA.ERS. Statistical Bulletin 376. Washington, D.C. p. 176. (1966).
2. USDA.ERS. Statistical Bulletin 489. Washington, D.C. p. 168. (1972).
3. American Soybean Association. Soya Blue Book, St. Louis, MO., p. 141. (1981).
4. American Soybean Association. Soya Blue Book, St. Louis, MO., p. 168. (1982).
5. American Soybean Association. Soya Blue Book, St. Louis, MO., p. 156. (1980).
6. American Soybean Association. Soya Blue Book, St. Louis, MO., p. 166. (1981).
7. U.S. Bureau of Census. Current Industrial Reports. (M20-K 81-13) Table 3B. (1982).
8. Cowan, J. C., "Soy Oil Bibliobraphy," American Soybean Association, St. Louis, MO. (1981).
9. "Handbook of Soy Oil Processing and Utilization", (D. R. Erickson et. al., eds), p. 17. AOCS Monograph 8, AOCS, Champaign, (1980).

10. USDA. SEA. Composition of Foods. Fats and Oils. Raw/Processed/Prepared. Ag. Handbook 8-4. Washington, D.C. (1979).
11. Becker, K. W., JAOCS 55, 755 (1978).
12. Ong, J.T.L., in "Proceedings of the Second ASA Symposium on Soybean Processing," The American Soybean Association. Brussels, Belgium, (1981).
13. Kock, M., Ibid. (1981).
14. Swift & Company. British Patent 1385303. January 19, 1973.
15. Lawhon, J. F. et al., J. Food Sci. 46, 391 (1981).
16. Lusas, E. W. et al., Oil Mill Gazetteer, p. 28 (1982).
17. Sullivan, D. A., Oil Mill Gazetteer, p. 24 (1982).
18. Friedrich, J. P. and G. R. List., J. Agric. Food Chem. 30, 192 (1982).
19. "Handbook of Soy Oil Processing and Utilization," (D. R. Erickson et al., eds), AOCS Monograph 8, AOCS, Champaign, p. 14 (1980).
20. Wiedermann, L. H., JAOCS 58, 159 (1981).
21. Allen, R. R., in "Bailey's Industrial Oil and Fat Products," 4th Ed. Vol. 2, p. 1. Wiley-Interscience, New York (1982).
22. Hastert, R. C., JAOCS 58, 167 (1981).
23. Latondress, E. G., JAOCS 58, 185 (1981).
24. Gavin, A. M., JAOCS 58, 175 (1981).
25. Zehnder, C. T., JAOCS 53, 364 (1976).
26. "Proceedings 2nd ASA Symposium on Soybean Processing." American Soybean Association. Brussels, Belgium (1981).
27. Dutton, H. J., JAOCS 58, 234 (1981).
28. Dutton, H. J. et. al., JAOCS 25, 384 (1948).
29. Erickson, D. R. and Falb, R. A., in "World Soybean Research Conference II: Proceedings." (F. T. Corbin, ed.), p 851. Westview Press, Boulder, (1980).
30. Holman, R. T., Amer. Jnl. Clin. Nutr. 35, 617 (1982).
31. Brongeest-Schoute, H. C. et. al., Amer. Jnl. Clin. Nutr. 34, 1752 (1981).

5

Use of Oils in Food Products

CLYDE E. STAUFFER

THE NUTRITIONAL ROLE OF OILS

Why are fats and oils part of our diet? Indeed, they are almost too large a part, if the recommendation that we decrease caloric intake derived from fat from 45% to 30% reflects our present nutritional situation. And yet, no one questions the desirability and necessity for having fat in some form in our daily food.

In the first place, fats and related lipids are essential structural components of our bodies. Thus cell membranes and nerve sheaths are high in lipid content, and depend on the presence of lipids with fairly narrow melting point and structural requirements for the proper functioning of the membrane. A degree of membrane fluidity is needed which is obtained by having a certain amount of unsaturation in the fatty acid chains. Further, it appears that the curvature of the cell membrane at the cell surface depends in part on having more polyunsaturated fatty acids present on the outside than on the inside of the membrane. Due to steric considerations, then, the membrane is enabled to have the small radius of curvature necessary for its dimensions with a favorable overall free energy.

The polyunsaturated fatty acids are also necessary to the lipid metabolism and functionality in our bodies. Human biosynthetic systems cannot make linoleic or linolenic (C18:2 and C18:3, respectively) acids *de novo*. These are important acids in and of themselves, and also they are the starting points for generating precursor fatty acids for prostaglandins. When fed a diet totally lacking in these two "essential fatty acids" (EFA) rats develop severe disease symptoms. However, because of the ubiquity of one or both of these EFAs in dietary food, this syndrome is almost never seen in everyday life.

The third physiological function of fats and oils is as a high energy calorie source. One gram of dietary lipid yields nine Kcal of energy, vs. four Kcal for carbohydrate or protein. While for the average over-nourished American this is a disadvantage, our nutritional status today might be termed a historical, geographical and evolutionary anomaly. For most of the people in our world today, and throughout most of human history, the high-energy food source represented by fats and oils when available has been nutritionally desirable.

Cereals and Legumes in the Food Supply, edited by Jacqueline Dupont and Elizabeth M. Osman © 1987 Iowa State University Press, Ames, Iowa 50010

Energy storage by the organism is more efficient using the energy dense, unhydrated lipid versus utilizing carbohydrate. Storage carbohydrate, e.g. glycogen, binds a considerable amount of water, so that our bodies carry about three pounds of weight in order to have one pound of stored energy source as carbohydrate. Thus on an equal body weight basis lipid is about seven times as efficient an energy storage medium as is carbohydrate. While storage fat can be (and is) synthesized from food material such as carbohydrate the overall process is energetically inefficient.

This consideration is germane when developing foods for calorie deprived parts of the population. Geriatric diets, for instance, often include a high proportion of fat. Sometimes older people don't eat enough calories for good health maintenance. Appetizing, calorie-dense food items are often part of sound overall nutritional planning for this subgroup of our population.

This is a quick and by no means exhaustive overview of the importance of fats and oils in our diet. The consequences of ingesting these materials are generally desirable, but they are not motivational. No one consciously thinks about cell membrane fluidity when deciding whether or not to eat a certain food.

We intentionally include fats and oils in our food because of improved palatability. Fats do things for the flavor, texture, mouthfeel and structure of foods which we have learned to prize in the things we eat.

FAT STRUCTURE AND FUNCTION

The mechanism by which fats and oils do these things may depend on a number of factors. Trace constituents found in oils may contribute to flavor. The physical state (solid or liquid) has much to do with the functionality of the lipid, while the crystal habit of the solid components present affects the way in which the fat contributes to textural aspects of food. Molecular structure also influences the impact which the fat has on our palate. Molecular interaction of the fat with other components of food (i.e., carbohydrate and protein) plays a large role in the contribution of fat to our enjoyment of food.

Since I will be talking about the uses of oils in foods, and relating mechanisms of fat functionality to the desired organoleptic results, I first need to describe these various factors. Based upon a few principles of structure, properties and reactivities, the function of fats and oils in foods will be easier to understand.

We start by considering the structure of a triglyceride, the molecule which makes up nearly 100% of a purified food-grade fat. It is made up of a molecule of glycerol which is esterified to three fatty acid molecules. In Figure 5.1 I show an example where the fatty acids are saturated (stearic), mono-unsaturated (oleic), and polyunsaturated (linoleic) acids. The middle, or #2 position, of glycerol is sterically distinct from the other two positions, #1 and #3. While this is usually of little significance, in certain cases such as cocoa butter (discussed later) the esterification of certain fatty acids at certain positions has a great deal to do with the properties of the fat.

The melting point of the fat is governed by two factors; the chain length and the degree of unsaturation of the fatty acids. This is

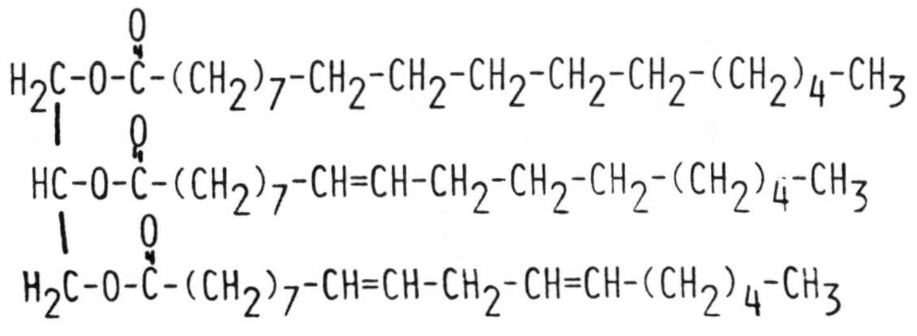

$$
\begin{array}{l}
\quad\;\; \overset{O}{\|} \\
H_2C\text{-}O\text{-}C\text{-}(CH_2)_7\text{-}CH_2\text{-}CH_2\text{-}CH_2\text{-}CH_2\text{-}CH_2\text{-}(CH_2)_4\text{-}CH_3 \\
\;| \quad \overset{O}{\|} \\
HC\text{-}O\text{-}C\text{-}(CH_2)_7\text{-}CH{=}CH\text{-}CH_2\text{-}CH_2\text{-}CH_2\text{-}(CH_2)_4\text{-}CH_3 \\
\;| \quad \overset{O}{\|} \\
H_2C\text{-}O\text{-}C\text{-}(CH_2)_7\text{-}CH{=}CH\text{-}CH_2\text{-}CH{=}CH\text{-}(CH_2)_4\text{-}CH_3
\end{array}
$$

FIG. 5.1. Chemical structure of a triglyceride. This particular structure represents stearoyl oleyl linoleyl triglyceride, reading the fatty acid chain from top to bottom.

illustrated by the comparisons in Table 5.1. Lauric acid is six carbons shorter than stearic, and melts 25 degrees lower. But the effect of unsaturation is even more marked. Within the family of 18-carbon fatty acids, the first double bond lowers the m.p. by 53 degrees, while the second double bond lowers the m.p. even further.

In the triglycerides these effects are maintained. Coconut oil is highly saturated but, being made up largely of C12 and C14 fatty acids, it is a liquid at room temperature. Soy oil contains mostly C18 fatty acids but is highly unsaturated and hence liquid at room temperature. As double bonds are removed by hydrogenation, the melting point increases, to give a fat which is quite hard at ambient temperatures. Since most commercial oils are based on soy, cottonseed and palm which contain primarily C16 and C18 acids, for our purposes we can relate melting point to degree of unsaturation; the more unsaturation, the lower the melting point.

In a given oil, such as soy, there are a large number of triglyceride species. Five different fatty acids comprise 99% of soy oil. If they are esterified on the three positions of glycerol according to random probabilities, over 50 different triglycerides may be formed. Since this represents a very "impure" material from an organic chemistry standpoint, we would expect that no sharp melting point would be observed. This is the case. While a melting point may be operationally defined for such a heterogeneous system, it has only limited usefulness.

TABLE 5.1. Factors Influencing Melting Point

Fatty Acid	Chain Length	Double Bonds	Melting Point, °C
Lauric	12	0	44.1
Stearic	18	0	69.6
Oleic	18	1	16
Linoleic	18	2	- 9.5

In practice we use a blend of fats and partially hydrogenated oils from different sources, and we make use of them at a variety of temperatures from refrigerator to warm room temperatures. Thus it is more useful to characterize a fat by a parameter which reflects its hardness over a range of working conditions. This parameter is called the Solid Fat Index (SFI). The SFI is the percentage of total lipid which is in the solid state at several specified temperatures, usually 50°F, 70°F, 80°F, 92°F, and 104°F.

In Figure 5.2 we can see the SFI curves for four different fats used industrially. The very shallow curve represents a pumpable shortening, sold for use in bread bakeries. It contains a small percentage of a highly hydrogenated soy oil (140°F m.p.) suspended in a liquid oil. The steepest curve is cocoa butter, which is quite hard even at 80°F, but which melts completely over a relatively narrow temperature range. The two intermediate curves are for a lard sample and for a partially hydrogenated soy oil, both of which are soft and workable ("plastic") at normal room temperatures, but which melt completely at temperatures above that of cocoa butter. The ratio of solid to liquid governs the functionality of oils in many food applications.

The crystal habit of the solid fraction of the fat also has a great deal to do with its functionality. There are three basic crystal types which may be formed in fats. The alpha crystal is the least regular structure. It is formed by rapidly chilling a completely melted fat, and has some characteristics of a glass. It is the least stable state thermodynamically, and rapidly changes to the more stable beta prime crystal. In fact, because of this instability the alpha crystal is seldom observed with a pure triglyceride fat.

The beta prime crystal is small and needle-like. Solid fat in this form will make a brush-heap structure which is very efficient at holding oil. It is usually made by quickly chilling melted fat and then tempering the fat under closely controlled conditions. This crystal form is the basis for most of the plastic fats used today in the food industry.

If tempering and storage conditions are not right (i.e., get too warm) the beta prime crystals will transform into the thermodynamically most stable form, beta crystals. These are large, dense plate-like crystals which form a solid aggregate. These structures don't hold oil as do the beta prime crystals.

An illustration of this principle is shown in Figure 5.3, where glass and water represent the solid and liquid phases of a shortening. In the funnel on the left is 12 g of glass wool with the fine, filamentous structure and high surface area characteristic of beta prime crystals. The funnel on the right contains 12 g of glass beads representing a dense, solid structure such as that of beta crystals. Over each one I poured 100 ml of water (colored with a dye). You can see that the glass wool retained about half the water, while the beads retained essentially none of it.

This phenomenon is utilized to make plastic shortenings with varied properties to give differing results in food applications. A fat based on beta prime crystals is a smooth, plastic shortening workable over a range of temperatures. If the crystals have converted to the beta form due to improper tempering or storage conditions much of the oil will bleed out and the shortening will have a hard lumpy consistency.

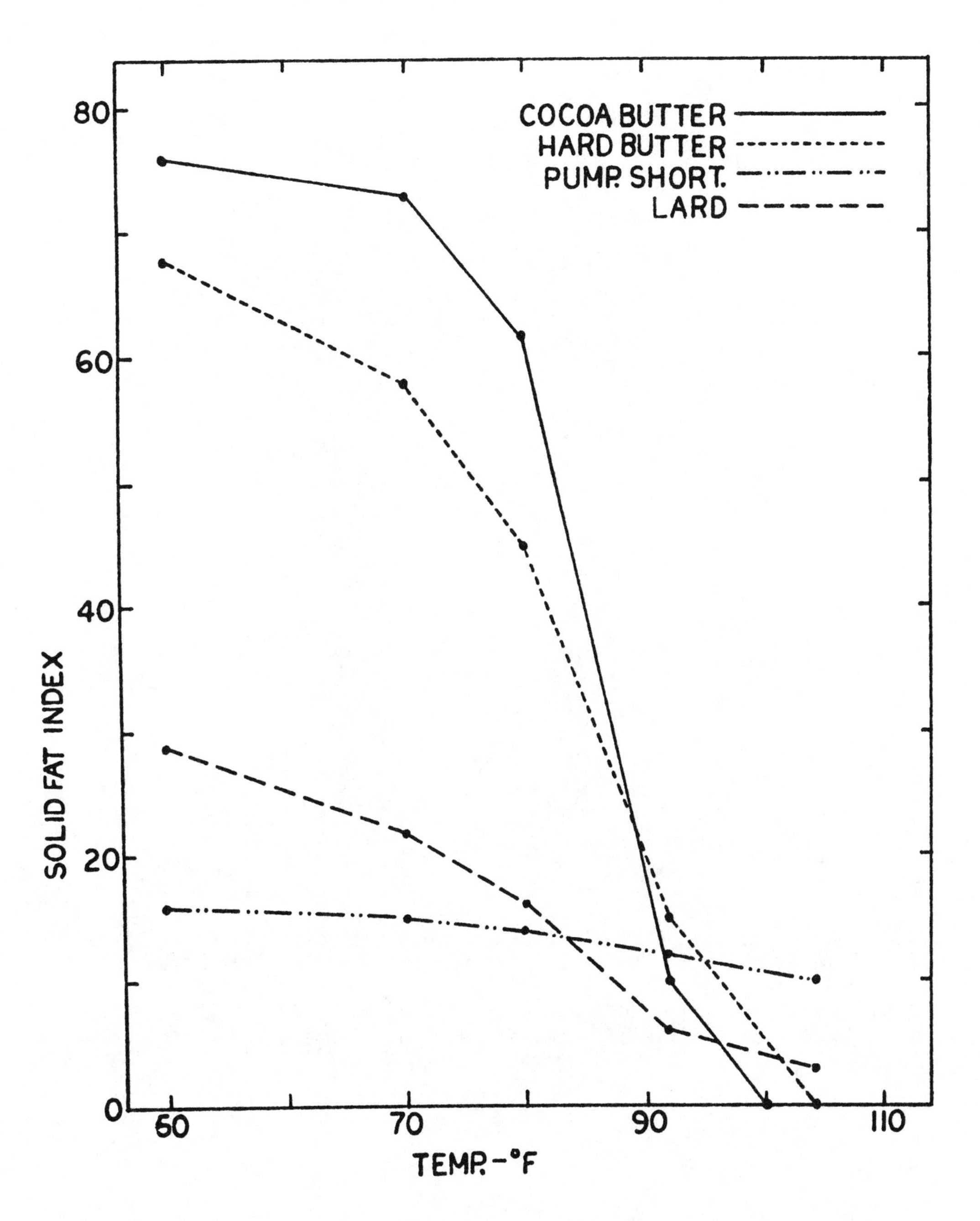

FIG. 5.2. Solid Fat Index of some typical fats. Representing the SFI for four commercially used fats, the curves are for cocoa butter; hard butter (soy oil hydrogenated to approximate cocoa butter characteristics); pumpable shortening (a slurry of high melting fat in soy oil); and a typical lard sample.

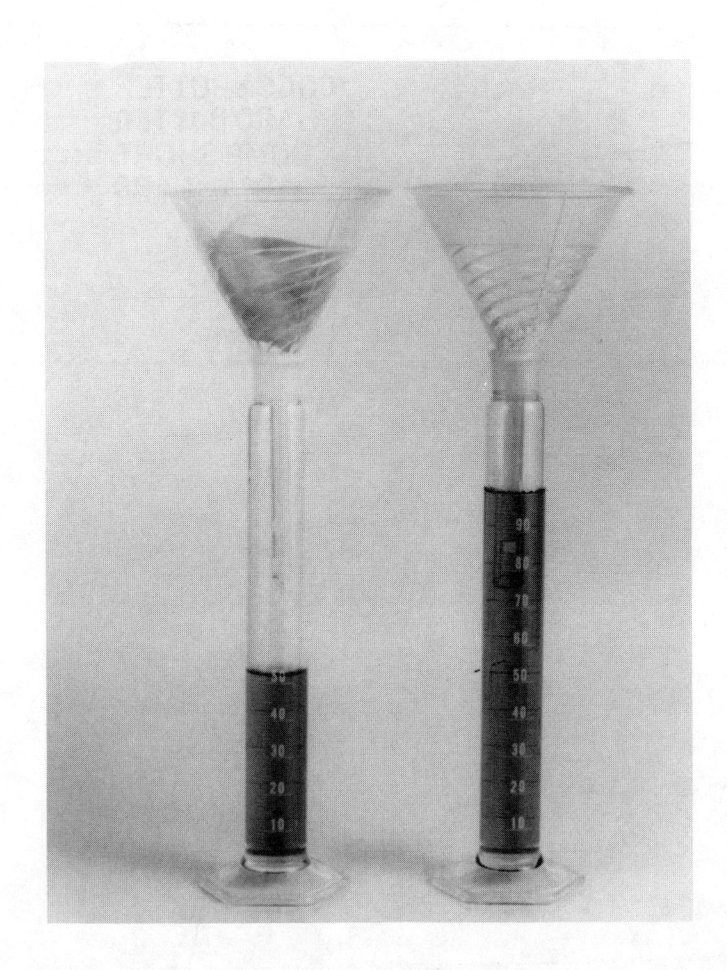

FIG. 5.3. Effect of physical structure on lipid-retention capacity. The funnel on the left contains 12 g of glass wool (high relative surface area) while the funnel on the right holds 12 g of glass beads (low relative surface area). Of 100 ml of colored water poured through each funnel, the glass wool retained about 55 ml while the glass beads retained 3 ml.

But if the fat is required to contribute strength and rigidity to the final food product, then the beta crystal would be desirable.

The factors of melting point, SFI, and (to some degree) crystal structure can be influenced by changing the fatty acid composition of the oil by hydrogenation. This process adds hydrogen to the double bonds, reducing the degree of unsaturation and raising the SFI curve. Use of selective catalysts and reaction conditions can affect the relative susceptibilities of the mono and poly-unsaturated acids to being hydrogenated and so skew the SFI curve in different ways. Also the degree of hydrogenation will influence the properties of the final fat.

Besides changing the hardness of the oil, hydrogenation also tends to react first with linolenic acid, C18:3. This is desirable, because linolenic is particularly reactive towards oxygen, contributing to oxidative rancidity in the fat. Table 5.2 shows the change in the ratios of the various fatty acids in soy oil during one particular experimental run. As is apparent, C18:3 goes quickly to C18:2, which in turn is hydrogenated somewhat more slowly to C18:1. This fatty acid, oleic, is rather unreactive under the conditions used, and accumulates.

The molecular structure can also be altered in ways which are not particularly desirable. The fat can undergo changes which lead to rancidity. There are two types of rancidity; hydrolytic and oxidative.

Hydrolytic rancidity is simply cleavage of ester bonds in the triglyceride by water, giving free fatty acids and diand monoglycerides. At low levels (up to 1%) the free fatty acids usually give a detectable but not objectionable flavor. An exception is with fats such as butter which contain significant amounts of short-chain fatty acids. Very low levels of butyric acid (a 4-carbon acid) gives a flavor which is quite repugnant to American palates. (But a taste for it can be acquired; ghee owes much of its characteristic flavor to the presence of short-chain fatty acids.)

Oxidative rancidity is another matter. Fatty acids with methylene-interrupted double bond systems such as linoleic and linolenic acids will react with oxygen to give peroxides. These in turn break down through a series of reactions to yield carbonyls of various sorts which are responsible for the off-flavor of rancid fat (see Figure 5.4).

The initial oxygenation of the reactive double bonds is catalyzed by traces of metal ions such as copper, iron, and manganese. They also catalyze the formation of free radicals from the scission of the

TABLE 5.2. Hydrogenation of Soybean Oil

	Hours of Reaction				
Acid	0	1	2	4	6
C16:0	11	11	11	11	11
C18:0	4	7	8	12	18
C18:1	25	41	52	63	70
C18:2	53	38	28	13	1
C18:3	6	3	1	0	0
Iodine Value	130	110	98	75	62

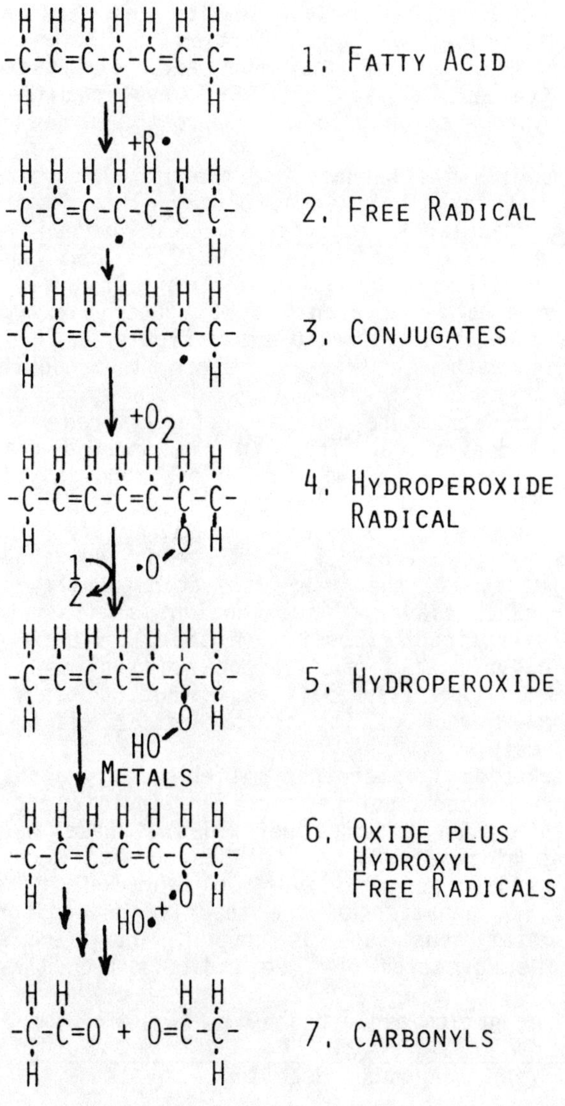

1. Fatty Acid

+R•

2. Free Radical

3. Conjugates

$+O_2$

4. Hydroperoxide Radical

5. Hydroperoxide

Metals

6. Oxide plus Hydroxyl Free Radicals

7. Carbonyls

FIG. 5.4. Some of the steps in autoxidation of polyunsaturated fatty acids. The multiple arrows between steps 6 and 7 indicate the presence of several poorly characterized reactions which lead to cleavage of the fatty acid chain and formation of carbonyl compounds.

hydroperoxide formed in the overall reaction. Thus citric acid, a chelating agent, is added at low levels to keep the metal ion concentration low and so hinder the development of peroxides. Also the process is auto-catalytic. The initial peroxide formed can catalyze oxygenation of more fatty acids via a free radical reaction mechanism. Addition of free radical scavengers--BHA, BHT, or propyl gallate--interrupts this process and slows down the development of rancid flavors. If the oil is to be used under severe conditions, such as frying doughnuts or snacks, selective catalytic hydrogenation is employed to reduce the linoleic and linolenic levels to as near zero as practicable.

While oxidation products can be objectionable at higher levels, traces of carbonyls give oils a flavor which is deemed desirable. In addition to the flavor of the carbonyls and carbonyl fatty acids, reaction with other food constituents such as sugars and amines gives a variety of compounds which have been identified as contributing to the flavor of foods cooked in oils.

A particular carbonyl compound, di-acetyl (2,3-butanedione) is found in butter, and is a major component of the characteristic butter flavor.

During the use of oils in food products, they interact with the other components of food, i.e. the carbohydrates and proteins. Hydrocarbon chains will interact with starch molecules to give inclusion products. This phenomenon has been known for many years. Fatty acids will also interact with starch, although this is usually thought of as being restricted to the interaction of something like monoglycerides with soluble amylose.

While there is no direct evidence, I think there may be some interaction of triglycerides with insolubilized starch molecules and granules. This hypothesis comes from observations of the effect of oils on mouthfeel and texture of various starch-based foods such as puddings, and also on some effects of shortening during the baking of bread. More will be said about this below.

The interaction of lipids with proteins, however, is a well-known reaction. Parts of protein chains are lipophilic due to the hydrophobic side chains of some amino acids in the chain. Many cereal and legume proteins are relatively high in their content of hydrophobic amino acids, and so interact more strongly with lipids than do the proteins of milk or eggs. One measure of the hydrophilicity of a protein is the ratio of polar to nonpolar amino acid residues. Using this indicator we can calculate the following ratios: gluten, 0.47; soy protein, 1.31; ovalbumin, 2.63. This ratio reflects the "water-friendliness" of the protein, or, conversely, the lack of compatibility with lipids. The relative lipophobic ratings given correlate well with observations made when oils are mixed with the respective proteins.

Gluten, the main protein of wheat flour, interacts strongly with fat added to a dough being mixed. In fact, the interaction is responsible for much of the characteristic texture of bread. Analytical results show a strong binding of lipid by the protein.

Soy protein also will interact with lipid, although apparently not quite so strongly as does gluten. Nevertheless, soy protein will readily emulsify fat, and is used for that purpose by many sausage manufacturers.

Ovalbumin (the main protein of egg white) is not very hospitable to lipid. An obvious expression of this tendency is the effect of traces of oil on the whipping of egg whites. The lipid inhibits the spreading of ovalbumin molecules at the air/water interfaces being generated, and so prevents the formation of the meringue. Also, egg whites are not particularly good emulsifiers. Attempts to make a mayonnaise based on white rather than egg yolk don't work.

OIL IN FOOD APPLICATIONS

Having discussed at some length how the molecular architecture of fats and oils influences their propeties, it is time to turn to a consideration of the ways in which these properties contribute to a pleasing palatability in our diets. I will use several specific examples to illustrate the ways in which oils enhance foods via: flavor, texture, mouthfeel, structure.

Flavor

The trace constituents in fat contribute to flavor. To illustrate this from your own experience, buy some tortilla chips which have only been baked, then compare them with the flavor of the same chips which have been briefly deep-fried. The difference is marked. Another comparison may be made by frying potatoes in a Teflon-lined pan using no fat, then fry potatoes in the same pan but with a little fat present. Again, there is quite a difference in flavor.

A few years back I demonstrated this for a group of students in a short course which I was teaching. I had made some cake doughnuts fried in the normal doughnut frying fat, and then some from the same batter fried in a completely bland foodgrade mineral oil. The students first took a bite of the normal doughnut, and found it to be acceptable. Then they took a bite of the second doughnut fried in mineral oil. Most of them chewed about twice and then spit the bite out. The flavor was very bland, and not at all something they wanted to swallow.

The flavor comes not only from trace constituents in the fat but also from compounds formed during cooking. Frying, especially deep frying, represents a high temperature anhydrous (or low water) reaction system. A wide variety of materials can be formed, including: imines and acetals involving carbonyls, amino groups and sugar hydroxyls; fatty amides and esters, from transacylation from the triglycerides; cyclic compounds, from dehydration reactions involving fats, proteins, and carbohydrates; a variety of other less-well characterized compounds. All of these reaction products contribute to our overall impression of how the food tastes.

Texture

Fats and oils contribute to texture in food. This factor is described in many ways; tenderness, flakiness, crispness, are only some of the terms used to describe desirable textures during taste testing of a food. These effects are readily seen in baked foods. A good example is Danish dough.

In making Danish the soft, chilled dough piece is rolled out into a rectangle. Shortening is spread over two-thirds of the piece, the uncovered third is folded towards the center, then the other end is folded on top. The result is a laminate of three layers of dough with two layers of fat interspersed. The laminate is rolled out again, folded, and this process is repeated twice more. The final result, if all goes well, is a sheet having some 60 to 80 layers of dough with a thin film of shortening between all the layers. This oil film prevents coalescence of the dough during subsequent make-up, proofing and baking, so that the final result is the flaky tender texture found in good Danish rolls.

The properties of the shortening are very critical to successful completion of this process. If the fat is too soft (too low SFI) the result will be isolated pockets of oil rather than thin layers of laminated fat. If it is too hard (too high SFI, or if some beta crystal formation has occurred) it will tear the dough during rolling, and poor-quality product will result. The proper plasticity of the fat is necessary so that during each rolling step it extends along with the dough and gives thin, continuous layers of fat in the laminated dough.

A somewhat similar "oil insulation" effect takes place in pie crusts. The ingredients--flour, water, and shortening--are combined in such a way that the hydrated flour forms mini-sheets and flakes of dough which are kept from coalescing by a coating of fat. If the dough is worked too much, the fat begins to combine with the gluten present and the whole mass develops into a continuous dough-piece. This gives a very tough pie crust. When the separation of the small dough elements is maintained and the fat is retained as a lubricating agent, the crust has the desirable tender, flaky texture.

Mouthfeel

While texture is sensed with the teeth (i.e., via resistance to chewing and the muscles involved in that operation) mouthfeel is related more to the response of the tongue and the other surfaces of the mouth. There are several descriptors which are used in characterizing mouthfeel: smoothness, grittiness, moistness, dryness, gumminess, etc. These relate to what the tongue and palate feels while the bite of food is being chewed.

An example of mouthfeel is the pleasing creaminess which is engendered in a starch-based pudding by the addition of a little oil. The pudding or pie filling as made would have had a slightly resilient, short texture if the oil were left out. Adding the fat at the last stage of cooking results in a final product which has a softer character. This is experienced as a creamy or smooth impression during eating.

Another aspect of mouthfeel is demonstrated by chocolate. Chocolate is made by grinding the roasted centers--the nibs--of the cocoa bean to give a suspension of finely-ground fibrous material in a fat matrix. The texture of chocolate depends largely on how small are the particles of the cell wall and intercellular lignin materials. The fat--cocoa butter which makes up nearly half the content of chocolate--influences how the product feels in our mouth and on our teeth.

If you remember the SFI curve for cocoa butter shown in Figure 5.2, it has a very sharp melting curve, being quite solid (SFI of 62) at 80°F and nearly liquid (SFI of 10) at 92°F. So at mouth temperature there is no appreciable solid fat content in chocolate. This means that it is quickly cleared from the surface of the teeth, leaving a clean, desirable impression.

This very sharp melting point is due to a unique character of cocoa butter; it has a large proportion of molecules which act as one functional species. This triglyceride contains only oleic acid at the #2 position of the glycerol molecule, and either palmitic or stearic (the C16 and C18 saturated fatty acids) at positions #1 and #3. To express this in lipid chemists shorthand, the major components of cocoa butter are: POP (1-Palmitoyl-2-Oleyl-3-Palmitoyl glycerol), 16%; POS, 40%; SOS, 26%; for a total of 82% of the butter.

Cocoa butter is expensive. Over the years companies which supply fats and oils to food manufacturers have been doing research to come up with a less costly "compound coating" based on modified vegetable oils. Recent products have properties quite similar to cocoa butter but it is very difficult to obtain a sharp melting point due to the inherently more heterogeneous molecular structures of the vegetable oils used as the starting materials. If the oil is hardened (hydrogenated) enough to keep the coatings from softening at room temperature then there is always a small residue of solids at body temperature, resulting in a waxy mouthfeel.

Another effect of oils in certain foods is to give the impression of moistness. Again, this depends on having a liquid fat at mouth temperature. An example is eating a cracker plain or with butter. The former will seem dry in your mouth. The latter will not, although no actual water has been added. This is why snack crackers typically are sprayed with about 15% of their weight of oil directly after baking; they have a more pleasing mouthfeel when eaten.

When we learned how to use the proper emulsifiers so that cakes could be made with oil instead of a plastic cake shortening, the cakes tasted moister to the consumer. This is so, even though analytical data show no difference in actual water content. The effect is one of lubrication. The oil keeps the food particles from sticking to the surfaces inside the mouth. It is easier to chew and swallow the food, making it more enjoyable to eat.

Structure

Speaking of cakes brings me to the fourth function of fats and oils structure. When cakes are made with a plastic shortening the first step is "creaming" the fat and sugar. The functional effect is to incorporate a lot of air in the form of very small bubbles. These air bubbles are the nuclei for the expansion of carbon dioxide and steam during baking which leads to the expansion of the whole cake and a light, porous final structure. With no air incorporation the gases tend to tunnel to the surface of the cake and escape. The result is a low-volume, dense cake of poor eating quality.

To incorporate and hold the air during batter preparation it is necessary to have the kind of fat matrix given by a beta prime crystal structure. The "brush-heap" not only holds oil but it also entraps air. The presence of emulsifiers then promotes the formation of many tiny bubbles rather than fewer large ones, giving a fine grain in the final cake.

A similar sort of phenomenon is responsible for the eating properties of a good creamy frosting. Again the shortening or butter is whipped to incorporate air, then the sugar and flavorings are added. If there is little aeration the frosting gives a dense, waxy, undesirable mouthfeel. Air entrapment, aided by beta prime crystal structure and emulsifiers, results in a pleasing frosting for the cake.

My last food usage example gets us back to the first love of all cereal chemists--bread. As mentioned above the gluten protein in wheat flour is quite lipophilic and reacts avidly with oils. The presence of the lipid helps the gluten to form elastic sheets of protein which entrap and hold the gases formed during yeast fermentation and during baking. Drs. Y. Pomeranz and O. K. Chung at the USDA Research Lab in Manhattan Kansas have directed a great deal of work towards elucidating the role of lipids in bread baking. The simplest and most direct demonstration is shown in Figure 5.5, where the final volume of the baked loaves is shown to have a definite dependence on the amount of shortening added to the dough during mixing.

But there is also another interaction which apparently involves the starch granules in the dough. Dr. Carl Hoseney and co-workers at Kansas State University have been studying the phenomena which occur during baking using a resistance-heater baking device. They monitor the volume of the bread throughout the baking cycle. As shown in Figure 5.6 during heating there is a smooth, continuous increase in volume due to gas expansion until the gelatinization temperature of the starch is reached. At that point gelatinization occurs, the structure sets, and no further volume expansion takes place.

The interesting point is that the presence of lipid apparently delays the gelatinization of starch in some way, allowing the bread to expand longer in the oven before the structure sets. We have no direct evidence for a physical interaction of lipid with the starch which would lead to such an inhibition of gelatinization. Nevertheless the experimental evidence certainly indicates that some sort of interaction, presently not understood, is taking place between lipid and starch.

SUMMARY

To recapitulate very briefly the uses of oils in foods, they are of two kinds:

Physiological, involving structural body elements, cell membrane fluidity, and as a dense calorie source; and

Palatability, for desirable flavor, texture, mouthfeel, and structure.

Used in the right ways, at appropriate levels, oils are not only a necessary but also an enjoyable part of our diet.

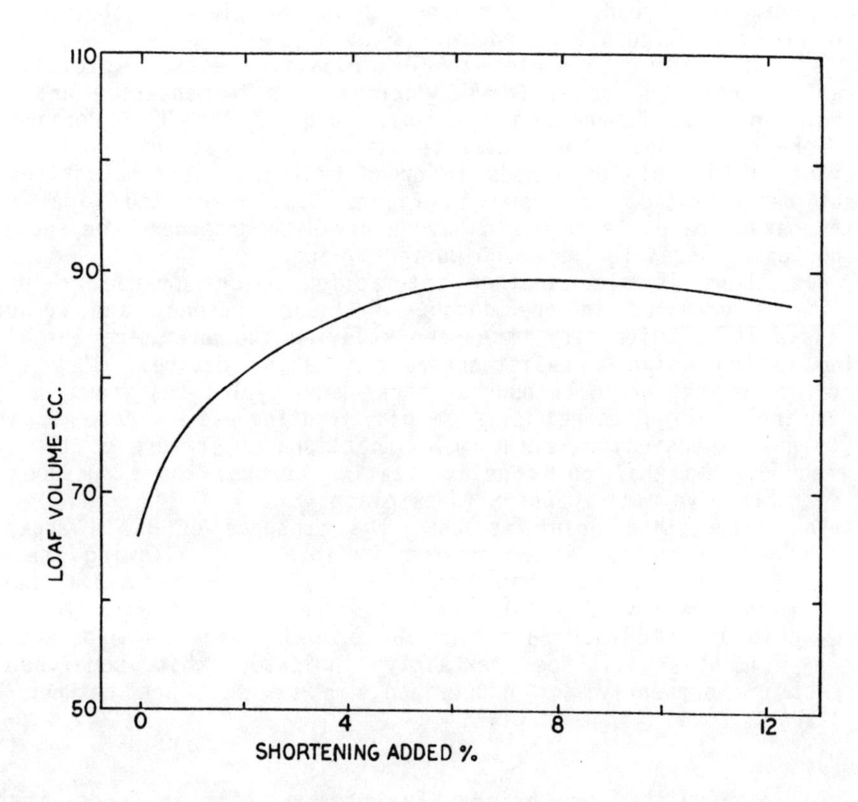

FIG. 5.5. The effect of adding shortening on loaf volume in experimental bread baking. Adapted from Fig. 6 in O. K. Chung, M. D. Shogren, Y. Pomeranz, and K. F. Finney, Cer. Chemistry, 58, 69, 1981. Used by permission of the American Association of Cereal Chemists.

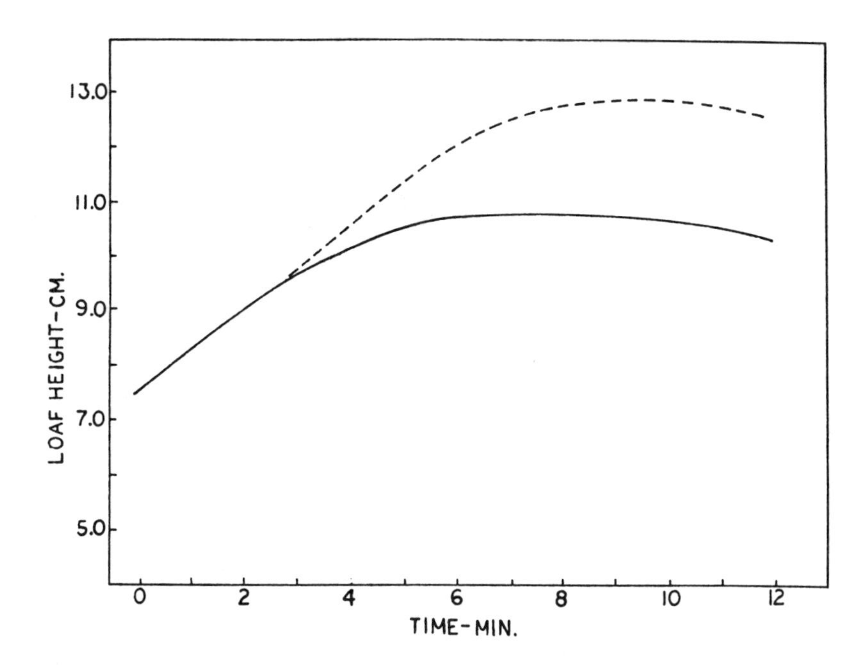

FIG. 5.6. The effect of shortening on "setting temperature" of bread. The temperature in the experimental baking apparatus used increases linearly with time, so the X-axis may be thought of as representing internal temperature of the bread dough. The solid line is control dough, with no added shortening; the dotted line is dough with 3% added shortening. Adapted from Fig. 5, R. C. Junge and R. C. Hoseney, Cer. Chemistry, 58, 408, 1981. Used by permission of the American Association of Cereal Chemists.

6

Value of Isolated Soy Protein in Food Products

STANLEY H. RICHERT and CHARLES W. KOLAR

INTRODUCTION

Isolated soy proteins are making significant contributions to the world's food supply. Advances in this ingredient technology have resulted in products which meet specific technical and/or dietary needs of food processors on a worldwide basis. As a primary food ingredient in a growing number of foods, these unique proteins provide quality, nutrition, and value to producers and consumers alike.

Isolated soy protein has the highest protein content of all the soybean-sourced ingredients and is the most versatile. It is defined as the major proteinaceous fraction of the soybean prepared from high-quality, sound, cleaned, dehulled soybeans by removing a preponderance of the non-protein components and shall contain not less than 90% protein on a moisture-free basis. Isolated soy protein is generally recognized as safe by the Food and Drug Administration.

Processing

The manufacturing process for isolated soy protein is an aqueous extraction process as outlined in Figure 6.1. The raw material is defatted soy flour or flakes with a high Protein Dispersibility Index. The soy flour or flakes are wetted or dispersed with the proper amount of water at controlled temperature, and mixed with the necessary amounts of high-quality food grade chemicals for a specified length of time. The water-soluble carbohydrates and proteins are dissolved. Then the residue, which is primarily the insoluble polysaccharide fraction or cell-wall material of the defatted soy flour, is removed by centrifugation.

The extract contains the water soluble carbohydrates and the major protein fractions. The pH of the extract is then adjusted to approximately pH 4.5 with a food grade acid to precipitate the major proteins. The precipitated protein is commonly referred to as soy protein curd, which is then washed to remove residual soluble carbohydrates, color and flavor components. The soy protein curd can be spray dried to yield products at the iso-electric point, or the pH can be increased with various bases to produce a number of neutralized products.

Isolated soy proteins are available in a number of physical forms dry powder, dry granules, and frozen fiber. These products are designed

Cereals and Legumes in the Food Supply, edited by Jacqueline Dupont and Elizabeth M. Osman 1987 Iowa State University Press, Ames, Iowa 50010. Reprinted by permission of the Ralson Purina Co., St. Louis, Missouri.

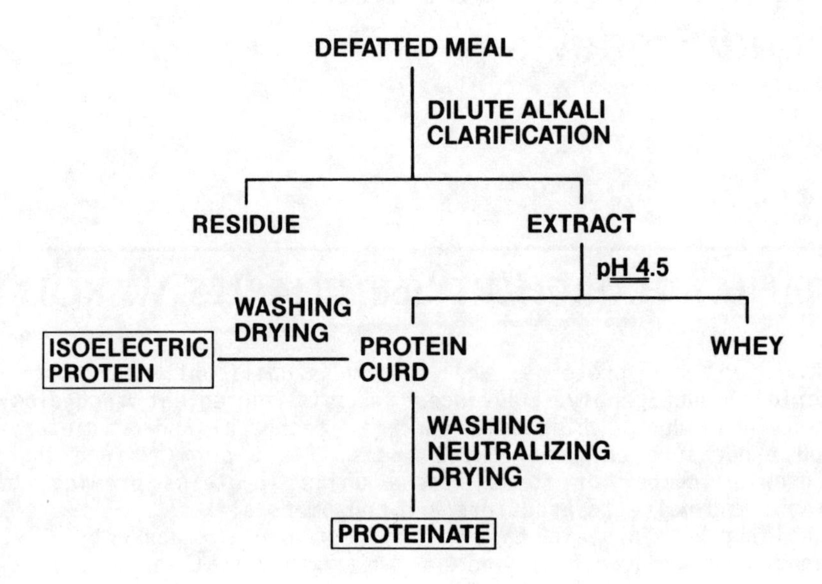

FIG. 6.1. Process for production of isolated soy protein.

to meet specific physical requirements for various food products or systems. Each of these forms will be discussed later.

The typical composition of isolated soy proteins on a moisture-free basis is shown in Table 6.1 (8). Based on this composition, isolated soy proteins are essentially pure proteins.

The typical mineral composition, shown in Table 6.2, illustrates that isolated soy proteins contain nutritionally significant minerals such as calcium, iron, copper, phosphorus, and zinc (8). Low sodium isolated soy proteins are made be neutralization by potassium or calcium hydroxide.

Protein Quality

From a nutritional standpoint, protein quality is generally determined by three factors: 1) essential amino acid composition of the protein, 2) digestibility, and 3) the amino acid requirements of the species consuming the protein. In the case of food proteins for humans, the essential amino acid requirements of humans are the critical parameters for assessing protein quality or value.

The Food and Nutrition Board (FNB) of the National Research Council, the National Academy of Sciences, a recognized scientific body in the United States for establishing dietary allowances, recently published recommended dietary allowances that contain the amino acid pattern for a high-quality protein (2). This pattern is based on the amount of essential amino acids per gram of protein required by the infant

TABLE 6.1. Typical Composition of Isolated Soy Protein (Moisture Free Basis)

Component	Percent
Protein	92.0
Fat	0.5
Ash	4.0
Total Carbohydrates*	3.5

*By difference

TABLE 6.2. Typical Mineral Composition of Sodium Hydroxide Neutralized Isolated Soy Protein

Element		Element	
Arsenic	<0.2 ppm	Lead	<0.2 ppm
Cadmium	<0.2 ppm	Magnesium	380 ppm
Calcium	0.18%	Manganese	17 ppm
Chlorine	0.13%	Mercury	<0.05 ppm
Chromium	<1 ppm	Molybdenum	<3 ppm
Cobalt	<1 ppm	Phosphorus	0.76%
Copper	12 ppm	Potassium	960 ppm
Fluorine	<10 ppm	Selenium	0.36 ppm
Iodine	<10 ppm	Sodium	1.1%
Iron	160 ppm	Zinc	40 ppm

and serves as a conservative guide for the protein requirements of adults.

The Food and Nutrition Board's pattern (2) for a high-quality protein and the essential amino acid content of the typical Ralston Purina isolated soy protein are shown in Table 6.3. The amino acid content of isolated soy protein compares favorably with the criteria established for a high-quality protein because the recommended pattern is met.

The nutritional value of Ralston Purina's isolated soy proteins has been established through extensive nutritional research programs conducted at Ralston Purina's laboratories and at research institutions around the world.

Digestibility studies have been conducted with children, young adults, and laboratory animals such as pigs and rats. A summary of these data shown in Table 6.4 indicates that the relative digestibility of isolated soy protein in comparison to milk is 98.9% for children (3); in comparison to egg it is 98.4% for adults (4); and in comparison to beef it is 99.5% for adults. Thus, isolated soy protein is comparable to other high-quality proteins such as meat, milk, and eggs in digestibility with both children and adults.

A series of human evaluations for protein quality was conducted with two- to four-year old children and young male adults using the

TABLE 6.3. Comparison of Amino Acid Profile of Isolated Soy Protein with FNB Pattern for High Quality Protein

Essential	FNB* pattern for high quality protein (mg/g)	Typical amino acid content of isolated soy protein (mg/g)
Histidine	17	28
Isoleucine	42	49
Leucine	70	82
Lysine	51	64
Total sulfur amino acids	26	26
Total aromatic amino acids	73	92
Threonine	35	38
Tryptophan	11	14
Valine	48	50

*Food and Nutrition Board, National Academy of Sciences, 1980.(2)

TABLE 6.4. Relative Digestibility of Isolated Soy Protein

Reference protein	Age	Relative digestibility of isolated soy protein
Milk	Child	98.9%
Egg	Adult	98.4%
Beef	Adult	99.5%

nitrogen balance technique. A summary of three studies on protein quality conducted with young male adults by Scrimshaw and Young (4, 5) is shown in Table 6.5. They reported that isolated soy protein is comparable in protein quality to milk and beef, and 80-90% of whole egg. Protein intake levels in these studies were below minimum requirements recommended by FNB.

The Food and Agriculture Organization/World Health Organization (FAO/WHO) (6), and the Food and Nutrition Board (2), recommended that the daily allowances for protein be 1.19 and 1.75 grams of protein per kilogram of body weight per day, respectively. A summary of three studies conducted with one- to four-year old children indicated that the minimum daily protein intake required from isolated soy protein for children is 0.81 to 1.00 grams per kilogram of body weight per day, as shown in Table 6.6. Thus, isolated soy proteins met young children's protein requirements at protein intakes below that recommended by these two widely recognized groups.

It can be concluded from these studies that isolated soy protein produced by Ralston Purina Company and tested with human volunteers meets the protein and amino acid requirements at the minimum recommended protein intake for humans. Thus, these proteins are comparable in nutritional quality to other high-quality proteins when consumed by humans.

Physical and Functional Properties

All foods have unique physical, chemical, and textural properties. The physical properties of meat, poultry, seafood, eggs, and dairy products are generally related to the proteins present in these products. In addition, proteins contribute functional properties such as gelation,

TABLE 6.5. Comparison of Nutritional Quality of Isolated Soy Protein with Other Proteins, Baseed on Human Studies

Reference protein	Relative protein quality of isolated soy protein
Egg	83%
Beef	100%
Milk	100%

Based on protein intake requirements for nitrogen balance

TABLE 6.6. Comparison of Protein Intake Level Required from Isolated Soy Protein with Recommended Levels (Children 1-4 Years of Age)

	g protein/kg body weight/day
FAO Recommendations	1.19 g
Food & Nutrition Board RDA	1.75 g
Isolated Soy Protein Feeding Studies	0.81-1.00 g*

*Includes a 30% increase over mean value to correspond with FAO recommendations.

viscosity, emulsification, water absorption, solubility, and textural characteristics.

These functional properties are also influenced by other variables in the food system such as pH, ionic strength, temperature, emulsifiers composition, processing conditions, etc. The functional properties of proteins are usually described by the measurement of various physical properties such as solubility, emulsion capacity, emulsion stability, gel strength, viscosity, water absorption, whipping properties, etc. These tests are generally conducted in an aqueous system. The variables mentioned above must be considered in the measurement of physical properties of proteins in order to predict performance in food systems.

Usually, these tests are performed under one set of conditions to develop comparative data. However, a protein may have superior properties under one set of conditions and poor properties under another set of conditions. Therefore, physical tests may not determine the functional performance of a protein in a food product or system. In most instances, two or more physical property parameters are needed to estimate the performance of proteins in a food product. The preferred method is to evaluate the protein in the final food where it will be used. Performance may then be correlated with physical property tests.

To successfully incorporate soy proteins in traditional food products, the protein ingredients must exhibit properties in the food product similar to the traditional protein. Isolated soy proteins are developed and produced to have different functional or physical properties to meet the requirements of various food products. Thus, a series of isolated soy proteins with a broad range of physical properties is available to food technologists. They can be described using the physical properties that follow.

The range of solubilities of isolated soy proteins is illustrated in Figure 6.2[1]. The solubilities range from about 25 NSI (Nitrogen Solubility Index) to 95 NSI. It is difficult to predict the performance of a protein in an application based only on the solubility property.

The emulsion capacities of isolated soy proteins can differ by a factor of nearly four as illustrated in Figure 6.3[1]. Emulsion capacity, or the ability of a protein to emulsify fat, is an important parameter of protein in many food products. Based on this emulsion capacity test, isolates can emulsify from 10 to about 35 milliliters of oil per 100 milligrams of protein.

Isolated soy proteins vary in their ability to form gels. Some are designed to form gels while others will not form gels at 14% solids content. The presence or absence of salt can affect the gel strength. Since salt is a common food ingredient, it was used in the preparation of the gels made with the various isolates. Gel strengths of isolates are shown in Figure 6.4[1].

The viscosities of isolated soy proteins are shown in Figure 6.5[1]. Isolates have thixotropic properties. Some isolates have the same viscosity at 18% solids as others have at only 10% solids concentration. Moreover, application of heat to the protein solutions can alter the viscosity.

---

1. Anon (1982). Unpublished data. Ralston Purina Company.

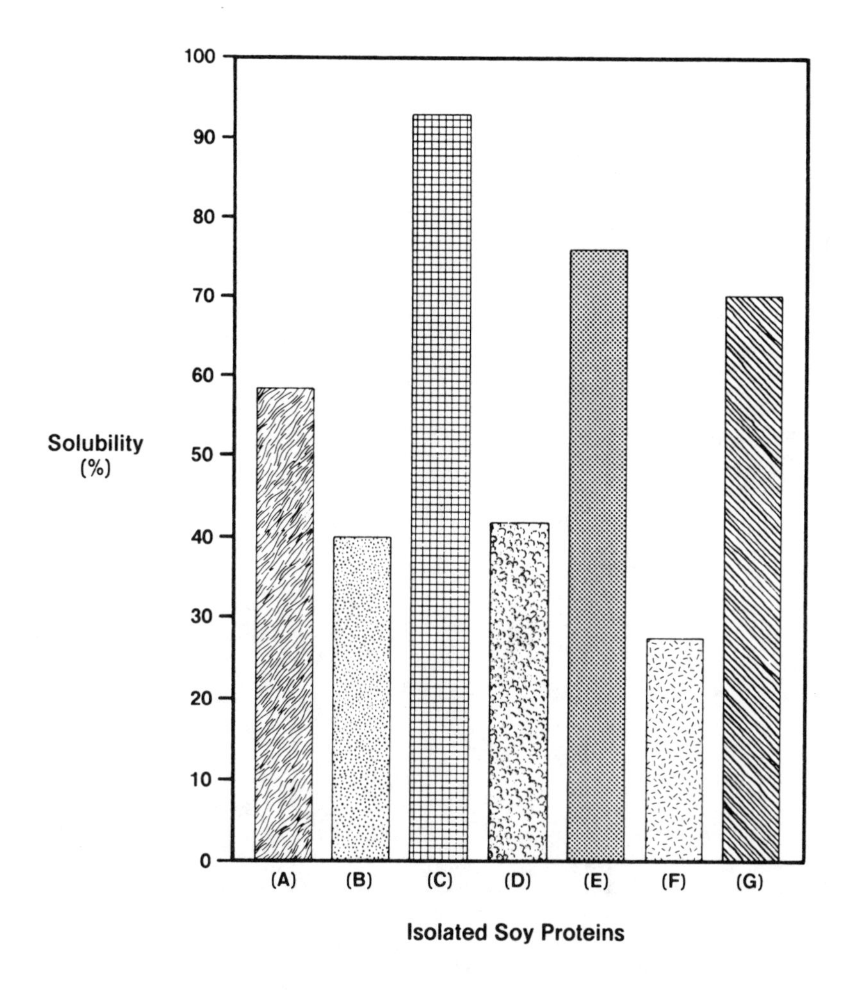

FIG. 6.2. Solubilities of commercial isolated soy proteins.

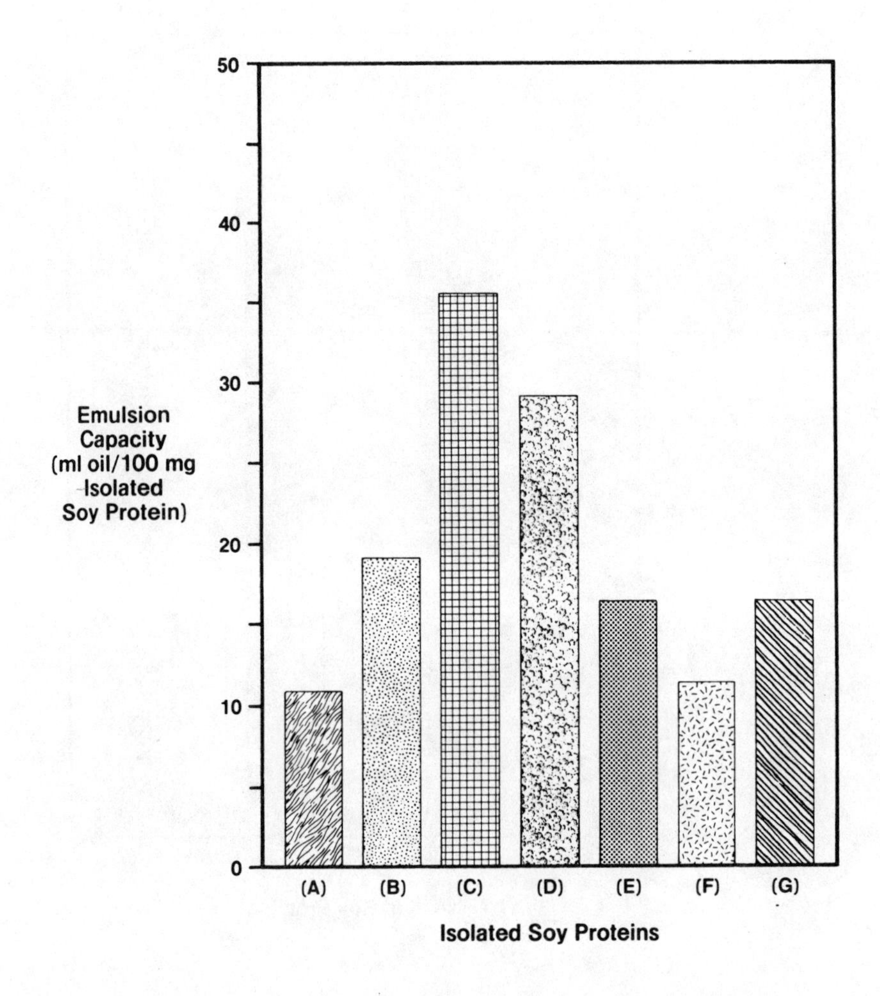

FIG. 6.3. Emulsion capacities of commercial isolated soy proteins.

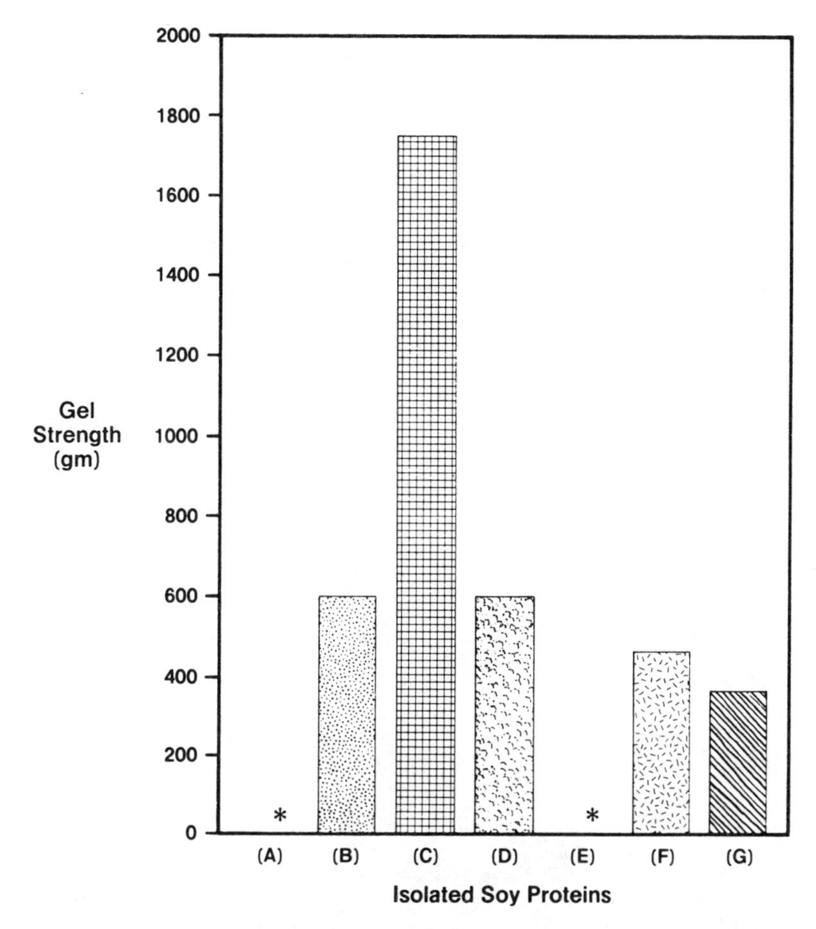

FIG. 6.4. Gel strengths of commercial isolated soy proteins (14% solids, 2% salt).

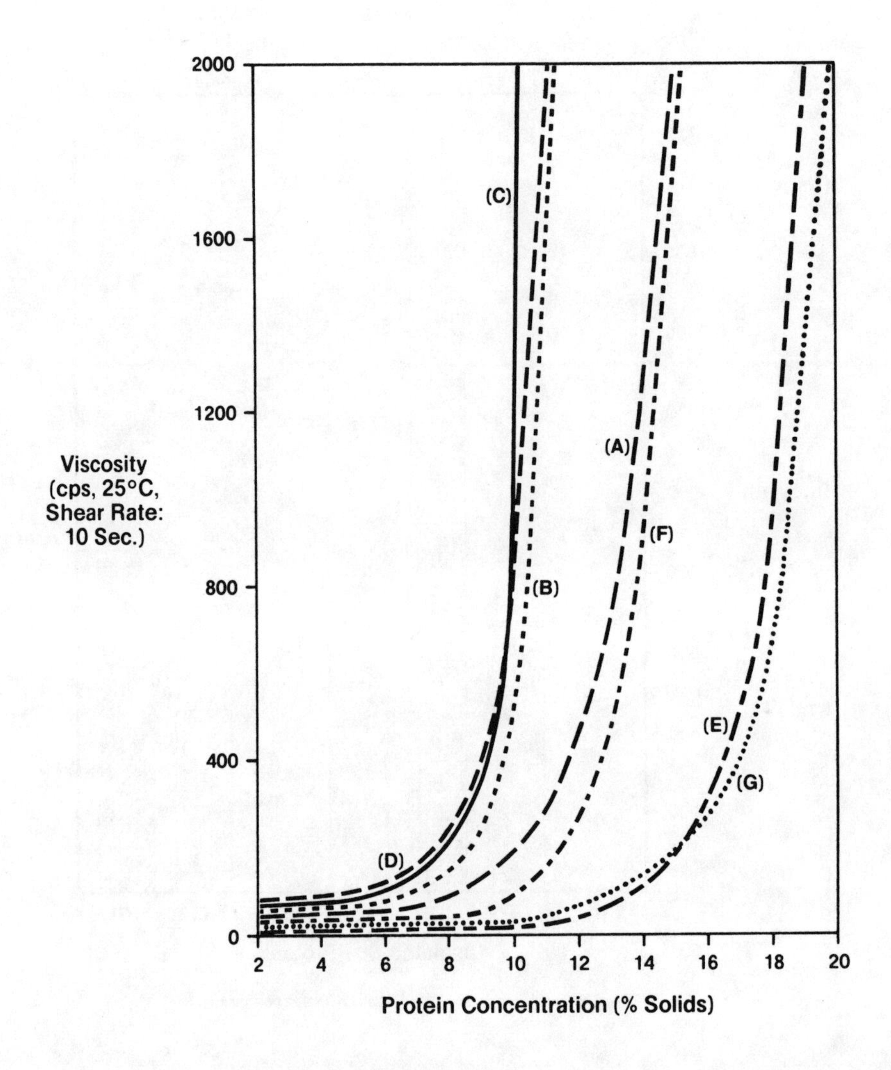

FIG. 6.5. Viscosities of commercial isolated soy proteins.

Water absorption is another property that can be used to describe proteins. This test is based on the Farinograph method for determining the water absorption of wheat flour. This method has been modified to measure the water absorption of the test protein in the presence of wheat flour. The water absorption of isolated soy protein, shown in Figure 6.6, ranges from about 150% to 250%[1]. Interactions between the isolated soy protein and wheat gluten and/or starch in wheat flour will have an effect on the water absorption value.

Forms of Isolated Soy Proteins

The physical properties described above were for several isolates available in powder form, which were designed to perform in specific food systems. One isolated soy protein in powder form was developed to have gelling and emulsifying properties similar to that of meat. This is an important characteristic in replacement of protein in traditional food products such as emulsified meats. It becomes a gel when hydrated with water and is compatible with emulsified meat products which have gel-like characteristics.

Another form of isolated soy protein is a structured product having a fibrous appearance. It has been designed to restructure poultry and red meats and, in some cases, augment the meat in the formula. An example of this application is a poultry roll. The fiber-like structure of this isolate is designed to add texture and mouthfeel to poultry roll products.

Another form of isolated soy protein is a granule. Since coarse ground meat products are particulate in nature, an isolated soy protein with granular characteristics was designed for this application. It can be used at levels as high as 20% on a hydrated basis while maintaining the traditional eating quality of ground beef. In addition to maintaining traditional eating quality, isolated soy protein can be used to alter the composition of traditional meat products. As an example, the fat level of ground beef can be reduced by 20% with the use of isolated soy protein.

Utilization of Isolated Soy Protein in Foods

There are two basic principles for utilization of isolated soy protein. One is for replacement of traditional protein ingredients in established food products. The second is as the primary protein ingredient (or as a functional ingredient) in new foods.

The successful use of soy proteins in traditional foods is based on reformulating traditional products in such a manner that the traditional quality of that product is maintained. This means identical color, flavor, texture, odor, overall eating quality, chemical composition, and nutrition. In new foods where quality expectations may not have been established, soy products must also contribute to the overall appeal of the product. It is important to note that economics and nutrition are not sufficient criteria for acceptance of new food ingredients. Past attempts to enter the food stream based only on these criteria have failed.

---

1. Anon (1982). Unpublished data. Ralston Purina Company.

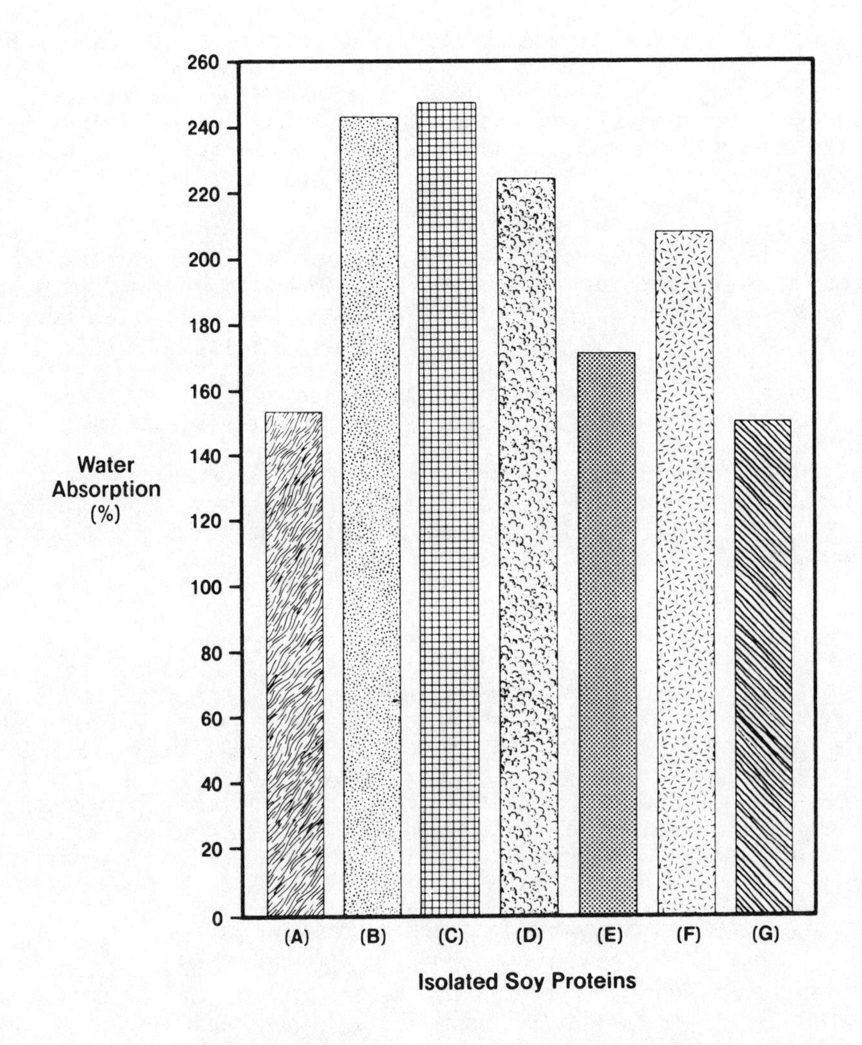

FIG. 6.6. Water absorption of commercial isolated soy proteins (contains 2% salt).

Today, we fully recognize that maintenance of traditional eating quality is the single most important factor in the utilization of isolated soy protein in established food products. Food is more than protein, fat, carbohydrate, vitamins, minerals, fiber and textural characteristics. It is part of every population's culture and tradition. Thus, when isolated soy proteins are being integrated with animal proteins in a given food system, it is critically important that traditional food characteristics and quality not be changed.

In addition to their use in established food products, isolated soy proteins are used in some foods as the sole source of dietary protein. Of course, to be successful, new foods must be quality products in their own right, even though quality standards have not been established in the consumer's mind.

Isolated soy proteins are being used in a wide variety of food products today. These include meat, poultry, seafood, baked goods, dairy-type products, infant foods, and dietary/healthy foods.

Meat Applications

A major application of isolated soy protein occurs in traditional meats where experience has shown that traditional quality must be maintained. The intact muscle products, such as hams and roasts, and comminuted products, such as frankfurters, bologna, and ground beef, are the principle forms in which meat is consumed. Comminuted products were one of the first meat applications for isolated soy protein.

Emulsified meats and coarse-ground meats are two important classes of comminuted products. In emulsified meats, non-meat proteins must perform the same functions as the salt-soluble meat proteins. These functions include emulsification, gelation, and fat and water binding. Levels of usage range from 1-4% in emulsified meat products.

In coarse-ground meats, textural properties are particularly important. Level of use in coarse-ground meats range from 1-5% of final product on a dry basis of isolated soy protein.

Whole cuts of meat can be augmented with isolated soy protein using techniques for cured meat preparation. A slurry of isolated soy protein, water, and salts can be injected into the muscle using a stitch pump or the slurry can be massaged into the muscle using other forms of cured meat technology. As in the case with emulsified meat products, the isolated soy proteins must perform the same functions as salt-soluble proteins. To reiterate an earlier point, they must perform these functions without interfering with the expected eating qualities of the finished product.

The following reformulation experiment is used to illustrate the concept of maintaining traditional quality (1). In this case, an emulsified meat product containing 90%--lean beef, 65%--lean pork, and pork backfat was reformulated with graded levels of commercial soy product (A) such that the fat and protein contents of the final products were constant and equal to those in the all-meat control. The color of the reformulated products was held constant using low levels of beef blood as a source of heme pigment.

A similar series was prepared with another commercial soy product (B). A trained sixteen-member panel compared the reformulations to the all-meat control for difference in color, flavor, odor, texture, and overall quality. Products were evaluated by panelists by comparing each sample to an identified all-meat control for degree of difference using a scale from 0 (no difference) to 10 (extreme difference). An unidentified control was included among the samples being evaluated.

As shown in Figure 6.7, a good linear relationship existed between the percentage of meat replacement and degree of overall quality difference for emulsified meats prepared from both soy protein products [(A), r = 0.997 and (B), r = 0.996]. About 16% of the meat could be replaced with soy protein product (A) before the test product became significantly different from the hidden all-meat control, whereas only 9% meat could be replaced with soy product (B) before the difference became significant. Because a color correction was made, the differences are primarily due to flavor and texture. These data indicate that soy products differ with regard to ability to replace meat and maintain traditional quality.

Preference or acceptance testing is often used to evaluate products reformulated with soy ingredients and it is not uncommon to find reports on reformulated products that are significantly preferred over the all-meat control. This type of testing misses the concept of maintaining traditional quality in which the goal is to be exactly like the control, not superior nor inferior to the control. Difference testing can help avoid erroneous and misleading conclusions when maintenance of traditional quality is of interest and, particularly, when the tests are conducted using a group of people who are culturally different from the ones for whom the information is intended.

In addition to maintaining the traditional quality of established products, new meat products are also important applications for isolated soy proteins. New products are frequently highlighted in trade journals and include formed and fabricated products, and convenience foods (such as frozen dinners, meat pies, and fabricated bacon-type products).

The use of isolated soy protein in the poultry industry is emerging today as the further processed segment of the industry grows. Isolated soy protein has proven to be a valuable ingredient to innovative poultry processors who have made great progress with new, highly successful further processed poultry products. In many cases, the isolated soy proteins are essential to the excellent quality of these products (7). Many new products, such as poultry rolls and poultry-based convenience foods, contain isolated soy protein. Poultry breasts injected with slurries of isolated soy protein, salt, and flavors are also becoming popular. These products are considered new products since only traditional whole cuts of poultry existed in the past.

The use of isolated soy protein in traditional seafood is best illustrated by the Japanese fish based products. Kamaboko, Chikuwa, and Agekama are traditional comminuted gel-like products that have been consumed in Japan for centuries. These products are based on a minced fish flesh ingredient called surimi.

The amount of surimi which can be replaced with isolated soy protein, while maintaining traditional quality, has been determined through systematic studies. Japanese fish sausage also contains surimi and has been successfully reformulated with isolated soy protein. Final products contain 1-3% isolated soy protein.

Outside of Japan, the principle applications are in new, further processed seafood products. Examples of these new products using isolated soy protein include crab salad and crab cakes which have been successfully introduced into the U.S. market. Increased use of isolated

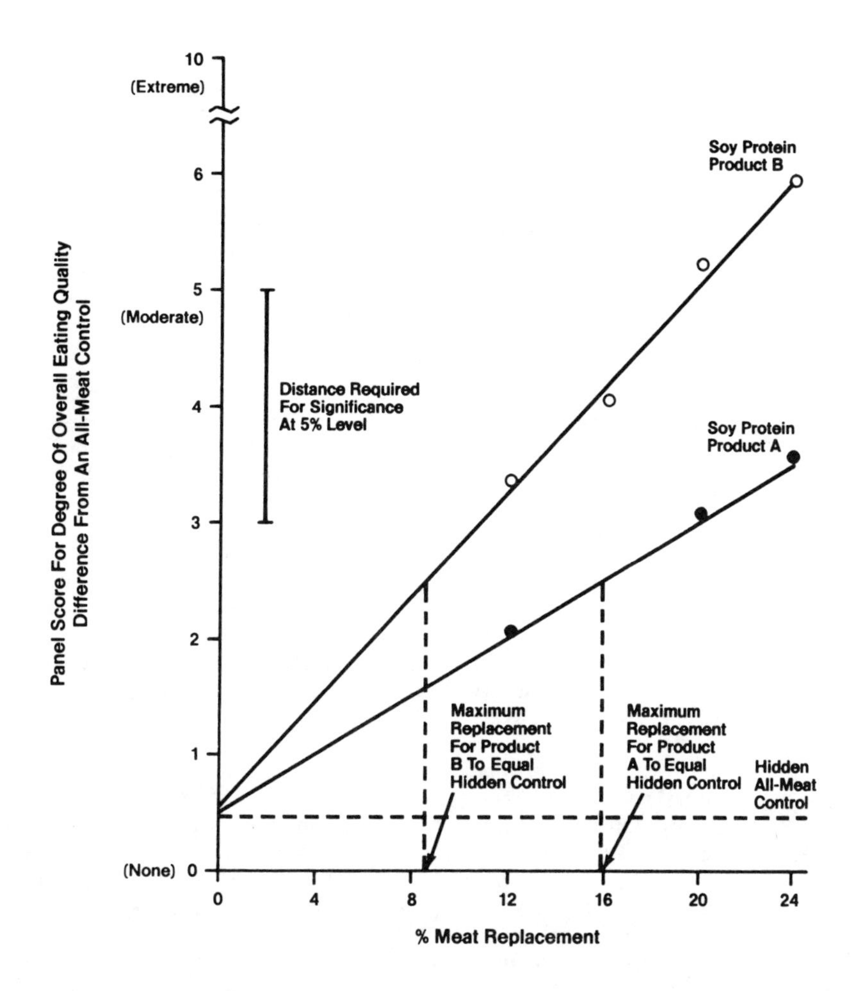

FIG. 6.7. Comparison of the effects of meat replacement by two soy protein products upon the difference in overall eating quality from an all-meat control.

soy protein in seafood is expected as the world becomes more efficient at utilizing its protein resources.

Bakery Applications

Isolated soy proteins are being used in cake mixes and cake donuts to simulate the functional properties of nonfat dry milk. Isolated soy protein, when combined with corn syrup solids, dried cheese whey, or other carbohydrates, functions as a replacement for nonfat dry milk in many of these baking applications.

Nonfat dry milk replacers in bakery products must control fat absorption, water absorption, flavor, film forming, and texture. These nonfat dry milk replacers are used at approximately a 2% level in retail mixes, commercial mixes, and by wholesale bakeries.

Dairy Applications

In addition to the replacement of nonfat dry milk in baked products, soy proteins are used to replace milk proteins in dairy-type products. Isolated soy protein has the highest protein content and lowest flavor profile of the soy products. This is a key factor in delicately flavored dairy products.

Isolated soy proteins are being used in emulsified products such as non-dairy coffee whitener and whipped toppings to replace sodium caseinate. The level of use ranges from 0.5-2.0%. Functional properties such as emulsification, emulsion stability, color, and flavor/aroma are critical factors in dairy applications.

Instant beverage mixes that are designed to be added to milk for use as a meal replacement use isolated soy protein as a primary protein source. In addition, isolated soy protein is being used in sour dressings to emulsify fat and to control viscosity and textural characteristics of the final product.

Infant formulas have traditionally been milk-based. Isolated soy protein-based formulas are being used successfully to replace milk-based formulas for the nutritional management of infants who are suspected to be allergic to milk or to have some type of milk intolerance. The first non-milk infant formulas contained full fat soy flours. These were of low quality, being dark in color and having a beany flavor. The carbohydrates in the soy flour were responsible for flatus and foul smelling stools. With the development of isolated soy protein, the manufacture of higher quality soy-based infant formulas became possible. These products have improved color, flavor, and aroma, and do not contain the indigestible carbohydrates found in soy flour products. Isolated soy protein supplies 100% of the protein present in these formulas. These formulas do not contain lactose, thus they can be used by infants or people who exhibit lactose intolerance because they are deficient in the enzyme lactase.

In addition to the milk-free or soy-based infant formulas, special formulas utilizing isolated soy proteins are designed and manufactured for older infants, geriatric, hospital and post-operative feeding. In addition, some isolated soy protein-based formulas are manufactured to be low in carbohydrate content to aid in the diagnosis of disaccharidase deficiency. In related products, soy proteins are used to increase the protein content of cereal products such as rice and wheat which are used as the first solid foods for infants.

Dietary/Healthy Foods

Another major use of isolated soy proteins is in the formulation of dietary/healthy foods. These foods have altered composition designed to provide a nutritional or health benefit, such as weight control, reduction of blood lipids or supplementation where certain foods cannot be tolerated. Some examples are protein supplements containing over 80% protein, meal replacement beverages, nutrition bars and low calorie entrees.

Isolated soy proteins are used in these products for several reasons. The isolates contain up to 90% protein on a dry weight basis and are very low in non-protein calories. Beyond the basic compositional and nutritional aspects, other properties of the protein must be considered for use in protein supplements and meal replacers. These include color, dispersibility, suspension, flavor, aroma, mouthfeel, and storage stability.

In addition to protein supplements and diet foods in dry powder form, there are a number of nutritionally complete foods on the market in the form of bars. These products contain a defined number of calories with the necessary vitamins and minerals for a complete meal and are marketed by major food companies. In some of these commercial diet food bars, the isolated soy protein is used to replace animal protein. In others, it is used as a major protein source. This is just one example of a growing number of weight control and healthy foods appearing on the market today.

In summary, isolated soy proteins are used in a wide range of products including meat, seafood, poultry, baked goods, dairy-type products, infant foods and dietary/healthy foods. They are intended to be used as a primary protein source, for functional purposes, and for formulating new foods to meet compositional requirements. Isolated soy proteins meet the criteria for a high-quality protein as defined by the Food and Nutrition Board of the National Research Council, National Academy of Sciences. Isolated soy proteins are being successfully utilized in traditional foods to add value while maintaining traditional eating quality and for the development of new food products. Experience has shown that it is important to maintain the traditional quality in these foods to meet the expectations of consumers. In new products, high quality is also an important factor even though quality standards have not been established as a result of a long history of consumption.

REFERENCES

1. Waggle, D. H., Decker, C. D., and Kolar, C. W., J. American Oil Chemists Soc. 58, 341 (1981).
2. FNB, "Recommended Dietary Allowances", Ninth Edition. Food and Nutrition Board, National Research Council/National Academy of Sciences, Washington, D.C., (1980).
3. Torun, B., Cabrera-Santiago, M. I. and Viteri, F. E., INCAP Annual Report. INCAP, Guatemala City, Guatemala, (1980).
4. Scrimshaw, N. S., and Young, V. R., in "Soy Protein and Human Nutrition" (H. L. Wilke, D. T. Hopkins and D. H. Waggle, eds.), pp. 121. Academic Press, New York, (1979).

5. Young, V. R., Rand, W. M., and Scrimshaw, N. S., Cereal Chemistry 54, 929 (1977).
6. FAO/WHO, Technical Report Series No. 522. World Health Organization, Geneva (1973).
7. Kardouche, M. B., Pratt, D. E., and Stadelman, W. J., J. Food Sci. 43, 882 (1978).
8. Waggle, D.H. and Kolar, C.W. in "Soy Protein and Human Nutrition" (H.L. Wilke, D.T. Hopkins and D.H. Waggle, edsr), pp 19. Academic Press, New York (1979).

7

Use of Whole Soybeans by the Consumer

RICHARD LEVITON

INTRODUCTION

The soybean is rightly called "the cow of China," a glowing reference to the impressive range of delicious foods fashioned from this 3,000 year old legume. With the bustle and excitement of entrepreneurial activity in the U.S. today with soyfoods--new companies, new products, growing consumer awareness--the soybean might also be dubbed "the golden cow."

The generic term "soyfoods" denotes the expanding family of high protein, lightly processed, Asian-styled, naturally-positioned, edible consumer products made from the soybean. These include tofu, tempeh, miso, soymilk, soy sauce, soy sprouts, soy flour, soy nuts, and an unlimited range of prepared secondary products, e.g. tofu burgers, tofu dips, tofu ravioli, tofu pizza, tempeh paté, soylami, tempehroni, miso dressings, soymilk ice cream. The term generally excludes soy protein ingredients intended for industrial reformulation, e.g. grits, isolates, concentrates, and extruded soy flour.

The rapid growth of the soyfoods industry since 1976 falls into four descriptive phases, described below.

A CAPSULIZED INDUSTRY HISTORY

The tofu industry was "invented" in early 1976 with the publication of The Book of Tofu by William Shurtleff and Akiko Aoyagi (1). This seminal work, which contained 500 recipes and information on the history, varieties, and commercial production of tofu, inaugurated the Caucasianization of tofu through the eventual entry of over 100 new young, middle class Caucasian American-run tofu plants. Prior to 1976 tofu slumbered in urban Oriental markets, being produced by about 50 Asian companies, but it was scarcely known even in the attentive natural foods market, and certainly nowhere else besides its own ethnic context. By May 1979, however, 120 tofu manufacturers were identified--nearly all of this increase was Caucasian proprietors bringing the healthful message of soyfoods to the general public.

The second phase, which began in 1979, was the advent of supermarket commodity retailing of tofu made possible through shelf-stable packaging. Tofu moved into mainstream, visible markets, somewhat divested of its

Cereals and Legumes in the Food Supply, edited by Jacqueline Dupont and Elizabeth M. Osman State University Press, Ames, Iowa 50010 low profile Asian connotations. Where anonymous bulk pails of beancurd had once prevailed, now branded water-filled plastic tubs lasting two to three weeks became the custom.

In the third phase--the soy delicatessen--we witnessed the proliferation of prepared convenience soyfoods in 1981. Numerous primary tofu and tempeh producers introduced attractively packaged, ready-to-eat products, thereby enhancing soyfoods consumption and renewed interest in the basic commodities.

The fourth phase, which began in 1983, involved brand name distinctions, price wars and market competition. About a dozen leading soy companies had grown sufficiently large to promote their products on an exclusive brand name basis, through advertisements, promotions, discounts, and demonstrations.

By mid-1983, then, in only seven years, tofu had achieved the status of household word among a significant portion of the American population. Casual references to tofu appear regularly in cartoons, articles, advertisements, and television talk shows. Tempeh-burgers have become the butt of a cantankerous journalist writing in a major men's magazine. Nearly every state has its own producer of tofu or at least a reliable supplier. Tofu is widely distributed in natural foods stores and supermarkets and is entering commercial food service such as restaurants and institutions. All of this presumes widespread awareness and consumption; and with tofu pioneering the field, tempeh and the other soyfoods will follow.

MAJOR SOYFOOD CATEGORIES: SALES AND VOLUME

In September 1981 the first per capita figures for U.S. soyfoods consumption were released (2). Lightly processed Asian-style soyfoods ($391 million, annual sales) combined with high technology soy ingredients ($616 million) represented $998 million in retail value. Per capita consumption of Asian soyfoods was 2.13 pounds, per capita expenditures, $1.77. Per capita consumption of high technology soy products was 6.45 pounds, expenditures, $2.80. The total per capita consumption was about 9 pounds, expenditures $4.57.

Examining these categories individually, we find that for tofu, the pack leader, in mid-1983 there were 182 manufacturers in the U.S. producing 27,500 tons annually from 11,000 tons of raw soybeans and for retail value of $55 million. By comparison, in May 1979, with 120 companies, 13,250 tons worth $23.2 million were produced. This was a 50% increase in manufacturers and over 100% rise in production in four years (3).

Tempeh (an Indonesian-derived fermented soybean patty, with a wonderful chewy, meaty texture, suitable for fish and poultry style dishes) is represented by 46 American companies. Annual production (as of May 1982) for the top 16 firms--92% of the industry--was 459 tons worth $1.8 million retail (3).

Miso is a Japanese fermented soybean condiment or paste, resembling peanut butter in consistency and used in soups, dressings, and dips. The U.S. industry is comprised of 8 companies with 1982 domestic (continental) production at 750 metric tons, up 525% from 1975, with another 640 metric tons from Hawaii alone. Total consumption (with imports of 959 metric tons included) for 1982 was 2,349 metric tons, up threefold from 1975 and growing 17% annually; total sales were $8.2 million (3).

Soy sauce (including wheat-free tamari and wheat-based shoyu) is an older, more established category with domestic annual production of 10.8 million gallons and imports of 2.3 million gallons; combined sales for 1981 were $203 million (3).

Soy nuts and soy nut butter, manufactured by 12 companies, represented sales of $4.6 million in 1981 while soy sprouts, with 5 companies, generated $250,000 (3).

One area of rapid growth and product proliferation is prepared convenience soyfoods. At least 30 companies (some primary tofu or tempeh makers, others reformulators) market innovative dairylike products, often frozen. Innovations include Tofu Slices (marinated, broiled slabs), Herb Garden and Spice Garden Tofu (preflavored with vegetables), Baked Tofu Slabs, Tofu Vegi-Dips, Tofu Lasagna, Soyettes, Ice Bean soy ice cream, and others. Such products capture the interest of nutrition and convenience-minded consumers while making "raw" soyfoods more accessible, less formidable, in the average kitchen. Tofu handily resembles dairy ingredients in secondary foods; tempeh assumes fish and poultry, even veal, qualities; and the range of product applications is probably unlimited.

SOYFOODS MARKET CHARACTERISTICS

In an almost breathless manner, soyfoods have grown in popularity in recent years largely due to a unique and highly favorable cultural context. By contrast, in the 1950s, when textured soy protein meat analogs were introduced, the climate was less favorable as Americans experienced rising affluence and meat-centered diets. Even yogurt, today phenomenally successful, began its plodding growth back in 1945 and had to wait until the late 1960s for its sales to finally skyrocket. In the late 1960s important new factors came into play in the United States which contributed greatly to the prominence today of soyfoods.

This distinctive milieu can be characterized in eight observations. First, the influential and often cited Baby Boom population cohort--the now maturing group of 72 million babies born between 1947 and 1965, now aged 20-38--has fueled, largely, the emergence of natural foods as the supportive context for soyfoods. This population segment--called "the pig in the python" in reference to its enormous numbers of educated, semi-affluent young Americans entering careers, starting families--has set the food and nutrition agenda for the 1980s.

Second, the natural foods industry is burgeoning with sales of $2.9 billion for 1982, up 21% over 1981. Supermarkets are pioneering "nutrition centers" within their general formats and now account for 1,500 centers out of the industry's total of 9,350 retailing units (4).

Third, the message of natural foods has entered the mainstream market as a clamoring for "light" foods, low calorie, non-fattening, dietworthy, cholesterol-free products. Reportedly 60% of households include someone trying to lose weight or making dietary alterations.

Fourth, the Asian Boom in population--Orientals are now 1.5% of the population, having doubled since 1970--has had a strong market influence. Japanese Americans have swelled their numbers by 19%,

Filipinos by 126%, Koreans 413%, Hawaiians 67% compared to the general population which grew by only 11% in 1981 (5). All of these Asian Americans are lifelong tofu and soyfoods customers. The new Asian demographics are influencing eating trends, generating an ethnic wave in prepared foods and restaurant fare.

Fifth, there is a heightened medical awareness of the connections between diet and disease, of proper eating and physical well-being. Suddenly millions of Americans have discovered the need to reduce animal fats and cholesterol intake while maintaining varied and nutritious diets.

Sixth, there is what Craig Claiborne terms "the gastronomic revolution." This is an upswell of interest in gourmet cooking, weekend home entertaining, the demand for exotic specialty foods, and the search for the chic in new foods.

Seventh, the rising cost of living has led consumers by the pocketbook to soyfoods for their comparative inexpense for high quality protein. The Food Marketing Institute in 1982 reported that 50% of consumers were changing their eating habits for economy rather than nutrition whereas in 1978, 75% had made changes for nutritional reasons (6).

The eighth factor summarizes the impact of the previous seven. Soyfoods today have the invaluable patina of "natural," "organic," "wholesome," and "simple" associated with them. These are connotations nearly impossible to impose artificially on products of lesser quality. The 1950s style meat analogs and soy protein ingredients tend to have an image of "imitation," "processed," and "high technology," connotations very hard to shake off and rendering them unacceptable in the emerging soyfoods consumer market.

These factors have synergistically propelled soyfoods into prominence and retail success and constitute the continuing foundation for market survival. Meanwhile, one becomes curious about the demographics of soyfoods consumers.

Regarding the typical natural foods shopper, the profile may be drawn as follows: women, aged 25-34, college educated, single, median income of $19,000, living in the Northeast and West (7). The average yogurt user is aged 18-44 years, college educated, employed full-time, female, professional or managerial, single, white, living in a metro suburban area, in a 3-4 person household, with income of $25,000 (8).

In 1981, the FIND/SVP consumer research group made the first survey of actual tofu consumers in an 800 person sample from 25 SMSAs. They found 17% of tofu consumers have an income between $10,000 and $16,000; 14% earn less than $10,000; 18% earn more than $40,000. While this profile covers both "upscale" and "downscale" consumers, the link is education and nutrition literacy. Further, they found 39% are aged 25-34, about 75% female, 50% living on the West Coast, 85% were regular yogurt purchasers, and 25% heavy yogurt users (9).

FACTORS EXPANDING THE MARKET FOR SOYFOODS

The soyfoods industry has entered a phase of deliberate market building through heightening both generic and brand name awareness of tofu and tempeh products. There are four principal aspects to this strategy.

The first method is through both purchased and invited media attention. In 1982 the first national advertising for tempeh appeared in consumer magazines. The ADM company began sponsoring soy protein commercials on Sunday morning news programs and referred to soyfoods (including tofu) while reaching 9.5 million households. A leading tofu company generated 8 million discount coupons for their branded tofu as part of full page color ads in national magazines in April 1983. The hugely popular Phil Donahue presented two natural foods chefs who demonstrated tofu recipes on television in May 1983.

Second, soyfoods companies are staging major promotions in natural foods outlets and supermarket chains. One company sponsored a successful Chinese New Year tofu promotion for three years. Several leading retailers sponsored multievent Soyfoods Weeks with demonstrations, discounts, prizes, and print attention. Another leading natural foods distributor has designated an entire month for soyfoods promotions and discounting.

Third, the flow of generalized consumer information on soyfoods is accelerating. The Soyfoods Association, the newly formed industry trade group (February 1983) has ambitious plans for a thorough soyfoods promotional campaign using all media channels. On a somewhat random though steady basis, a constant flow of soy cookbooks appear in the market. As of mid-1983 about 45 are in print and available in the U.S. Many manufacturers provide recipes with their products while many communities have individuals presenting soyfoods cooking classes.

The fourth, and probably most significant, contribution to an expansion of the use of soyfoods is the interest exhibited today by full service restaurants and institutional food service. The estimated 600 natural foods restaurants in the U.S. consume 611.5 tons of tofu annually, or about 2.25% of industry production. Asian style formats use another 3,822 tons; combined restaurant usage is at least 4,433.5 tons or 16.25% of current volume (10).

Institutional food service, including hospitals, colleges, correctional, elderly meal sites, are experimenting with soyfoods. Their interest in tofu, particularly, focuses on its low cost (An estimated $9,204.00 per year could be saved with a 50/50 ratio of tofu/hamburger in patties served six times weekly), versatility, adaptability, and convenience (11).

Typical uses for tofu in restaurants run the ethnic gamut from Italian, French, and Greek to Chinese and Japanese, while recipe presentations include wok-fried with vegetables, blended into dairy-like dips, dressings, and sauces, as a cheese replacer, and in desserts like cheesecakes and puddings. In institutional food service tofu is often used as a beef or poultry extender in patties and casseroles or as an enhancer in prepared dressings, or as an extender for eggs or cheese.

FUTURE PROSPECTS

Various industry observers have prognosticated on the future prospects for soyfoods in the United States and the view is consistently positive. The FIND/SVP report predicted that tofu sales, then (1981) at $50 million, would grow by 300% by 1986 to $200 million, or at the rate of 32% yearly. The statement was offered that tofu might become "the yogurt of the 1980s." The report concluded, "The potential market for tofu, positioned as a tasty, versatile food, is at least as great as the current market for yogurt," now $575 million (9).

Tofu will enter the American mainstream diet in three ways, suggests William Shurtleff (3). This will be as a vacuum-packed firm, cheeselike product, retailed in cheese sections; as a line of low calorie tofu dressings, chip dips, and mayonnaise; and as a line of convenience, heat-and-serve, meatless entrees using tofu. Moreover, one of America's large food processing firms is likely to introduce a tofu or soyfood product; similarly, one of Japan's major tofu or soymilk processors is likely to enter the American market. In either case, the national marketing and large scale promotions that will ensue will broaden consumer awareness of soyfoods significantly.

My own view of prospects runs as follows. The market for whole soybean products, or soyfoods, will be enhanced by these coming developments: 1) a tightening up in the manufacturing community with little increase in new plants; 2) a deeper penetration of existing markets by present manufacturers; 3) a widening of selected product distribution to a 1,000 mile radius, with some companies becoming national in scope; 4) extended shelf life capabilities of one to two months for tofu and tempeh, less soured or spoiled products in stores; 5) a further diversification of prepared convenience soyfoods; 6) the entry of "soyfoods" into the vernacular; 7) widespread and well-reported restaurant menuing of tofu; 8) tofu products that closely resemble dairy items like yogurt and parfaits; tempeh products that resemble prepared beef patties or veal cutlets; 9) continuing support from the diet-health-fitness-soyfoods link.

The 1980s will be viewed by historians as a remarkably supportive period for the rise of soyfoods which became daily staples for millions of consumers.

REFERENCES

1. Shurtleff, W. and Aoyagi, A. The Book of Tofu, (insert publishers) (1976).
2. Leviton, R., and Shurtleff, W. Soyfoods 6, 6 (1982).
3. Shurtleff, W. Soyfoods Industry and Market--Directory and Databook. Soyfoods Center, Lafayette, CA (1983).
4. Second Annual Market Overview. Natural Foods Merchandiser 45, (1983)
5. Leviton, R. Soyfoods 7, 61 (1982).
6. Economics Overshadows Nutrition. Natural Foods Merchandiser 116 (1982).
7. Health Food Store Shoppers. Prevention Magazine, Emmaus, Penn. (1981).
8. Simmons Market Research Bureau, (1980).
9. The Tofu Market: Overview of a High Potential Industry. FIND/SVP Information Clearing House, New York (1981).
10. Leviton, R. Soyfoods 9, 18 (1983).
11. Leviton, R. Soyfoods 8, 18 (1983).

8

Technical Aspects of Whole Soybean Use

HWA L. WANG

INTRODUCTION

Whole soybeans have been used as food in the Orient since ancient times. To a great extent, they have provided the Oriental population with needed protein. People outside of the Orient have had the idea that soybeans are not suitable for human food. Undoubtedly, cultural traditions dictate food customs, but the inherent characteristics of the beans and the aversion of people to anything new also curb the popularity of soybeans.

In addition to the beany flavor, disagreeable taste, flatulence, and antinutritional factors, one other notable problem that limits the use of whole soybeans as food is the cooking time required. These problems are not really unique to soybeans, but rather are common to many beans and other pulses consumed around the world. Because of the high oil and low carbohydrate content, lack of starch and compact texture, soybeans do not cook as soft or as readily as many other beans. Traditionally, soybeans are soaked in water overnight before cooking. The addition of a small amount of sodium bicarbonate (baking soda) in soaking or cooking water has been suggested as a means of reducing cooking times of dry legumes. The recommended amount of sodium bicarbonate varies between 0.07% (1, 2) and 0.5% (3). However, beans prepared with greater than 0.2% sodium bicarbonate solution were darker and had a mealy texture and an alkaline flavor (1). Another method suggested for reducing cooking time is to fry soaked beans in oil prior to boiling (4). When soybeans are cooked properly, consumers may find them as acceptable in their diets as other beans.

TRADITIONAL USES OF WHOLE SOYBEANS AS FOOD

To make soybeans as palatable as possible, many methods of preparation have been devised in the Orient. Not only are whole soybeans

1. The mention of firm names or trade products does not imply that they are endorsed or recommended by the U.S. Department of Agriculture over other firms or similar products not mentioned.

Cereals and Legumes in the Food Supply, edited by Jacqueline Dupont and Elizabeth M. Osman 1987 Iowa State University Press, Ames, Iowa 50010

cooked and consumed, they are also simply processed to make such products as soybean sprouts, soybean milk, protein-lipid film, soybean curd, and soybean flour, as shown in Table 8.1 (5).

When the developing soybeans are at about three-fourths maturity, the beans reach a maximum fresh weight and are green and firm; the pods are greenish yellow. The Orientals refer to the beans at this stage of development as Mao-tou, edamame, or fresh green soybeans; they are used as an important seasonal vegetable. The green beans are cooked in the pods or they can be shelled and cooked like peas. The immature soybeans cook to a tender stage much more quickly than the dry beans.

Fermentation processes are widely used in the Orient for preparation of soybean foods. The well-known ones are listed in Table 8.2 (5). Soybean fermentations are neither strictly fermentation of soybeans alone nor carried out by only one kind of microorganism; they may involve a substrate consisting of both cereals and soybeans and an inoculum consisting of bacteria, yeasts, and molds, such as the soy sauce and miso fermentations. Some of these fermented products are flavoring agents, and others like tempeh and natto are staples.

Because of the recent popularity and interest in tofu and tempeh, this presentation will be limited to the technology of preparing these two products. Traditional fermentation processes have been reviewed elsewhere (6-8).

HYDRATION OF WHOLE SOYBEANS

Regardless of process used and the end product desired, the first and most important step in making soybean foods from dry beans in the Orient is to thoroughly soak and dry beans. The beans first are washed and rinsed to clean, and then are soaked in fresh water. Depending on air temperature, beans are soaked from several hours to overnight. When the soaked beans are drained and rinsed, they are ready for use.

The soaking process long has been believed to increase wholesomeness of the product, as well as to reduce cooking time; however, it is not always clear whether the improvement is in nutritional value, texture, or flavor of the product. Modern soybean processing technology, on the other hand, employs no soaking, so as to reduce processing time and loss. However, Lo et al. (9) have shown that despite a loss in solids during soaking, soybeans soaked before water extraction yielded milk with higher solids than did unsoaked beans or bean flour, because the soaking process made possible a better dispersion and suspension of bean solids during wet extraction.

The conditions for hydrating whole soybeans, the changes that take place during hydration, and the effects of hydration on cooking quality have been investigated by Wang et al. (10). Data (Fig. 8.1) indicate that soybeans absorb water rapidly for the first 2 hr at 20°C to 37°C and then at a declining rate until they become saturated. Also, the rate of water absorption increases as the temperature increases. The soybeans absorbed an equal weight of water after steeping 5.5 hr at 20°C and 2.5 hr at 37°C. When the weight of the hydrated beans is about 2.4 times that of the original beans, complete hydration is approached. Saturation is reached at different times, depending on the temperature: 6 hr at 37°C, 12 hr at 30°C and 25°C, 16-18 hr at 20°C. As expected, the loss of solids is greater with longer soaking

TABLE 8.1. Oriental Nonfermented Soybean Foods (5)

Foods	Local names	Description	Uses
Fresh green soybeans	Mao-tou, edamame	Picked before mature, plump, firm, bright green	Steamed or boiled in the pods and shelled or shelled before cooking, served as fresh green vegetable
Soybean sprouts	Huang-tou-ya, daizu no moyashi	Bright yellow beans with 3-5 cm sprouts	Steamed or boiled, served as vegetable or in salad (parboiled)
Soybean milk	Tou-chiang	Water extract of soybeans, resembling dairy milk	Boiled, served as breakfast drink
Protein-lipid film	Tou-fu-pi, yuba	Cream-yellow film formed over the surfact of simmering soybean milk, moist and firm, or dried and brittle in the form of sticks, sheets, or flakes	Cooked and used as meat
Soybean curd	Tofu, tou-fu, tubu, tahoo, touhu, tau-foo, dou-fu, dan-fu	White or pale yellow curd cubes precipitated from soybean milk with a calcium or magnesium salt or vinegar, soft to firm	Served as main dish with or without further cooking
Soybean flour	Tou-fen, kinako	Ground roasted dry beans, nutty flavor	Used as filling or coating for pastries

TABLE 8.2. Oriental Fermented Soybean Foods (5)

Foods	Local names	Organisms used	Substrate	Nature of Product
Soy sauce	Chiang-yu, shoyu, toyo, kanjang, ketjap, see-ieu	*Aspergillus*, *Pediococcus*, *Torulopsis* and *Saccharomyces*	Whole soybeans or defatted soy products, wheat	Dark reddish brown liquid, salty taste suggesting the quality of meat extract, a flavoring agent
Miso	Chiang, doenjang, soybean paste	*Aspergillus*, *Pediococcus*, *Saccharomyces*, *Torulopsis* and *Streptococcus*	Whole soybeans, rice or barley	Paste, smooth or chunky, light yellow to dark reddish brown, salty and strongly flavored resembling soy sauce, a flavoring agent
Hamanatto	Tou-shih, tao-si, tao-tjo	*Aspergillus*, *Streptococcus* and *Pediococcus*	Whole soybeans, wheat flour	Nearly black soft beans, salty flavor resembling soy sauce, a condiment
Sufu	Fu-ru, fu-ju, tou-fu-ju, bean cake, Chinese cheese	*Actinomucor*, *Mucor*	Soybean curd (tofu)	Cream cheese-type cubes, salty, a condiment, served with or without further cooking

Tempeh	Tempeh kedelee	*Rhizopus oligosporus*	Whole soybeans	Cooked soft beans bound together by mycelium as cake, a clean fresh and yeasty odor. Cooked and served as a main dish or snack
Natto		*Bacillus natto*	Whole soybeans	Cooked beans bound together by and covered with viscous, sticky polymers produced by the bacteria, ammonium odor, musty flavor, served with or without further cooking as main dish or snack

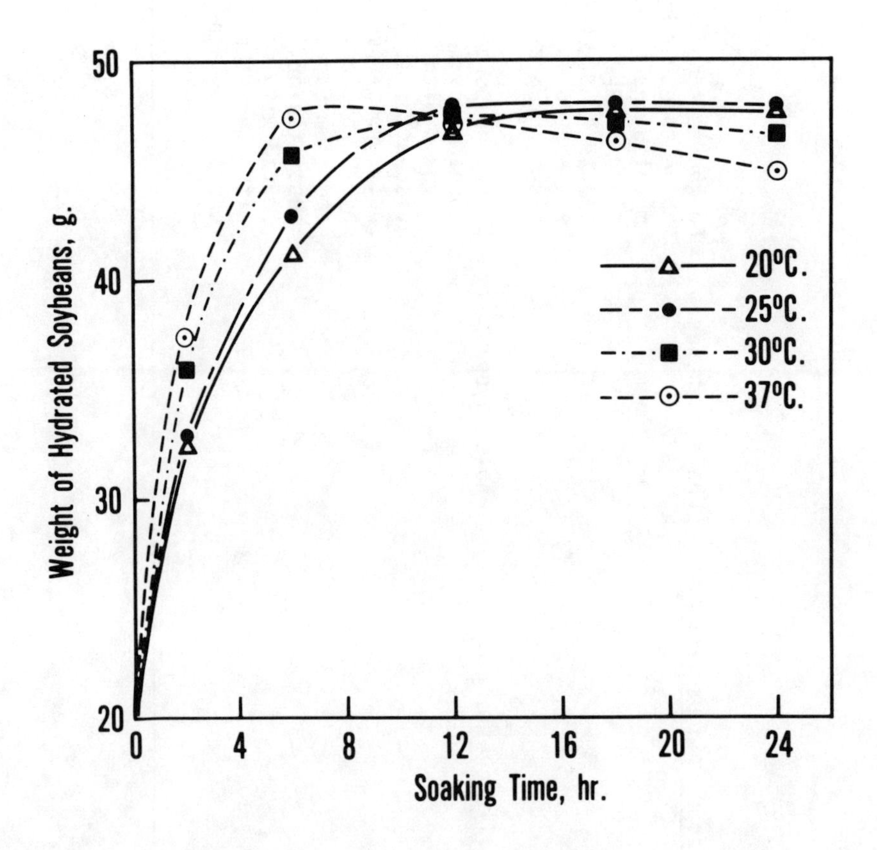

FIG. 8.1. Rate of water uptake by soybeans with time as affected by temperature (10).

time and higher temperature. The amount of solids leached out during the time required to reach complete hydration at different temperatures is about 4.5% of bean dry weight, but the percentage of protein loss is greater at higher temperature (25-37°C) than at lower temperature (20°C). At lower temperature, the rate of solids losses beyond the saturation point is much slower than at higher temperatures. Thus, soaking time becomes more critical at higher temperatures. It was also found that, at complete hydration, about 30% of raffinose and 20% of stachyose are leached out of the soybeans. Raffinose and stachyose are the oligosaccharides that cause gas formation in the digestive tract.

Very little quantitative information is available on the effect of hydration on cooking rate of the beans. By using the Instron Universal Testing Machine, it was found that unsoaked soybeans required 1.5 hr of cooking to achieve the same degree of tenderness as soaked beans cooked for 1 hr (Table 8.3). Furthermore, longer cooking of the unsoaked beans does not give tenderness approaching that of the soaked beans. Complete hydration, on the other hand, does not further increase the tenderness as compared to partially hydrated beans, but it does increase the weight. Also, the maize-yellow color of the soaked cooked beans is lighter and brighter than that of the unsoaked cooked beans.

The soaking process, therefore, can reduce flatulent factors, improve cooking quality, and facilitate extraction. For convenience and to keep the soaking loss at a minimum, soaking the soybeans at an ambient temperature around 20-22°C for 16-18 hr is most suitable. It is also recommended that the beans be washed before soaking to reduce surface microbial contaminations. The presence of heat-resistant, sporeforming bacteria observed on soybeans (unpublished data) suggests that bacterial contamination could occur. This could shorten the shelf-life of such foods as soybean milk, tofu, and tempeh, which are made from whole soybeans and have a short cooking time during processing.

TOFU

Tofu is the most important soybean food in the Orient. It has much the same importance to the people of the Orient that meats, eggs, and cheese have for the people in Western countries. Tofu is closely associated with soybean milk, because the initial step in making tofu is to make soybean milk, a protein-oil emulsion extracted from whole soybeans. When a mineral salt or vinegar is added to soybean milk, coagulation occurs that is quite similar to cottage cheese produced from dairy milk.

Tofu is a highly hydrated, gelatinous product and, depending on the water content, an array of tofu with different characteristics can be produced. In the Orient, the typical type has an approximate composition of 85% water, 7.5% protein, and 4.3% oil. This type of tofu has a soft pudding-like texture but is firm enough to retain its shape after slicing. Japanese prefer tofu having a smooth, soft, fragile texture, which contains 90% or more water. In addition to this soft tofu, the Chinese produce many types of firm tofu products with water content as low as 50-60%. This type of tofu product is usually flavored with soy sauce, sugar, and spices and has a meat-like texture. Tofu found in the U.S. supermarkets contains about 75 to 80% water. According

TABLE 8.3. Effect of Hydration and Cooking Time on the Weight and Tenderness of Cooked Soybeans (10)

Soaking time (h)	H_2O uptake (%)	Cooking time (h)	Weight of cooked beans[a] (g)	Inverse of tenderness TEA[b] (kg-cm)
0	0	1.0	45.4	0.269±0.067[1]
		1.5	46.8	0.169±0.067[2]
		2.0	46.8	0.162±0.052[2]
		2.5	46.6	0.146±0.052[2]
3.5	100	1.0	46.9	0.148±0.049[2]
		1.5	47.5	0.106±0.038[3]
		2.0	47.8	0.0.94±0.027[3]
18	142	1.0	48.6	0.155±0.049[2]
		1.5	48.6	0.114±0.051[3]
		2.0	48.8	0.089±0.030[3]

[a]From 20 g of dry soybeans.

[b]TEA, an instron reading, indicates total energy absorbed. Data are means ± S.D. of 50 beans from two experiments. Means without a superscript number in common are significantly different.

to the tofu producers, Western consumers prefer tofu with a firm and chewy texture.

Traditionally, three main steps are involved in making tofu: preparation of soybean milk, coagulation of protein, and formation of tofu cubes in a mold (Fig. 8.2). Dry soybeans are washed and soaked in water overnight or until the beans are fully hydrated. The soaked beans are drained, rinsed, and ground with water. The slurry is brought to a boil and kept at boiling temperature for 15 min. The boiled slurry then is filtered through cheesecloth, yielding a milk-like product known as soybean milk. When bittern or nigari (the bitter liquid that remains after salt is crystallized from sea water), gypsum, magnesium, calcium salts, or vinegar is added to the milk, a curd forms. This curd is pressed to remove excess whey, forming a highly hydrated curd called tofu. The curd is cut into cubes and sold in a water-packed container. The tofu process is a very simple one; however, making a high-quality and reproducible product is another matter, especially for the Japanese type of soft tofu. Many factors, from the dry beans to pressing the curd, could affect yield and quality of the resulting tofu. Several studies have been made on tofu processing in recent years in an attempt to understand the process and to optimize processing conditions.

The amount of water needed to make the soybean milk is important. By experience, the Orientals have found the most suitable ratio of water to dry beans to be 8:1 to 10:1. Watanabe et al. (11) reported a significant reduction in the amount of protein and total solids extracted when the amount of water used is reduced to less than 6.5 times that of the dry beans. Increasing the amount of water above 10:1 increases the amount of solids and protein extracted. However, excess water would result in a soybean milk with a protein concentration too low to obtain proper curd formation. Therefore, a 10:1 ratio of water to dry beans, including that absorbed during soaking, is preferred.

Grinding or blending the soaked beans in water facilitates the extraction and also the formation of protein and lipid emulsion. Hot grinding (12) et al., and heating of the soaked beans before grinding (3) to prevent formation of off-flavor due to lipoxygenase have been suggested and practiced at some of the Western tofu production plants.

The heat treatment is essential not only for protein denaturation to attain proper curd formation (13) but also to improve nutritional value and reduce off-flavor. Tofu often is consumed without further cooking; therefore, sufficient heat treatment is essential to destroy the antinutritional factors and to obtain the maximum nutritional value of the soybean milk. In vitro digestibility (Fig. 8.3) and amino acid composition indicate that maximum nutritive value of soybean milk can be ensured by boiling for 10-15 min, but excessive heat may adversely affect the nutritive value. Watanabe et al. (11) reported that boiling the soybean slurry for more than 20 min not only reduces the total solids recovery, and thus reduces the tofu yield, but also affects the tofu texture. Therefore, boiling the slurry for 15 min is recommended.

Coagulation is the most significant step in terms of yield and texture of tofu, but it is the least understood. In the Orient, making of tofu has been considered an art; even today, the relationship between the ion binding to the soybean protein and the coagulation phenomenon of the protein is not completely understood.

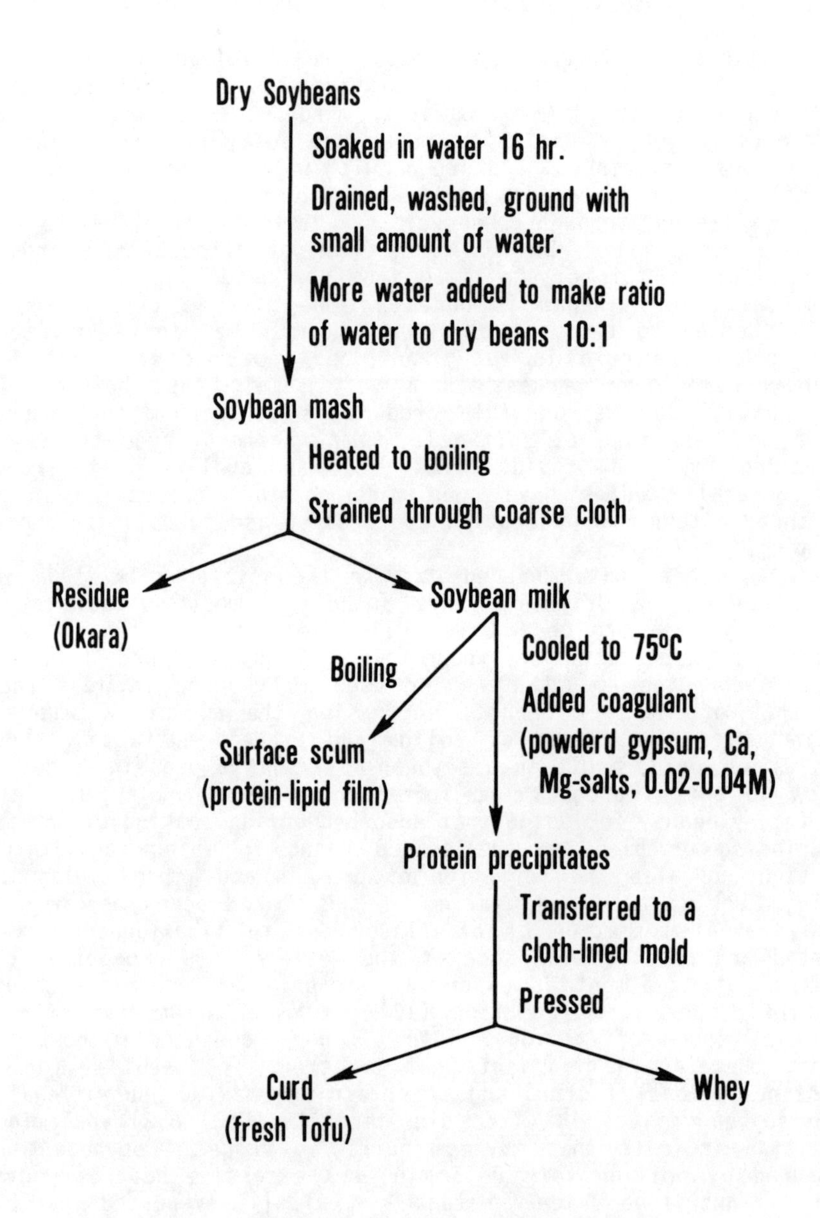

FIG. 8.2. Flow diagram for the preparation of tofu (5).

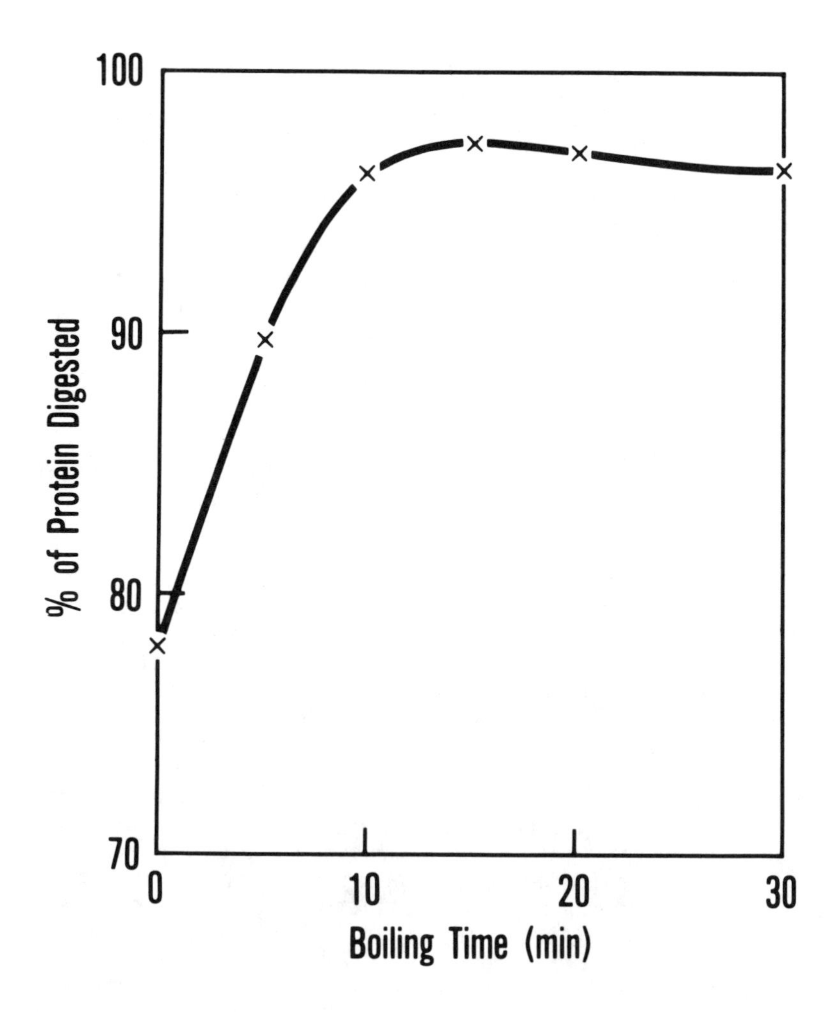

FIG. 8.3. *In vitro* digestibility of soybean milk as affected by duration of boiling (14).

Data in Figure 8.4 (14) show that both ionic concentration and type of coagulant affect the gross weight and moisture content of the final product, as well as total solids and nitrogen recoveries. Except when calcium sulfate is used, gross weight, moisture content of tofu, and total solids recovery decrease as the salt concentration increases from 0.01 to 0.02 M, remain about the same between 0.02 to 0.04 M, and then steadily increase at higher concentrations. No curd is noted when the concentrations of coagulant are lower than 0.008 M or higher than 0.1 M, although in some cases thickening of soymilk occurs. The percentage of nitrogen recovery, on the other hand, increases as the concentration of salt increases, remains the same at 0.02-0.04 M, and then decreases at higher concentrations. In studying the binding of unfractionated soybean proteins with calcium ion, Appurao and Rao (15) observed that, at higher concentrations of calcium ion, the extent of precipitation decreases and the protein becomes soluble again, which explains our results of decreasing nitrogen recovery at higher salt concentrations.

Because of the limited solubility of calcium sulfate, the actual ionic concentration at each level is uncertain, and the concentration gradient is less than that indicated. This could partly account for the smaller variation noted in tofu made with calcium sulfate as compared to that made with the other three salts. This study shows that salt concentrations between 0.02 to 0.04 M have the least effect on the four quantities investigated, and also result in the highest nitrogen recovery. Therefore, use of salt at a level between 0.02 to 0.04 M is more likely to yield reproducible product with high nitrogen recovery. For the same reason, calcium sulfate is preferred.

The texture characteristics of the curds also are influenced by concentration and type of coagulant (Fig. 8.5). When the concentration of the coagulant is increased from 0.01 to 0.02 M, significant increases in hardness, brittleness, cohesiveness, and elasticity are noted. No significant effect is observed at concentrations between 0.02 to 0.04 M, but above that range these measurements of the curds decrease steadily. Calcium chloride and magnesium chloride result in curds with much greater hardness and brittleness than do calcium sulfate and magnesium sulfate, suggesting that anions have a greater effect than cations on these two parameters. These data again indicate that the use of salt at a level between 0.02 to 0.04 M is likely to yield reproducible firm products. If all the conditions are the same, the use of calcium chloride or magnesium chloride would produce harder tofu than the sulfate salts.

Watanabe and coworkers (11) made a comprehensive study on tofu making. They found that the soybean variety used had an important effect in making tofu. However, their study was made at a salt concentration below 0.02 M, because the soft tofu most suitable for Japanese taste is made at salt concentrations between 0.01 to 0.011 M. From our results, one can see that the quality and yield of tofu made at salt concentrations below 0.02 M are greatly affected either by the amount of salt used or by the protein concentration of the soybean milk when salt concentration is kept constant. Therefore, without adjusting the salt concentration used for each batch of milk or adjusting each batch of milk to a constant protein content, reproducible and uniform products are difficult to obtain. Protein content of soymilk is related to the protein soluble index and protein content of the variety used.

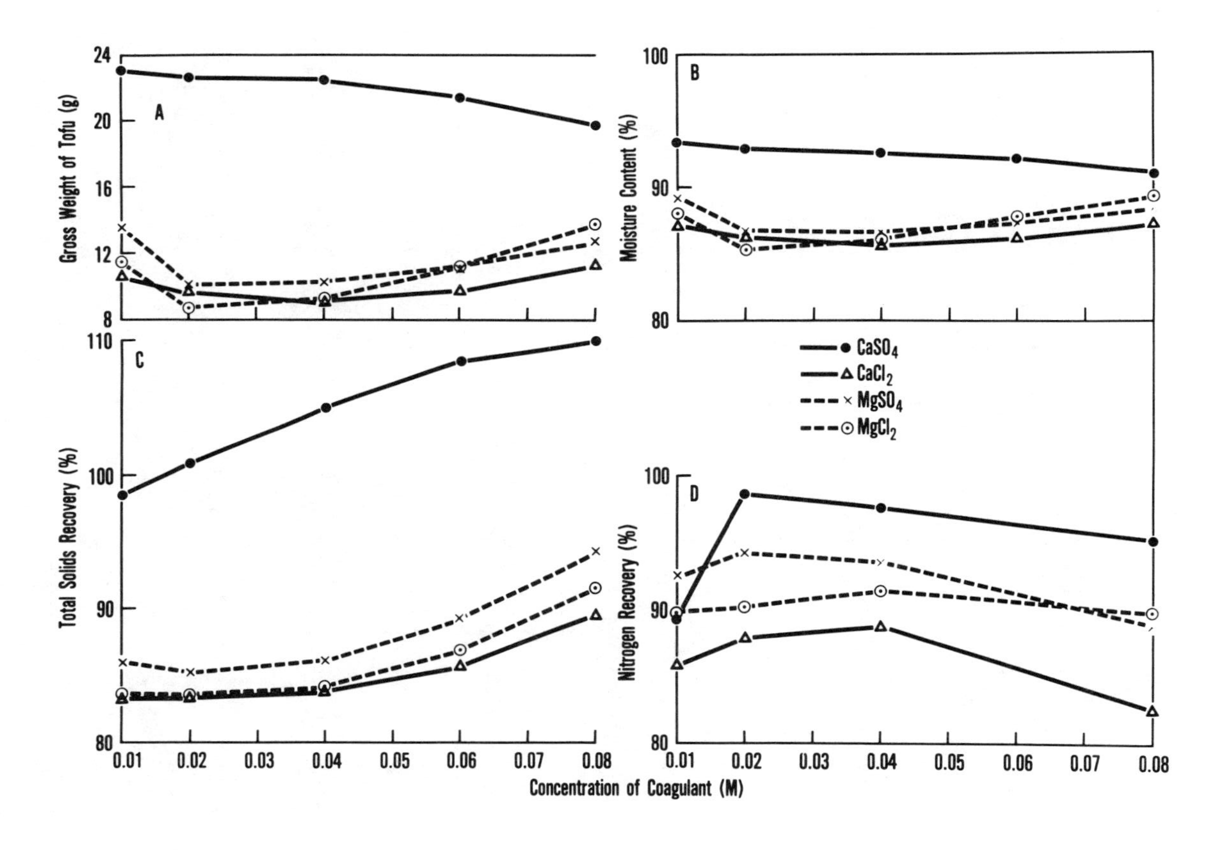

FIG. 8.4. Relationship of concentration and type of coagulant to the yield of tofu (14).

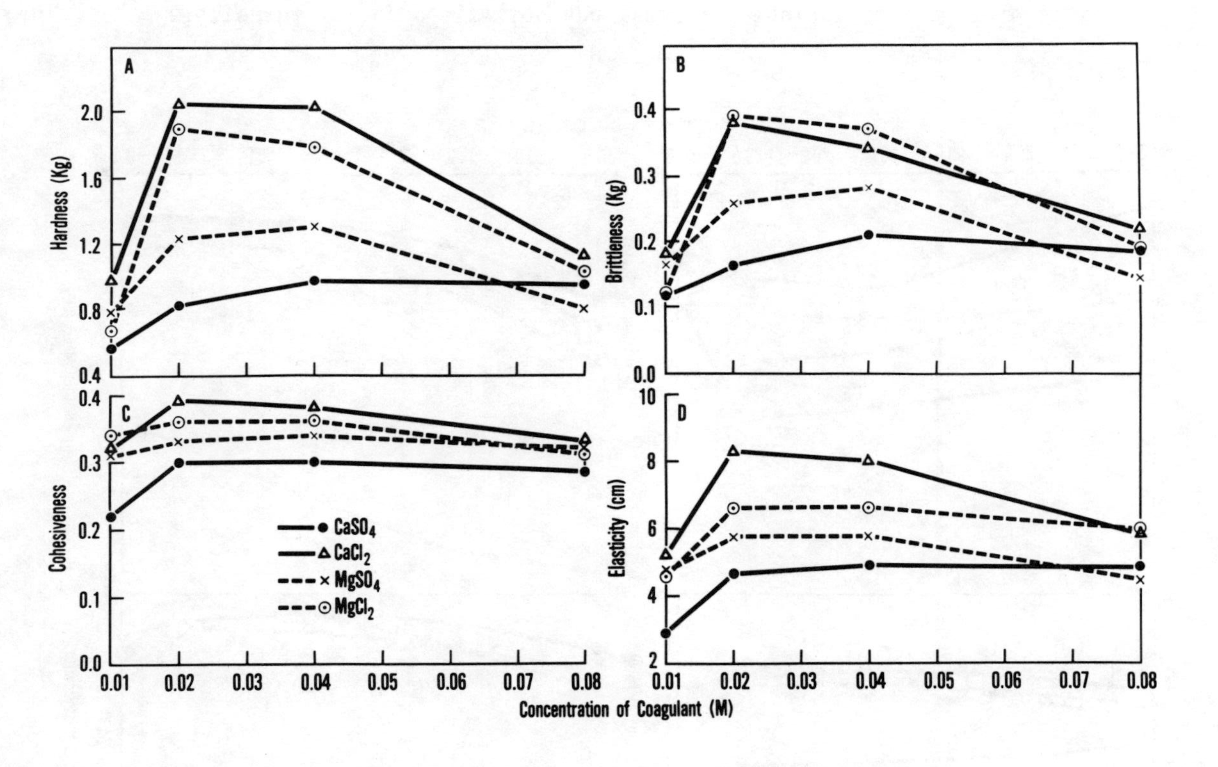

FIG. 8.5. Relationship of concentration and type of coagulant to the texture characteristics of tofu (14).

The temperature of soybean milk at the time the coagulant is added and the mode of mixing greatly affect the yield and texture of the tofu. As the temperature increases, the gross weight and moisture content of the curd decrease and its hardness increases (14). Increased mixing decreases tofu volume and increases its hardness (14, 16).

Thus, many factors come into play during the process, and each factor affects the final product. By knowing the effects exhibited by varying each factor, one can choose and establish a set of conditions to reproduce the desired type of tofu.

Chemical composition of soybeans has been reported to affect tofu texture. Saio et al. (17) found that gel made from 11S protein isolated from defatted meal is much harder than that made from 7S protein, and they also noted increasing tofu hardness as the amount of phytic acid added to soybean milk increased. Since the ratio of 7S to 11S protein and the phytic acid content of the beans may vary with the variety, Saio and coworkers speculated that soybean variety could have an effect on tofu texture. Skurray et al. (18) compared 15 varieties and found no significant correlation between the ratio of 7S to 11S protein or phosphorus content and the quality of tofu, but they did find that the quality of tofu is greatly affected by the amount of calcium ion added. Perhaps such chemical variations may not be great enough to have a significant effect when compared with the processing variables. It has also been suggested (11) that Japanese soybean varieties are better suited for tofu making than are U.S. soybeans. Therefore, soybean varieties originating from U.S. and Japan grown in the same location and the same environmental conditions were used to determine varietal variability in making tofu (19); it was found that composition and color of tofu are affected by soybean variety but that yield and texture are not significantly affected. Tofu made from a variety with a high protein content has a higher protein/oil ratio than tofu made from a variety with less protein (Table 8.4).

Thus, varieties with high protein content are preferred and varieties with dark hilum are not desirable. To develop a new variety for high tofu yield, desirable characteristics would be an increased protein and oil content and an increased nitrogen solubility index.

TEMPEH

Tempeh, originating in Indonesia, is made by fermenting dehulled and briefly cooked soybeans with a mold, *Rhizopus*; the mycelium binds the soybean cotyledons together into a firm cake. The raw tempeh has a clean, fresh, and yeasty odor. When sliced and deep-fat fried, it has a nutty flavor, pleasant aroma, and texture that are highly acceptable to almost all people around the world.

Tempeh fermentation is characterized by its simplicity and rapidity. Making tempeh in Indonesia is a household art. Soybeans are soaked in tap water overnight until the hulls can be removed easily by rubbing between the hands. After dehulling, the beans are boiled with excess water, drained, and surface-dried. Small pieces of tempeh from a previous fermentation, wrappings of the previous tempeh, or ragi tempeh (commercial starter) are mixed with the soybeans, which then are wrapped in wilted banana leaves or other large leaves and incubated in a warm place for 24-48 hr until the beans are covered with white mycelium and bound together as a firm cake. The cake is either sliced into

TABLE 8.4. Ratio of Protein to Oil Content of Tofu and Soymilk as Affected by Protein Content of Soybeans (19)

Soybean variety	Soybean protein %	Protein/oil Tofu	Protein/oil Milk
Wase-Kogane	45.2	2.07	2.49
Vinton	45.1	2.01	2.50
Toyosuzu	44.1	1.87	2.13
Coles	43.2	1.78	2.11
Yuuzuru	42.3	1.89	2.30
Tokachi-Nagaha	41.8	1.88	2.12
Weber	40.9	1.57	1.75
Hodgson	40.9	1.67	1.90
Corsoy	40.8	1.69	1.95
Kitamusume	40.8	1.57	1.86

thin strips, dipped in a salt solution, and deep-fat fried in coconut oil or it is cut into pieces and used in soups much as we use chunks of meat. In this country tempeh is used the same way as hamburger patties.

Studies carried out at our Center have identified Rhizopus oligosporus as the major organism responsible for the tempeh fermentation (20); subsequently, we developed a pure culture fermentation as shown in Figure 8.6. The preparation of soybeans for pure culture fermentation is similar to that for the traditional method: soaking, cooking, cooling, and surface-drying. In 1964, Martinelli and Hesseltine (21) introduced full-fat soybean grits (soybean cotyledons that have been mechanically cracked into four or five pieces) for tempeh fermentation. Since soybean grits absorb water easily, the soaking time can be reduced from more than 20 hr to 30 min. Furthermore, because the hulls are removed mechanically in producing grits, much labor can be saved. Steinkraus and his coworkers (22) suggested soaking the beans in 0.85% lactic or acetic acid solution to prevent bacterial contamination. This practice may reduce bacterial growth during soaking, but it does not help destroy the contaminating heat-resistant spores nor to prevent their germination during fermentation.

Starter containing spores of Rhizopus oligosporus NRRL 2710 was developed at our Center (23) and now is available commercially. Petri dishes were found to be the most convenient laboratory container for tempeh fermentation. Some commerical producers now are using petri dishes for making tempeh patties. Shallow aluminum foil or metal trays with perforated bottoms and perforated plastic film covers may be used to replace banana leaves. Perforated plastic bags, which are now available commercially, also have been used successfully. Like many other molds, R. oligosporus requires air to grow, but it does not require as much aeration as many others. In fact, too much aeration will cause spore formation and also may dry up the beans, resulting in poor mold growth. Therefore, it is important to perforate the containers and pack the beans properly for fermentation. Thickness of the package and size and distance of the perforations are important. Containers 2.5-3.0 cm deep with holes about 0.6 mm in diameter and about 5 mm apart are most desirable.

Tempeh fermentation can be carried out at temperatures ranging from 25-37°C; the time required for fermentation decreases as temperature increases. At higher temperatures the soybeans tend to dry out; consequently, the mold growth is suppressed. At an inoculum level of 1×10^6 spores per 100 g of cooked substrate, tempeh fermentation can best be carried out at 32°C for 20-22 hr (23). When fermentation is complete, the beans are covered with and bound together by white mycelium. Thus, raw tempeh looks like a firm white cake and has an attractive and slightly yeasty odor. Prolonged fermentation, on the other hand, often causes the product odor to become obnoxious due to the enzymatic breakdown of proteins.

Tempeh-like products can also be made by fermenting whole cereal grains such as wheat, oats, barley, rice, or mixtures of cereals and soybeans with the same mold, R. oligosporous (24). Wheat tempeh, soybean tempeh and wheat-soybean tempeh are commercially available in the U.S.

Tempeh has a short shelf-life. Shelf-life can be extended by freezing or by steaming for a few minutes to kill the mold and to inactivate the enzymes and then freezing.

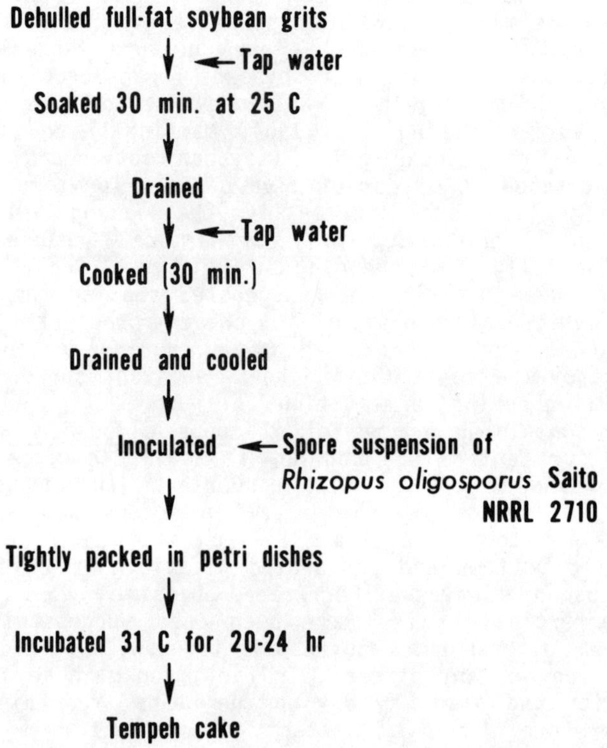

Dehulled full-fat soybean grits
↓ ← Tap water
Soaked 30 min. at 25 C
↓
Drained
↓ ← Tap water
Cooked (30 min.)
↓
Drained and cooled
↓
Inoculated ← Spore suspension of *Rhizopus oligosporus* Saito NRRL 2710
↓
Tightly packed in petri dishes
↓
Incubated 31 C for 20-24 hr
↓
Tempeh cake

FIG. 8.6. Flow diagram for tempeh fermentation (5).

Traditionally, tempeh is sliced and deep-fried for consumption. But it also can be cooked by roasting, baking, or sautéing just like meat. In the West, tempeh burgers and chips are popular.

Although the enzymes produced by the mold used in fermentation have acted upon the substrate and partly hydrolyzed its constituents into small molecules, the digestibility coefficient of tempeh tested by rat assay is not significantly different from that of heat-treated soybeans (25). Because the fermentation process does not significantly change total nitrogen content and amino acid composition, it is not surprising to find that the protein efficiency ratio (PER) of tempeh is no better than unfermented but properly heat-treated soybeans (25, 26). On the other hand, the PER value of wheat is greatly improved by R. oligosporus: 1.78 vs 1.28 (27). The improved PER value of wheat by fermentation is partly attributed to the increase in availability of lysine in wheat. Possibly, the proteolytic enzyme produced by the mold attacked the protein in such a way that more lysine could be made available by the digestive enzymes. Also, the protein quality of tempeh can be improved by making tempeh from mixtures of cereals and soybeans.

Microorganisms differ in their ability to synthesize vitamins. Some microorganisms are unable to synthesize any of the vitamins needed, whereas others may be able to synthesize some or all of the vitamins. Therefore, the vitamin content of various fermented foods may be greater or less than that of the unfermented substrate. As shown in Table 8.5, niacin and riboflavin contents of wheat and soybeans are greatly increased after tempeh fermentation, whereas thiamin and pantothenic acid are decreased (28-30); The increase in vitamins by fermentation is of great nutritional significance to people of those countries where food industries do not enrich foods with vitamins.

Rhizopus oligosporus and many other molds used in Oriental food fermentation synthesize antibiotics of a glycopeptide nature (31-33). The antibiotic compound does not exhibit broad spectrum activity, but it is very active against some gram-positive bacteria, including intestinal Clostridia. This finding supports the view expressed by natives and also by some scientists that those people who eat tempeh daily have few intestinal infections.

CONCLUSION

Two simple processes converting soybeans into interesting and attractive foods are discussed. Both tofu production and tempeh fermentation are low-technology processes. These processes may not improve the nutritive value of soybean protein, but they reduce the cooking time, improve the organoleptic characteristics, and increase the versatility of soybean uses. These protein-rich foods offer excellent possibilities for improving diets of people around the world.

REFERENCES

1. Perry, A. K., Peters, C. R., and Van Duyne, F. O., J. Food Sci. 41, 1330 (1976).
2. USDA, Home and Garden Bull. No. 105, U.S. Department of Agriculture, Washington D.C. (1971).

TABLE 8.5. Effect of *Rhizopus oligosporu* Fermentation on the Vitamin Content of Substrate

Substrate	Niacin		Riboflavin		Thiamin		Pantothenic acid		References
	B[a]	A[b]	B	A	B	A	B	A	
				μg/g					
Soybeans	9.0	60	3.0	7.0	10.0	4.0	4.6	3.3	28
Soybeans	17.5	65	2.6	8.6	7.8	5.8			29
Wheat	46.0	135	0.4	3.2	3.2	3.0			30

[a]Before fermentation.
[b]After fermentation.

3. Nelson, A. I., Steinberg, M. P., and Wei, L. S., J. Food Sci. 41, 57 (1976).
4. Onate, L. U., Aguinaldo, A. R., and Eusebio, J. S., Department of Home Technology, U. P. College of Agriculture, Los Banos, Philippines (1972).
5. Wang, H. L., in "Handbook of Processing and Utilization of Agriculture" Vol. II, part 2, "Plant Products" (I. A. Wolff, ed.), p. 91. CRC Press, Inc., Boca Raton, Florida (1983).
6. Hesseltine, C. W., and Wang, H. L., in "Soybeans: Chemistry and Technology, Vol. 1, Proteins" (A. K. Smith and S. J. Sidney, eds.), p. 389. AVI Publishing Co., Westport, CT (1972).
7. Wang, H. L., and Hesseltine, C. W., in "Microbial Technology," Vol. II (H. J. Peppler and D. Perlman, eds.), p. 96. Academic Press, New York (1979).
8. Wang, H. L., and Hesseltine, C. W., in "Prescott and Dunn's Industrial Microbiology," 4th edition (G. Reed, ed.), p. 492. AVI Publishing Co., Inc., Westport, CN (1982).
9. Lo, W. Y-L., Steinkraus, K. M., Hand, D. B., Wilkins, W. F., and Hackler, L. R., Food Technol. 22, 1322 (1968).
10. Wang, H. L., Swain, E. W., Hesseltine, C. W., and Heath, H. D., J. Food Sci. 44, 1510 (1979).
11. Watanabe, T., Fukamachi, C., Nakayama, O., Teramachi, Y., Abe, K., Suruga, S., and Mivanage, S., The Report of Food Research Institute, 14 B, p. 1, Ministry of Agriculture and Forestry, Japan. (1960).
12. Wilkens, W. F., Mattick, L. R., and Hand, D. B., Food Technol. 21, 1630 (1967).
13. Fukushima, D., in "ACS Symposium Series 123, Chemical Deterioration of Proteins" (J. R. Whitaker and M. Fugimaki, eds.), p. 211. ACS, Washington, D.C. (1980).
14. Wang, H. L., and Hesseltine, C. W., Process Biochem. 17, 7 (1982).
15. Appurao, A. G., and Narasingo Rao, M. S., Cereal Chem. 52, 21 (1975).
16. Saio, K., Cereal Foods World 24, 342 (1979).
17. Saio, K., Kamiya, M., and Watanabe, T., Agric. Biol. Chem. 33, 1301 (1969).
18. Skurray, G., Gunich, J., and Carter, O., Food Chem. 6, 89 (1980).
19. Wang, H. L., Swain, E. W., Kwolek, W. F., and Fehr, W. R., Cereal Chem., 60, 245 (1983).
20. Hesseltine, C. W., Smith, M., Bradle, B., and Ko Swan Djien, Dev. Ind. Microbiol. 4, 275 (1963).
21. Martinelli, A. F., and Hesseltine, C. W., Food Technol. 18, 761 (1964).
22. Steinkraus, K. H., Hwa, Y. B., Van Buren, J. P., Providenti, M. I., and Hand, D. B., Food Res. 25, 777 (1960).
23. Wang, H. L., Swain, E. W., and Hesseltine, C. W., J. Food Sci. 40, 168 (1975).
24. Hesseltine, C. W., Smith, M., and Wang, H. L., Dev. Ind. Microbiol. 8, 179 (1967).
25. Hackler, L. R., Steinkraus, K. H., Van Buren, J. P., and Hand, D. B., J. Nutr. 82, 452 (1964).
26. Smith A. K., Rackis, J. J., Hesseltine, C. W., Smith, M., Robbins, D. J., and Booth, A. N., Cereal Chem. 41, 173 (1964).

27. Wang, H. L., Ruttle, D. I., and Hesseltine, C. W., J. Nutr. 96, 109 (1968).
28. Steinkraus, K. H., Hand, D. B., Van Buren, J. P., and Hackler, L. R., in "Proc. Conf. on Soybean Products for Protein in Human Foods," September 13-15, 1961, p. 83. USDA, NRRC, Peoria, IL (1962).
29. Roelofsen, P. A., and Talens, A., J. Food Sci. 29, 224 (1964).
30. Wang, H. L., and Hesseltine, C. W., Cereal Chem. 43, 563 (1966).
31. Wang, H. L., Ruttle D. I., and Hesseltine, C. W., Proc. Soc. Exp. Biol. Med. 131, 579 (1969).
32. Wang, H. L., Ellis, J. J., and Hesseltine, C. W., Mycologia 69, 218 (1972).
33. Ellis, J. J., Wang, H. L., and Hesseltine, C. W., Mycologia 66, 593 (1974).

Cereal Products: Factors Affecting Physical, Nutritional, and Flavor Characteristics

9

Factors Involved in the Quality of Bread Wheats

BERT L. D'APPOLONIA

INTRODUCTION

The United States is unique in wheat production in that it produces five classes of wheat. Each of the wheat classes differs in quality and as a result differs in terms of end-product usage. We immediately realize, therefore, that wheat is not simply wheat; that is, each type or each class of wheat because of its properties makes it more or less desirable for a specific type of end-product. Table 9.1 lists the five main classes of wheat, the possible range in protein one might obtain for the different classes and their general end-product usage. Two of the five classes listed are considered as bread wheats. Hard red spring (HRS) and hard red winter wheat because, primarily, of their protein quantity and quality, are used for bread type products. The ability of the protein found in the flour derived from these two wheat classes to form gluten when mixed with water is the key factor in these wheats being used for bread products. Due to differences in terms of functionality and certain quality parameters between these two classes of wheat, their exact method of usage may differ.

Since I am more familiar with the quality factors associated with hard red spring wheat, my discussion will center primarily on this wheat class. It should be kept in mind, however, that most of what I will say can also be applied to hard red winter wheat.

It is difficult to give a straightforward definition of quality since the word quality may have different meanings to different people.

TABLE 9.1. Range in Protein Content and Uses of the Five Main Classes of Wheat Grown in the United States

Wheat class	Protein range %	Uses
Hard red spring	11-18	bread, blending
Hard red winter	10-15	bread products
Soft red winter	8-12	cakes, cookies, crackers
White	7-10	crackers, noodles
Durum	11-16	pasta products

To the farmer or grower, wheat generally has high quality: (a) if the plant matures properly and lends itself readily to a typical harvesting process without shattering, (b) it threshes easily and remains clean, and (c) most importantly it produces a good yield. To the plant pathologist a high quality wheat may be one which has disease resistance, to the agronomist a high quality wheat may be related to yield, heading date and lodging, while to the cereal chemist a high quality wheat is considered one which will produce the most acceptable end product for that class of wheat. It will be in terms of the cereal chemists definition of quality that I will center my discussion.

From a cereal chemist's viewpoint one will also find differing opinions as related to quality. The social, environmental and cultural traditions within a society will result in varied definitions for quality in wheat. I will, therefore, confine my discussion to those factors which we consider to be of paramount importance at North Dakota State University in assessing quality in our bread wheats.

It is not possible to define quality by means of one single criterion. I prefer to define quality by dividing its importance into several different areas. These areas include: (a) kernel quality, (b) milling and flour quality, (c) dough quality and (d) bread quality.

KERNEL QUALITY

Some of the quality factors of concern in this area include: test weight, vitreous kernel content, 1000 kernel weight, kernel size, kernel damage, protein content, ash and moisture content, amylase activity and grade.

One of the simplest and oldest criteria of quality is the weight per unit volume (test weight) of the wheat. In North America, this is expressed in terms of pounds per bushel. In most of the countries of the world where the metric system is employed, the term kilograms per hectoliter is used. Confusion arises at times over the fact that in the U.S. the Winchester bushel, having a capacity of 2,150.42 cu. in., is the unit of volume employed, whereas in Canada the Imperial bushel, having a unit volume of 2,219.36 cu. in. (larger volume), is used. As a result, if wheat is purchased from Canada it may appear to have a test weight approximately 3% higher than comparable wheat purchased from the U.S.

Test weight is an important factor in the grading of wheat and

TABLE 9.2. Wheat Kernel Quality Factors

1. Test weight
2. Vitreous kernel content
3. 1000 kernel weight
4. Kernel size
5. Kernel damage (sprout, heat, frost damage)
6. Protein
7. Ash
8. Moisture

can also be an indicator of flour extraction. It is related to variety as well as environmental conditions.

Vitreous kernel content is a grading factor for hard red spring wheat and may be an indicator of protein content. This factor also is related to variety and environmental conditions.

Weight per 1,000 kernels is another measurement made on wheat. Some people regard this as a better indication, perhaps, of flour extraction than test weight. The higher the value the better.

Kernel damage, which could include sprout damage, heat damage or frost damaged wheat, is of extreme importance in wheat quality. These factors will not only affect the quality of the resultant endproduct but may also result in an economic loss to the producer.

Besides the factors I have mentioned regarding kernel quality there are additional ones such as shrunken and broken kernels and foreign material that are of importance in determining the grade of wheat.

Protein content is a quality factor of primary importance. In addition to protein amount the quality of the protein is very important. In the assessment and development of new varieties of hard red spring wheat very often it is observed that two samples may contain the same protein content (say 14%) but the baking quality of the two samples may be very different. The primary reason why the two classes (HRW and HRS) can be used to produce bread-type products is because of the type of protein present and more specifically the gluten proteins. It is this gluten that allows us to make a loaf of bread by entrapping and holding the CO_2 gas produced as a result of the fermentation process.

The classical method for measurement of protein has been the Kjeldahl technique. Within the past few years a nondestructive method for measuring protein has been introduced, namely the nearinfrared (NIR) techniques. The NIR techniques simply involve grinding the wheat sample and, after standardization of the instrument, direct measurement of protein based on reflectance.

MILLING AND FLOUR QUALITY

Having considered some of the factors associated with kernel quality let us now turn to milling and flour quality. Table 9.3 lists some of the quality factors that we must consider in this area. In the development of either new hard red spring or hard red winter wheat varieties very often only small amounts of wheat are available for

TABLE 9.3. Wheat Quality Factors Associated with Milling and Flour Properties

1. Conditioning properties
2. Ease of milling
3. Flour extraction
4. Flour ash
5. Flour color
6. Protein content
7. Amylase activity
8. Wet gluten content
9. Starch damage

evaluation since the plant breeder has limited seed in the early generation. Experimental mills such as the Quadrumat Jr, Quadrumat Sr, Buhler and Miag are available for testing purposes. The particular mill used is dictated by the sample size. In most instances, the smaller the mill the cruder the separation that will take place. With a larger mill more flour streams are obtained and a flour with properties similar to that obtained in a commercial operation results. The miller is concerned with the conditioning properties of the wheat which may be related to its hardness. The hardness of the wheat may affect not only the manner in which water is taken up by the wheat kernels during the tempering operation but flour properties as well. The amount of flour obtained from the wheat, or flour extraction also can be affected as well as the amount of starch damage which will have an effect on the flours absorption.

Some of the flour properties that we are concerned with include flour ash and flour color. The ash content of a flour is a good indicator of the milling efficiency of a particular wheat sample under test. Protein content of the flour is very important as already discussed for wheat protein. In the milling operation there is a reduction in protein content from wheat to flour. A reduction of more than one percent is considered undesirable in experimental milling in the evaluation of potentially new wheat varieties. Wet gluten content is another factor that is often measured in bread flours as an indicator of flour quality.

A factor of main concern in flour analysis is the amount of amylase enzyme present. In sound wheat the amount of beta-amylase is generally sufficient, however, the amount of alpha-amylase is limited. A certain amount of amylase is necessary to obtain optimum bread baking performance. For this reason either barley malt or fungal amylase is added at a flour mill to provide this enzyme. If the natural alpha-amylase activity in a flour is excessive, dough and baking properties can be adversely affected. High amounts of alpha-amylase in flour normally result from wheat that has sprouted during harvesting. This occurs as a result of very wet rainy weather. Various methods are available to measure this amylase activity. Certain methods are based on starch gelatinization followed by measurement of the change in viscosity of which the falling number and amylograph techniques are examples. Procedures based on colorimetry and nephelometry are also used.

Starch damage, which I have already alluded to, is another quality parameter of importance in a bread flour. This starch damage is a physical damage to the starch granule during the milling of the wheat into a flour and can be controlled in the milling operation. For bread baking purposes a certain amount of starch damage is desirable as it will result in higher flour absorption and a greater percentage of fermentable sugars. Excessive amounts of starch damage will, however, result in sticky doughs and inferior bread quality.

PHYSICAL DOUGH PROPERTIES

Another important aspect in assessing the quality of a bread wheat such as hard red spring pertains to the measurement of the rheological or physical properties of the dough as a result of mixing flour and water together. Several different types of instruments are available for this purpose among which are the farinograph, mixograph, extensigraph and alveograph. Each of these instruments provides certain information on the physical dough properties of a flour.

The farinograph is a dynamic type dough testing instrument in that a measurement is made while the dough is in motion. Basically a curve is traced out as the flour and water is mixed in the mixing bowl of the instrument. The force required for mixing blades to move at constant speed through a dough of fixed initial consistency is graphically recorded in the form of a band and as the test proceeds this force varies according to the nature of the flour and, as a result, different shaped farinograms are produced. Table 9.4 illustrates some of the information one can obtain from a farinogram obtained from the farinograph instrument. These include absorption, dough development time, mixing tolerance, mechanical tolerance index, and valorimeter value.

Another physical dough testing instrument used in wheat quality evaluation is the mixograph. In general, similar type information is obtained as with the farinograph (Table 9.5). The mixograph is very useful in a plant breeding program for screening purposes to eliminate the poor quality wheat samples from the good ones. A smaller sample size than the farinograph can be used and more samples can be analyzed per a given time interval. A third physical dough testing instrument that is available is the extensigraph. The extensigraph differs from either the mixograph or farinograph in that it is a static dough testing instrument. It makes a measurement on a dough while the dough is at rest. The extensigraph gives a good indication of gluten strength.

Table 9.6 shows the type of information one can derive from the extensigram obtained from the extensigraph instrument. The measurements include extensibility, resistance to extension, area under the curve and proportional number. A flour producing a weak extensigram has gluten properties with little resistance to extension.

The final and most important parameter in a wheat quality breeding program is the baking test. Table 9.7 indicates some of the factors of concern in the bread baking evaluation. These include absorption, mixing time, dough characteristics, fermentation tolerance, and internal

TABLE 9.4. Quality Factors Measured with the Farinograph

1. Absorption
2. Dough development time (mixing time)
3. Mixing tolerance (stability)
4. Mechanical tolerance index
5. Valorimeter value

TABLE 9.5. Quality Factors Measured with the Mixograph

1. Absorption
2. Mixing time
3. Mixing tolerance

TABLE 9.6. Quality Factors Measured with the Extensigraph

1. Extensibility
2. Resistance to extension
3. Area under curve
4. Proportional number

TABLE 9.7. Quality Factors Involved in Bread Baking

1. Absorption
2. Mixing time
3. Dough characteristics
4. Fermentation tolerance
5. Baked loaf:
   a) volume
   b) external appearance
   c) crumb grain and texture
   d) crumb color

and external characteristics of the finished loaf of bread. In the experimental baking evaluation of wheats for quality it is most common to use "pup" loaves. These loaves can be made from 10, 25 or 100 grams of flour.

What I have presented so far refers primarily to the testing of bread wheat samples in a wheat breeding program for the release of new cultivars to the farmer.

Many of the quality factors I have talked about are the same ones that a commercial baker is concerned about. This is illustrated by the list in Table 9.8. One must keep in mind, however, that the type of bread product, formulation and method of processing will all affect the quality factors the baker desires in a bread flour.

TABLE 9.8. Flour Factors of Concern to the Commercial Baker

Moisture
Protein
Ash
Color
Starch damage
Mixing properties
Absorption
Amylase activity
Fermentation tolerance

10

Factors Involved in the Quality of Soft Wheat Products

WILLIAM T. YAMAZAKI

INTRODUCTION

For marketing purposes, five wheat classes are recognized by the Federal Grain Inspection Service (1). These are the hard red winter, hard red spring, durum, soft red winter, and white wheats, differentiated through kernel characteristics such as shape, size, color, and texture, as well as through region of cultivation. The hard red winter and spring wheats are grown in the Great Plains region, extending from North Dakota to the Mexican border. Durum wheat, spring in growth habit, is produced mostly in North Dakota and neighboring states. Soft red winter wheat is cultivated generally east of a line between Milwaukee and Dallas. White wheat, found mostly in the Pacific Northwest, is predominantly soft in kernel texture. It is also grown in the states of Michigan and New York. Soft wheat kernels are often distinguishable from hard wheat through the plumper shape, opaque rather than vitreous appearance, and lighter seed coat color among the red wheats. There are many exceptions to these general descriptions, however, and visual examination of a sample may be inadequate in classifying it properly.

A second basis for classifying wheat, especially for domestic consumption purposes, lies in the use to which a wheat class is normally put. Hard wheats, both winter and spring, are milled to flour in the United States almost exclusively for the production of bread, rolls, and related yeast-leavened baked goods. Hard wheat flour is also used in the production of certain sweet goods, such as Danish pastry. On the other hand, soft wheat is little used in bread production. Its flour is utilized in the baking of a variety of products such as cookies, cakes, saltines, wafers, pretzels, cones and similar goods (2). Most of them are chemically leavened, although a few, such as saltines and pretzels, are yeast-raised. Because hard and soft wheat flour products differ from each other in their ingredient content, texture, preparation, and consumption, the wheats and flours themselves can be expected to differ in a number of respects. Many of these differences reflect inherent cultivar or class distinctions. In the present discussion we shall speak of soft wheat mostly in terms of these inherent differences and the relation of these differences to the quality of products baked from the flours.

Cereals and Legumes in the Food Supply, edited by Jacqueline Dupont and Elizabeth M. Osman

Many years ago, in the early history of the United States, soft wheat, brought from western Europe, was the only wheat class grown. All food preparation requiring flour was carried out using soft wheat. The soft wheat cultivars grown in the east were not, however, well adapted to the Great Plains. With the introduction of hard wheat cultivars to the region, production increased greatly. The changing baking habit of the American homemaker, the evolution of the baking industry toward increased mechanization, the consequent greater specificity in raw material requirement, and advances in genetics and plant breeding, all combined to bring the utilization quality of wheat to the forefront as an important component of the wheat production industry.

The breeders in the hard wheat region are today working closely with cereal chemists in the development of new cultivars with increased field yield potential for the farmer while simultaneously incorporating good inherent milling and bread baking properties in the grain. In similar fashion, soft wheat breeders and cereal scientists are cooperatively improving this class of wheat for confectionery purposes. Because of this difference in expected applications, however, the quality objectives of the hard and soft wheat breeding programs differ greatly from each other.

SOFT WHEAT USES

In addition to the domestic utilization of soft wheat for confectionery purposes, soft wheat finds application in other parts of the world. For many years, a considerable portion of the soft white wheat production of the Pacific Northwest has been exported, as much as 85% in recent years. Its white seed coat color (actually a pale yellow) and finely granulated flour, together with the traditionally low protein content in the grain, have been factors in the overseas popularity of this wheat class. It finds particular application in the flat bread of the near and middle east. Western white wheat is consumed in Japan as specialty cakes, as noodles, tempura, sweet buns, and other confectionery products. The People's Republic of China utilizes soft red winter wheat in the production of steamed bread, noodles, and various confections.

Western Europe is not a significant market for our soft wheat. It does grow a large crop of soft wheat which, in their nomenclature, includes the spectrum of utilization quality ranging from the high protein strong gluten bread wheat to the low protein finely granulated biscuit type which resembles our soft wheat. Their term "hard wheat" refers to our durum wheat. Thus, in the context of their definition, soft wheat is used in the production of bread (baguette), regional breads, brioches, croissants, puff pastry, tortes, cakes, and biscuits (3). Although the full range of baked goods are produced from their domestic wheats, considerably greater emphasis appears to be placed on bread quality in their variety improvement programs because of their greater need for this type of wheat.

INHERENT QUALITY CHARACTERISTICS IN SOFT WHEATS

The early adaptation of soft wheat in the eastern United States and in the Pacific Northwest as well as that of hard wheat in the Great

Plains have been heavily reinforced through breeding programs. There is good reason for this. Point of origin, tied to utilization potential, is a principal guide in domestic (and export) wheat marketing. The introduction in a region of a new cultivar with unrepresentative utilization quality and its subsequent acreage increase would tend to lower the average quality picture through unavoidable mixing of cultivars after harvest. Therefore, an integral part of a breeding program within a region is the monitoring of lines to ensure that any new release would have satisfactory quality.

The breadbaking test is the means whereby this monitoring is carried out for hard wheat lines. In the same sense, baking tests are appropriate for soft wheat breeding line quality potential evaluation. However, as noted above, a great diversity of products are baked from soft wheat, each of which may depend for its quality on a different compositional factor. Moreover, many soft wheat products of commerce are not baked from a straight grade flour, but from those resulting from split milling or stream selection (4). Further, air-classified fractions may be used for certain baked goods (5). In view of these variables, soft wheat quality tests have tended to promote those characteristics present in commercially available cultivars, the assumption being that such cultivars have satisfactory quality under present processing conditions.

In contrast to quality standards for bread wheats which emphasize high protein content, strong gluten properties, and suitable dough mixing properties including moderate mixing time, good mixing tolerance, and high water absorption, wheat and flour properties considered advantageous in soft wheat include low protein content, low water absorption, and fine flour granulation.

Consensus prevails in the soft wheat processing industry that, in general, high protein in flour is to be avoided because it is thought that it promotes product toughness and that low protein is therefore to be desired to impart tenderness to baked goods. Product tenderness appears to be the trait favored for products in which soft wheat is commonly used, from the Arabic flat breads to the chappati of India, to the noodles, cakes, and other confections of the Orient, to the cookies, cakes, and saltines in the United States. I shall refer to this tenderness factor later.

Two baking tests approved by the American Association of Cereal Chemists have been employed as a means of evaluating soft wheat quality. These are the cookie baking and white layer cake baking tests. For variety testing in the soft wheat breeding program, we use a sugar snap cookie test which is similar to the official test (6). A minimally mixed dough disc of given dimension is baked. The extent to which the dough expands laterally is the measure of flour quality, a flour producing larger diameter cookies being considered to have better quality than one which expands to lesser extent.

Research has indicated that the spread of cookies is inversely associated with the water absorptive property of the flour (7) When a dough is mixed, it appears that a partition of the limited water in the system takes place between the major hydrophilic ingredients, sugar and flour. With a flour of lesser absorptive capacity, the partition favors sugar more than if a flour of greater hydration capacity were present. Thus, relatively more sugar is dissolved to a syrup. When such a dough is heated, the viscosity of the syrup is lowered such that the entire internal dough becomes quite fluid. Resistance to leavening gases is lowered and the dough expands both vertically and laterally. Baking is completed when the dough sets and no further dimensional changes take place. In the presence of flour of a relatively high absorption, however, the partition favors flour more than if a lesser hydration capacity flour were present, relatively less syrup is formed, and the heated dough then offers greater resistance to leavening gases because internal dough viscosity is not as low. Expansion is therefore restricted, resulting in cookies of smaller diameter than may otherwise be the case.

A physicochemical test called the Alkaline Water Retention Capacity (AWRC) test, measures the ability of a flour or sifted meal to absorb a dilute solution of sodium bicarbonate (to control pH) against centrifugal force (8). Data from the test have been found to correlate highly with cookie quality as measured by spread. The relationship is inverse, low AWRC denoting superior cookie quality potential. The test is useful in a breeding program when there is insufficient sample for baking.

The official cake test is essentially a high ratio white layer cake formulation (9). In variety evaluation, a short patent flour is milled and chlorinated. The quality criteria are cake volume and associated attributes such as internal texture. High cake volume is normally associated with superior grain texture.

It has been found that, other factors being equal, cake volume is associated significantly with flour granularity where granularity is due to inherent or cultivar differences for wheats milled in a standardized way (10). A flour with lower mass median diameter usually bakes a larger cake with superior internal textural characteristics than a coarser flour. Since flour granulation reflects kernel hardness, results of a test for kernel hardness correlate with volume of cakes baked from such wheats. This relationship is utilized in estimating cake potential in breeding lines (11).

The chlorine reaction converts flour to one which produces cakes with significantly increased volume together with light and finely textured crumb. The exact mechanism of the reaction is unknown, but the reaction is immediate and surface-dependent (12). It thus appears that a flour with fine granulation would respond more effectively to the treatment than one with coarse granulation. There is some evidence to indicate alteration in the starch and/or lipid fraction of flour (13). However, the effect of chlorine on the protein fraction has not been sufficiently studied to permit an appraisal of the importance of this reaction in cake improvement.

A cake baking experiment using air-classified fractions of similar mass median diameter obtained from wheats with widely different inherent granulation properties produced cakes with significantly differing volumes (14). The results indicated that particle size may not be the sole determinant of cake quality. A qualitative compositional or varietal effect may be operating. Flour protein content was found not to correlate significantly with volume in the experiment.

PRODUCT TENDERNESS, AN OPEN QUESTION

Soft wheat is usually lower in protein content than hard wheat. This condition appears to be the result of two basic causes: first, soft wheat adaptation to areas of adequate to plentiful rainfall and moderate temperature during the growing season which are conducive to high yield, and secondly, in more recent times the emphasis on inherently low protein content as a quality factor in variety development.

For years the thought has prevailed in the domestic soft wheat milling and baking industries that a principal factor in consumer acceptance of confectionery products has been its tenderness compared to those prepared from other wheat classes, and further that the low protein in the flour was *per se* the primary contributor to that tenderness. This belief may have developed from the time most higher protein flours were milled from hard wheat cultivars, and full consideration was not given to the possible role of heritable factors in tenderness.

There is evidence to indicate that within a soft wheat cultivar, high protein content may not necessarily be associated with hard kernel texture (15). Soft wheat cultivars grown under conditions promoting high protein content may produce kernels vitreous in appearance and thus resemble hard wheat. However, such grain still mills like a soft wheat, producing finely granulated flour. Conversely, low protein hard wheats often are opaque but mill and bake like hard wheat.

Informal trials of consumer acceptance of flat breads and confectionery goods seemed to show a preference for such goods baked from soft wheat over products from other wheat classes. It is not known however whether such tests were conducted under conditions which permitted direct comparison of products made from hard and soft wheats of equal protein content. The aforementioned study (10) relating flour particle size to cake volume also adds indirectly to uncertainty of relative importance of protein content.

It has been stated by cereal chemists experienced in commercial saltine production that detectable textural differences in favor of soft wheat are found in saltines manufactured from hard and soft wheat flours of equal protein content.

It thus seems that the final verdict is not yet in as to whether product tenderness in soft wheat products is due to low flour protein content regardless of variety or wheat class, or conversely whether soft wheat cultivars possess inherent compositional factors manifest even in flour of higher protein content thus contributing to tenderness.

REFERENCES

1. Federal Grain Inspection Service. The official United States standard for grain. U.S. Government Printing Office, Washington, D.C. (1977).
2. Loving, H. J. and Brenneis, L. J. in "Soft wheat; production, breeding milling, and uses" (W. T. Yamazaki and C. T. Greenwood, eds.). Am. Assn. Cereal Chem., Inc., St. Paul, Minn. p. 169 (1981).
3. Greenwood, C. T., Guinet, R., and Seibel, W. in "Soft wheat: production, breeding, milling, and uses" (W. T. Yamazaki and C. T. Greenwood, eds.) Am. Assn. Cereal Chem., Inc., St. Paul, Minn., p. 209 (1981).
4. Nelson, C. A. and Loving, H. J. Cereal Sci. Today 8, 301 (1963).
5. Pratt, D. B., Jr. Baker's Dig. 37(4), 40 (1963).

6. Finney, K. F., Morris, V. H., and Yamazaki, W. T. Cereal Chem. 27, 42 (1950).
7. Yamazaki, W. T. Cereal Chem. 31, 135 (1954).
8. Yamazaki, W. T. Cereal Chem. 30, 242 (1953).
9. Yamazaki, W. T. Cereal Sci. Today 15, 262 (1970).
10. Yamazaki, W. T. and Donelson, D. H. Cereal Chem. 49, 649 (1972).
11. Yamazaki, W. T. and Donelson, J. R. Crop Sci. 12, 374 (1972).
12. Wilson, J. T., Donelson, D. H., and Sipes, C. R. Cereal Chem. 41, 260 (1964).
13. Kissell, L. T., Donelson, J. R., and Clements, R. L. Cereal Chem. 56, 11 (1979).
14. Chaudhary, V. K., Yamazaki, W. T. and Gould, W. A. Cereal Chem. 58, 314 (1981).
15. Yamazaki, W. T. and Donelson, J. R. Cereal Chem. 60, 344 (1983).

11

Factors Involved in the Quality of Rice

BILL D. WEBB

INTRODUCTION

Rice constitutes one of the most important food sources for mankind and is, in fact, the primary food staple for over half of the world's population. Consequently rice milling, cooking, eating, processing, and nutritional qualities are, as they have been for centuries, factors of primary importance in rice eating areas of the world. It is these aspects of quality that establish the food and economic value of the rice grain. Worldwide, rice is consumed in hundreds--perhaps thousands--of different ways and with this diversity of consumption of rice in the diet it is little wonder that the manner of cooking rice and the form in which it is served differ widely. This means that the qualities expected of rice vary considerably with different countries and even between regions or areas within countries.

In the United States high grain quality is an absolute prerequisite for new varieties of rice. If a variety does not have satisfactory quality it simply won't make it commercially. The United States rice industry is highly diverse and innovative in the areas of rice production, domestic uses and consumption, export rice programs, and the qualities needed for each. A brief overview of the scope of the industry's complexity, diversity, and quality requirements is illustrated by the following information obtained from reports and fact sheets published by the Rice Council for Market Development, 6699 Rookin Street, Houston, Texas 77274, U.S.A.

United States Rice Production

The major rice-producing states (Fig. 11.1) are Arkansas, California, Louisiana, Mississippi, Missouri, and Texas. Arkansas produces about one-third of the U.S. rice crop (Table 11.1). Florida and a few other states also produce a limited amount of rice. In the southern states both long-grain and medium-grain rice is grown. In California nearly all production is medium and short-grain rices. In production value, rice ranks sixth among the major field crops and is grown on less than 2% of U.S. cropland acreage. More than 95% of the rice eaten in the United States is grown in the United States. Only small quantities are imported and these are generally specialty rices for specific dishes and recipes. About 1/3 of the U.S. rice crop is consumed domestically and 2/3 exported to over 100 countries (2).

Cereals and Legumes in the Food Supply, edited by Jacqueline Dupont and Elizabeth M. Osman

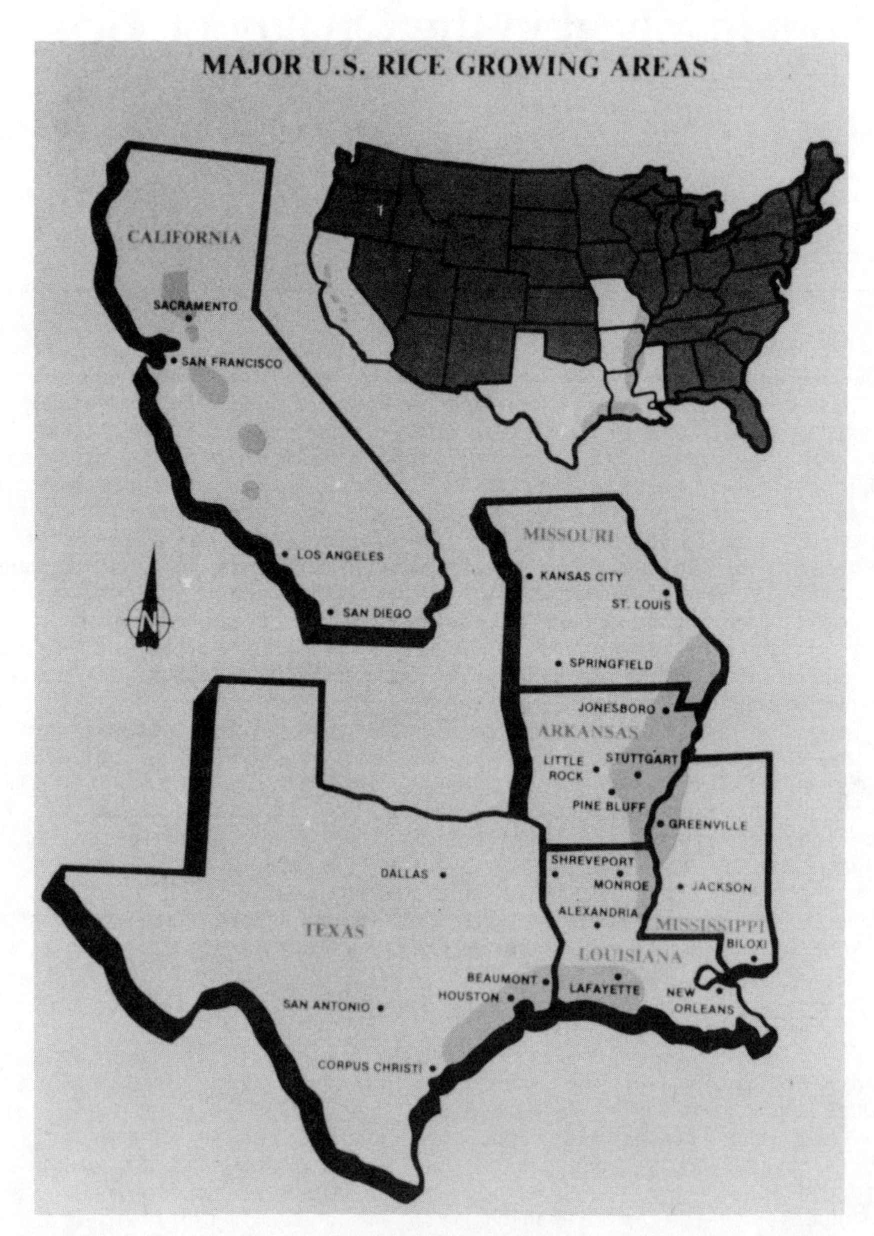

FIG. 11.1. Major U.S. rice growing areas. Courtesy of Rice Council for Market Development, Houston, Texas.

TABLE 11.1. U.S. Rice Production (1)

	Production		Acres Harvested	
	1981	1982	1981	1982
State	1,000 cwts[a]	1,000 cwts	1,000 acres	1,000 acres
Arkansas	65,463	56,843	1,423	1,353
California	42,004	35,749	595	563
Louisiana	27,306	23,489	650	584
Mississippi	15,065	9,859	344	280
Missouri	3,311	3,141	81	77
Texas	26,404	22,056	581	476
Florida	414	496	10	12
U.S. Total	179,967	151,633	3,684	3,345

[a] cwt = 100 pounds

United States Rice Consumption

Rice consumption in the United States has increased significantly in recent years and according to Rice Council reports (2) indications are that this trend will continue. Present per capita consumption of rice for direct food use and processed foods is about 13 pounds. If rice for brewing is included per capita consumption approaches 17 pounds. Direct food use includes regular-milled, parboiled, pre-cooked, and brown rices. Processed food use includes cereals, packaged mixes, etc. There was an increase of more than 70% in per capita rice consumption from 1975 to 1981. Rice Council reports (2) also indicates that over 31 million hundredweights of rice are now consumed in the U.S. each year; of which, 61% is for direct food use, 25% for brewing, and 14% for processed food use. The types and forms of rice used for direct food use and the percentage of each consumed in the U.S. are shown in Table 11.2. Traditionally regular-milled long-grain white rice is by far the most popular type and form of rice consumed in the United States.

United States Export Rice

According to Rice Council figures (2) about 2/3 of the U.S. rice crop is exported to over 100 countries and U.S. rice usually represents 20 to 25% of all rice moving in international trade. In most recent years the United States has been one of the top exporters even though it produces less than 2% of the total world rice crop. All three grain types of rice grown in the United States are exported. About 2/3 of the export rice is long grain, nearly 1/3 medium-grain, and the remainder short-grain. While exports fluctuate each year depending on world demand, regular-milled rice has accounted for about 60% of U.S. export rice in recent years. European importers of U.S. rice purchase principally brown rice and parboiled brown rice which European processors then mill, package, and distribute.

CONCEPTS OF RICE QUALITY

United States rice has many diverse uses both domestically and exportwise, consequently its quality is evaluated according to its suitability for a specific end-use for a particular consumer. Because nearly all the U.S. rice crop is milled to a relatively high degree the quality of rice is closely related to the quality of its milled endosperm. Also most rice, unlike many other cereals, is processed and consumed in whole kernel form, so the physical properties (Fig.

TABLE 11.2. Types and Forms of Rice Consumed in the U.S. (2,4)

Types/Form Rice	% Consumed in U.S.
Regular-milled long grain	57.8
Regular-milled medium grain	21.0
Regular-milled short grain	1.5
Parboiled	10.4
Pre-cooked	5.4
Brown	2.0

11.2) of the intact kernel such as size, shape, uniformity, and appearance are particularly important in assessing rice quality.

Much information is available on rice quality, rice processing, and related areas such as nutritive properties, enrichment, snack items, rice oil, hulls, etc. Most of this information is consolidated in five books or handbooks each of which contains numerous original references (5-9).

Varieties of rice in the United States are classed for marketing purposes as long-, medium-, and short-grain types (Fig. 11.3) and each grain type is associated with specific cooking and processing behavior (10). High quality U.S. long-grain varieties cook dry and fluffy, and the cooked grains tend to remain separate. On the other hand, cooked kernels of high quality medium- and short-grain varieties are more moist and chewy than those of long-grain varieties and the kernels tend to cling together. All three grain types and their textural qualities are in widespread demand by the domestic and export trade. Different ethnic and cultural groups prefer specific and varied textures in homeboiled rice. In the United States most consumers prefer the dry, fluffy and separate texture characteristics of the long-grain types while others prefer the more moist and clingy texture of the short- and medium-grain rices.

Processors and reprocessors of rice need a variety of grain types and textural qualities for use in various kinds of prepared and convenience-type food products. An ever-increasing amount of the domestic rice crop is annually processed and reprocessed into numerous kinds of prepared products such as parboiled rice, quick-cooking rice, dry breakfast cereals, canned rice products, packaged mixes, baby foods, snack items, frozen dishes, etc. Rice is ranked high as a brewing adjunct and there is a strong demand for the broken grains in beer brewing. Rice flour is preferred for dietetic formulations and in non-allergenic baking. In some of these processed, convenience, and specialty foods the textural properties and grain size of U.S. long-grain types are preferred, whereas in others the properties of the short- and medium-grain varieties are required.

Both the domestic and world trade associates United States long-, medium-, and short-grain rices with specific cooking and processing qualities of each grain type, and for this reason the traditional standards of high milling, cooking, and processing behavior must be maintained in all new varieties. It is primarily in the United States that grain type, through planned breeding, is associated with specific rice cooking and processing behavior (10-13). Worldwide, there are short-grain varieties that have the cooking and processing qualities of U.S. long-grain types and long-grain varieties that have the characteristics of U.S. short- and medium-grain types.

The grains of nearly all rice varieties produced commercially in the United States are translucent, nonscented, nonwaxy types which contain various ratios of amylose and amylopectin type starch. One of two exceptions is the limited production of a specialty waxy (glutinous) rice. This rice, also called "sweet-rice", is characterized by a chalky white (opaque) endosperm containing virtually 100% amylopectin starch. When cooked, it loses its shape and is very glutinous. It is used primarily for specialty products by ethnic groups and in commercial product formulations. Another exception is the scented

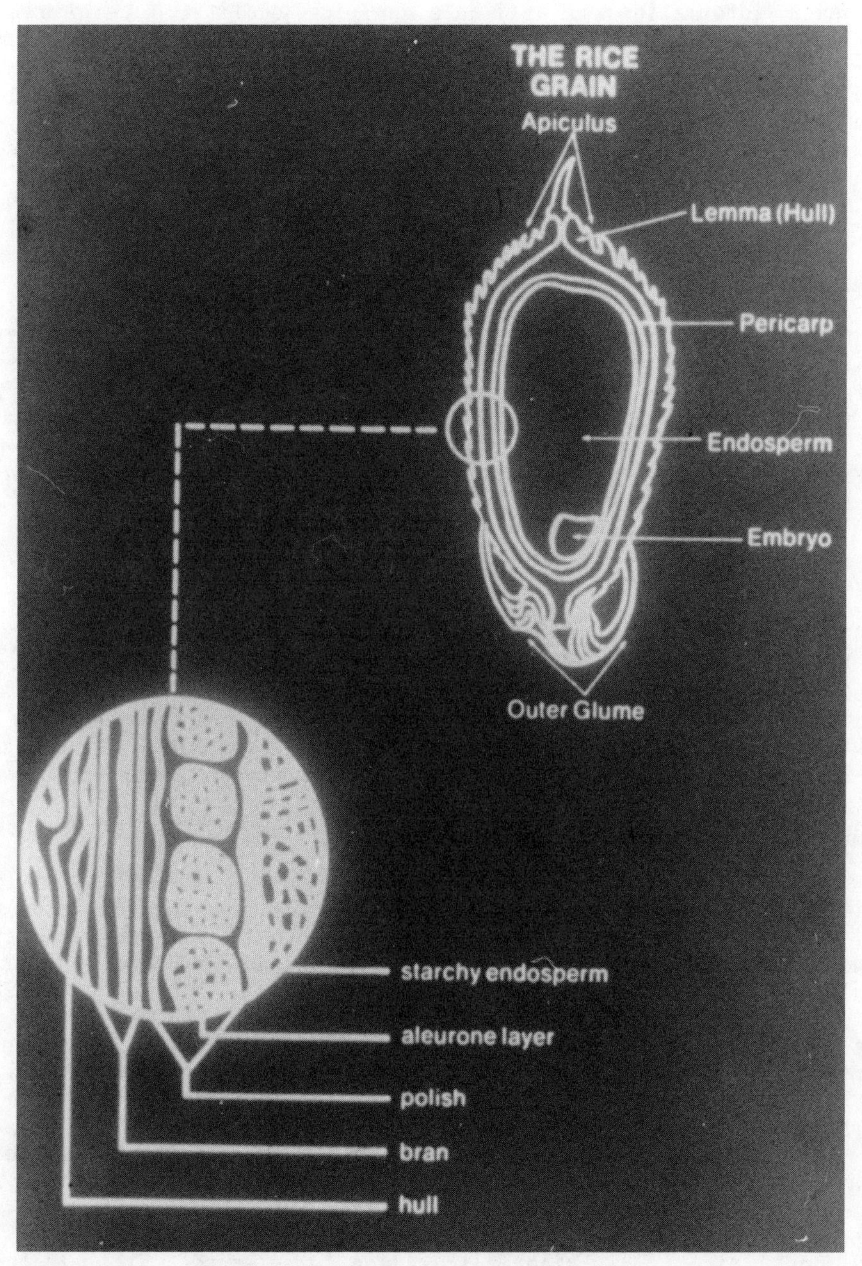

FIG. 11.2. Structure of the rice grain.
Courtesy of Riviana Foods, Inc., Houston, TX.

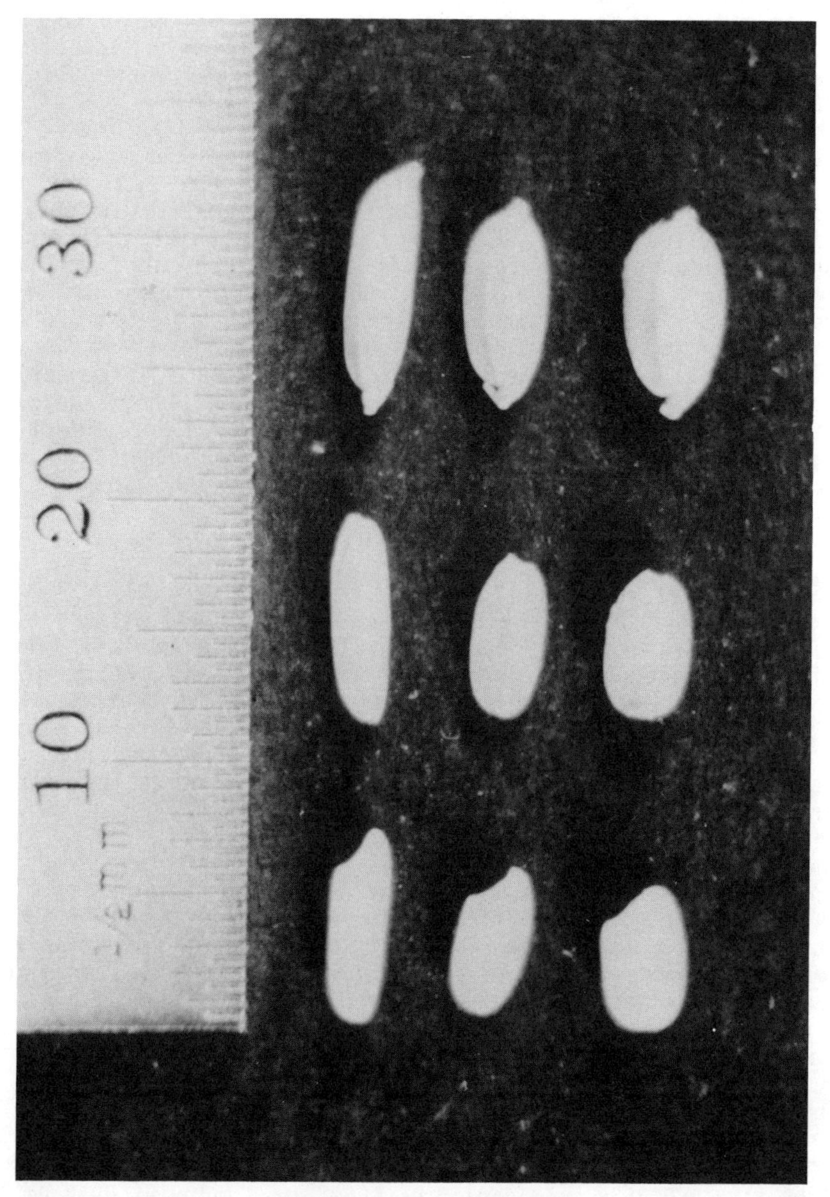

FIG. 11.3. United States rice grain types
Top (L to R) rough (paddy) rice long-, medium-, and short-grain
Middle (L to R) brown rice long-, medium-, and short-grain
Bottom (L to R) milled (head) rice long-, medium, and short-grain.

(aromatic) rice produced on small acreages as a specialty product. While cooking, this rice gives off an aroma similar to that of roasted popcorn or nuts and has a flavorful nutty taste. A major constituent responsible for its unique flavor has recently been identified as 2-Acetyl-1-pyrroline present in the volatile oil fraction of cooked rice (14).

The major factors influencing quality of rice, like that of other cereals, are the genetic makeup of the grain and the environment in which rice is grown. Modern breeding programs strive to refine the genetic characteristics of rice that influence quality to obtain the attributes required by the U.S. rice industry. Wide differences in rice quality also exist due to variations in environmental and cultural practices during growth. Many times quality differences due to these factors may be greater than differences due to genetic factors. Additional factors influencing rice quality include those associated with handling, storage, processing, and the incidence of objectionable and foreign material. For example, rice stored under unfavorable conditions may contain objectionable odors and flavors which would render it unsuitable for human consumption. On the other hand a short period of "aging" under favorable conditions is considered desirable to permit more uniform grain stabilization.

FACTORS INVOLVED IN THE QUALITY OF RICE

There are four broad rice quality categories: (1) milling quality; (2) cooking and processing quality; (3) nutritive quality; and, (4) qualities associated with cleanlines, soundness, and purity of rice. All are important collectively, in assessing the suitability of rice for a particular end use. The rice quality factors (12,13) considered to be the most important to the United States rice industry in producing, processing, and utilizing the U.S. rice crop for domestic consumption and for export are summarized in Table 11.3.

Hull and Bran Color

United States varieties are either light- (straw) or dark- (gold) hulled types (Fig. 11.4). In the manufacture of parboiled rice in the United States light colored hulls are usually preferred because they tend to produce a lighter colored finished parboiled product than dark-hulled varieties processed under comparable conditions. Most users prefer the lighter colored parboiled products although certain consumers prefer a darker parboiled rice. In the latter instance, gold-hulled varieties are selected when available.

Bran (pericarp) color, also shown in Figure 11.4, is a factor in parboiling where undesirable dark, non-uniform, or damaged and discolored bran colors may be imparted to the parboiled endosperm during the parboiling process. It is also a factor in milling of both parboiled and regular white milled rice where increased milling pressure (more often resulting in lower milling yields) must be applied to remove or minimize the dark colored areas. The bran appearance should be a uniform characteristic light brown (nonpigmented) color.

Grain Size and Shape, Weight, and Uniformity

The physical dimensions, weight and uniformity, of rice grains are of vital interest to those involved in the many facets of the rice

TABLE 11.3. Rice Quality Factors

Hull and Bran Color
Kernel Size, Shape, Weight and Uniformity
Kernel Translucency
Kernel Chalkiness
Milling Quality
Whole Kernel (Head) Rice Yield
Total Milled Rice Yield
Milling Uniformity
Cooking, Eating, Processing and Nutritive Quality Indices
Amylose/Amylopectin Ratio
Alkali Spreading Value
Protein Content
Gelatinization Temperature (BEPT)
Water Uptake Capacity
Amylographic Paste Viscosity
Parboil-Canning Stability
Additional Grading Factors Included in U.S. Standards for Rice
Moisture Content
Test Weight
Degree of Milling
Color
Dockage
Damaged Kernels
Heat Damaged Kernels
Odors
Red Rice

FIG. 11.4. Hull and bran types of United States rices (Top left) light (straw) hulls; (top right) dark (gold) hulls; bottom (L to R) light (non-pigmented) bran.

industry. These physical attributes are primary quality factors in breeding, handling, drying, marketing and grading, and processing of end-use products. In the U.S. rice is marketed under three grain (Fig. 11.3) types known as long-grain, medium-grain, and short-grain. Varieties within each grain type are expected to conform to rather narrow limits regarding size and shape specifications for their respective grain type. Average and ranges of values for grain size, shape, and weight of rough, brown, and milled forms of traditional U.S. long-, medium-, and short-grain types are shown in Table 11.4.

Intensive genetic selection is carried out to eliminate undesirable kernel abnormalities that would detract from or decrease milling quality, cooking and processing quality, and general appearance. These grain defects shown in Figure 11.5 include deep furrowed creases which leave bran streaks on milling, irregularly shaped kernels, sharp pointed extremities that break easily on milling, and oversized or deep seated germs that also decrease milling quality.

Kernel Translucency and Chalkiness

Virtually all aspects of rice quality are affected by kernel translucency and chalkiness. Clear, vitreous, translucent kernels (Fig. 11.6) are characteristics required by most segments of the rice industry. United States varieties are required to possess this trait to a high degree.

Chalkiness (Fig. 11.7) is undesirable in essentially all forms of rice because it detracts from the all-important general appearance and most often results in lower milling quality. Chalkiness is undesirable for many processes because of the potential for nonuniformity in manufactured products. Numerous processors specify the amount and kind of chalky kernels permissible for their products. Rice kernel chalkiness is called "white belly," "white core," "white back," "germ tip," or "immature" depending largely upon its location on or within the endosperm. Several factors cause chalkiness in rice. Immature chalk occurs in rice which is harvested at too high a moisture content and/or in rice of non-uniform maturity. Adverse weather conditions and cultural practices also influence chalkiness in rice. Specific types of chalkiness such as "white belly," "germ tip," etc. are highly heritable and intense breeding and genetic selection is required for its elimination. Translucent kernels inherently free of chalk is one of the first quality factors sought in United States rice varieties.

Milling Quality

Milling outturn is the quality factor of the most direct economic importance. No multipurpose rice variety is commercially successful unless it has high whole kernel (head) rice and total milled rice yields. Whole kernel (head) rice yield is the amount or percentage of whole milled kernels that can be obtained from a unit of rough rice. Total milled rice yield includes the whole kernel (head) rice and all sizes of broken kernels obtained from a unit of rough rice. The objective of rice milling is the removal of hulls, bran, and germ with a minimum breakage of the endosperm and this process consists of four broad operations: (1) cleaning the field run rough (paddy) rice to remove foreign material; (2) shelling the cleaned rough rice to remove hulls to produce brown rice; (3) milling the brown rice to remove the outer and inner bran layers, aleurone layers, and germ; and, (4) sorting

TABLE 11.4. Range of Average Grain Size and Shape Measurements among Typical U. S. Commercial Long-, Medium-, and Short-Grain Rice Types (13)

Grain Type	Grain Form	Average Length (mm)	Average Width (mm)	Aver. Length/ Width Ratio	Average Thickness (mm)	Average 1000 Grain Wt. (g)
Long		6.7 to 7.0	1.9 to 2.0	3.4:1 to 3.6:1	1.5 to 1.7	15 to 18
Medium	Milled[a]	5.5 to 5.8	2.4 to 2.7	2.1:1 to 2.3:1	1.7 to 1.8	17 to 21
Short		5.2 to 5.4	2.7 to 3.1	1.7:1 to 2.0:1	1.9 to 2.0	20 to 23
Long		7.0 to 7.5	2.0 to 2.1	3.4:1 to 3.6:1	1.6 to 1.8	16 to 20
Medium	Brown[b]	5.9 to 6.1	2.5 to 2.8	2.2:1 to 2.4:1	1.8 to 2.0	18 to 22
Short		5.4 to 5.5	2.8 to 3.0	1.8:1 to 2.0:1	2.0 to 2.1	22 to 24
Long		8.9 to 9.6	2.3 to 2.5	3.8:1 to 3.9:1	1.8 to 1.9	21 to 24
Medium	Rough[c]	7.9 to 8.2	3.0 to 3.2	2.5:1 to 2.6:1	1.9 to 2.1	23 to 25
Short	(Paddy)	7.4 to 7.5	3.1 to 3.6	2.1:1 to 2.4:1	2.1 to 2.3	26 to 30

Data based on measurements of fully developed mature kernels of typical varieties within each grain type.

[a]Whole milled kernels with hull, bran and germ removed.
[b]Grain with hull removed.
[c]Unhulled grain.

FIG. 11.5. Undesirable rice kernel defects and abnormalities.

FIG. 11.6. Desirable clean translucent rice kernels (Left) long-grain type; (right) medium-grain type.

FIG. 11.7. Undesirable chalky rice kernels.

the mixture of whole and broken milled kernels according to size classes known as whole kernel (head) rice, second head (larger pieces of broken milled kernels), screenings (smaller pieces of broken milled kernels), and brewer's rice (very small pieces of broken milled kernels).

Whole kernel (head) rice yield is the product of highest economic value. Yields of whole kernel (head) rice vary widely due to many factors such as: inherent characteristics of the variety, grain type, cultural practices, and other environmental factors affecting the chalkiness and plumpness of the grains, and the conditions of handling, drying, storing, and milling. The yield of total milled rice is also economically important and this yield is further affected by the hull percentage and the amount of fine endosperm particles unavoidably included in the bran fraction during milling.

Milling quality is reported as the percentage of whole kernel (head) rice and total milled rice obtained from a unit of rough rice. The range and average milling yields for U.S. long-, medium-, and short-grain types are given in Table 11.5.

Cooking and Processing Quality

Cooking and processing characteristics and milling outturn are the factors of quality that collectively establish the economic value of the rice grain. United States multipurpose rice varieties must possess both high milling quality and satisfactory cooking and processing behavior to be commercially acceptable.

There are several ways to assess rice cooking and processing qualities. In the United States the most useful way has been to characterize rice qualities in terms of chemical and physical properties which serve as indices of rice cooking and processing behavior. These indices and test methods include: (a) amylose content (15) or as modified by Juliano, (16); or Webb, (17); (b) alkali spreading value of whole milled kernels in contact with dilute alkali--an indicator of gelatinization temperature (18); (c) amylographic gelatinization and paste viscosity characteristics (19); (d) water uptake capacity at 77°C (19); (e) birefringence-end-point temperature, BEPT (20); (f) protein content (6); and, (g) parboil-canning stability (21).

Among these quality indicators, amylose content is considered to be the single most important characteristic for predicting the type of rice cooking and processing behavior (15, 22-24). Alkali spreading value--used to classify rices according to gelatinization temperature

TABLE 11.5. Range of Average Milling Yields among Typical U.S. Commercial Long-, Medium-, and Short-Grain Rice Types (13)

Grain Type	Average Milling Yields (%) Whole Kernel (Head Rice)	Total Milled Rice
Long	56-61	68-71
Medium	65-68	71-72
Short	63-68	73-74

Data based on clean, mature, rough rice samples of varieties within each grain type.

type--is of comparable importance for U.S. rice varieties. Virtually all current U.S. varieties have been developed by selecting for these two characteristics as indices of cooking and processing qualities.

Some rice cooking and processing quality indices for typical long-, medium-, and short-grain United States rices are given in Table 11.6. Typical U.S. long-grain varieties which cook dry and fluffy are characterized by: relatively high amylose content, slight-to-moderate alkali spreading reaction of whole kernel milled rice in dilute alkali, moderate water uptake capacity at 77°C, and intermediate gelatinization temperature. Amylographic paste viscosity characteristics of the typical long-grain varieties usually show an intermediate peak and a relatively high cooled paste viscosity. Canning stability of parboiled kernels of typical long-grain varieties in terms of solids loss during canning are relatively low (desirable), and the canned kernels show a minimum amount of splitting and fraying of edges and ends.

By way of comparison, typical U.S. short- and medium-grain varieties which cook moist and clingy are characterized by: comparatively low amylose content; extensive spreading reaction of whole kernel rice in contact with dilute alkali; relatively low gelatinization temperature; and relatively high water uptake capacity. Amylograms of the typical medium- and short-grain varieties usually show comparatively low cooled paste viscosities. The parboil canning characteristics of the typical short- and medium-grain types show high (undesirable) solids loss on canning and the canned kernels show extensive splitting and fraying of edges and ends.

Nutritive Quality

The nutritional qualities of rice are of utmost importance and are of necessity an integral part of all aspects of rice quality. Some common compositional and nutritional characteristics of the basic forms of raw and cooked and enriched and unenriched rices are given in Table 11.7 for ready comparison (3). For a more complete and detailed review of the many aspects of the nutritional qualities of rice, the reader should consult (5, 25-27).

RICE GRADES AND QUALITY FACTORS

United States Standards for Rice

The U.S. standards for rough rice, brown rice for processing, and milled rice provide a means of marketing rice by grade which takes into account those quality factors associated with the cleanliness, soundness, and purity of rice. However, because of the many diverse end-uses and quality requirements of various processors, the classification system does not and was not intended to meet all requirements of all phases of the rice industry. For this reason many processors require that rice meet their own additional specific requirements for particular products such as baby foods, breakfast cereals, parboiling, instant rices, and brewing. The official United States rice designations as to type, kinds, classes, and grades of rice are given in Table 11.8. Rice grading factors are adequately described and defined in the U.S. Rice Standards (28) and the Rice Inspection Manual (29) so they will be discussed here only briefly.

TABLE 11.6. Range of Average Chemical and Physical (Quality) Characteristics among Typical U.S. Commercial Long-, Medium-, and Short-Grain Rice Types (13)

Endosperm Characteristics	Grain Type		
	Long	Medium	Short
Amylose content (%)	23 to 26	15 to 20	18 to 20
Alkali spreading value (aver.)	3 to 5	6 to 7	6 to 7
Gelatinization temperature (BEPT)	71 to 74	65 to 68	65 to 67
Gelatinization temperature (class)	Intermediate	Low	Low
Water uptake at 77°C (ml/100g)	121 to 136	300 to 340	310 to 360
Protein (N x 5.95)(%)	6 to 7.5	6 to 7	6 to 6.5
Parboil-canning stability: solid loss %	18 to 21	31 to 36	30 to 33
Amylographic past viscosity:			
peak-brabender units (B.U.)	765 to 840	890 to 980	820 to 870
cooked 10 min at 95°C (B.U.)	400 to 500	370 to 420	370 to 400
cooled to 50°C (B.U.)	770 to 880	680 to 760	680 to 690

Data based on measurements of fully developed mature kernels of typical varieties within each grain type.

TABLE 11.7. Composition of Basic Forms of Rice (3)

	100 Grams									
	Brown		Regular White (Unenriched)		Regular White (Enriched)		Parboiled (Enriched)		Pre-Cooked (Enriched)	
	Raw	Cooked	Raw	Cooked	Raw	Cooked	Dry Form	Cooked	Dry Form	Ready To-Serve
Water (percent)	12.0	70.3	12.0	72.6	12.0	72.6	10.3	73.4	9.6	72.9
Food energy (kcal)	360	119	363	109	363	109	369	106	374	109
Protein (gms.)	7.5	2.5	6.7	2.0	6.7	2.0	7.4	2.1	7.5	2.2
Fat (gms.)	1.9	.6	.4	.1	.4	.1	.3	.1	.2	Trace
Carbohydrate:										
Total (gms.)	77.4	25.5	80.4	24.2	80.4	24.2	81.3	23.3	82.5	24.2
Fiber (gms.)	.9	.3	.3	.1	.3	.1	.2	.1	.4	.1
Ash (gms.)	1.2	1.1	.5	1.1	.5	1.1	.7	1.1	.2	.7
Calcium (mgs.)	32	12	24	10	24	10	60	19	5	3
Phosphorus (mgs.)	221	73	94	28	94	28	200	57	65	19
Iron (mgs.)	1.6	.5	.8	.2	**2.9	**.9	**2.9	**.8	**2.9	**.8
Sodium (mg.)	9	***	5	***	5	***	9	***	1	***
Potassium (mgs.)	214	70	92	28	92	28	150	43	---	Trace
Thiamine (mgs.)	.34	.09	.07	.02	**.44	**.11	**.44	**.11	**.11	**.13
Ribofalvin (mgs.)	.05	.02	.03	.01	****	****	****	****	****	****
Niacin (mgs.)	4.7	1.4	1.6	.4	**3.5	**1.0	**3.5	**1.2	**3.5	**1.0
Tocophorol (mgs.) (Vitamin E)	29	8.3								

*The information in this table was taken from the "Composition of Foods," Agriculture Handbook No. 8, Agricultural Research Service, U.S.D.A. (Revised December 1963) except for tocophorol which was taken from "The Chemical Composition of Rice," U.S.D.A., Agricultural Research Administration, Bureau of Agricultural and Industrial Chemistry (1951).

**Values for iron, thiamine, and niacin are based on the minimum levels of enrichment specified by the U.S. Government.

***Varies with sodium ion content of water and the addition of salt in cooking.

****Minimum and maximum requirements for riboflavin have not been specified as yet by the U.S. Government.

TABLE 11.8. Official U.S. Rice Designations (9)

TYPES

Long grain rice
Medium grain rice
Short grain rice

KINDS

Rough Rice

SPECIAL GRADES	CLASSES	GRADES
Parboiled	Long grain rough rice	U.S. No. 1
Smutty	Medium grain rough rice	U.S. No. 2
Weevily	Short grain rough rice	U.S. No. 3
	Mixed rough rice	U.S. No. 4
		U.S. No. 5
		U.S. No. 6
		U.S. sample grade

Brown rice for processing

SPECIAL GRADES	CLASSES	GRADES
Parboiled	Long grain brown rice for processing	U.S. No. 1
Smutty	Medium grain brown rice for processing	U.S. No. 2
	Short grain brown rice for processing	U.S. No. 3
	Mixed brown rice for processing	U.S. No. 4
		U.S. No. 5
		U.S. sample grade

Milled rice

SPECIAL GRADES	CLASSES	GRADES	CLASSES	GRADES
Coated	Long grain milled rice	U.S. No. 1	Second head milled rice	U.S. No. 1
Granulated brewers	Medium grain milled rice	U.S. No. 2	Screenings milled rice	U.S. No. 2
Parboiled	Short grain milled rice	U.S. No. 3	Brewers milled rice	U.S. No. 3
Undermilled	Mixed milled rice	U.S. No. 4		U.S. No. 4
		U.S. No. 5		U.S. No. 5
		U.S. No. 6		U.S. sample grade
		U.S. sample grade		

Moisture Content

Sound, dry, rough rice can be maintained in good condition for years if properly stored but high moisture rice can spoil in only a few days. Moisture content of 13% is commonly accepted as a safe level for storage of rough rice for 6 months or less, but 12% or less is recommended for long term storage. Moisture content in excess of 14% is designated as sample grade under the U.S. standards for rough rice.

Equally important is the effect of moisture content on milling yields in rice. To approach optimum milling quality, rice must be harvested at the proper moisture levels, dried carefully to suitable moisture levels and stored and milled under moisture and humidity conditions recommended for maximum milling yields.

High moisture levels in rice may adversely affect some cooking and processing quality characteristics since changes in these characters during storage usually occur more quickly in high than in low moisture rice. The moisture content of rice, like that of other commodities is of direct economic importance.

Weight per Unit Volume

Test weight is used as an indicator of potential total milled rice yields by the industry. It also provides an indirect estimate of dockage and/or foreign materials present, and of the proportion of unfilled, shriveled, and immature kernels. Average test weight of U.S. rough rice is 58 kg/hl (45 lb/bu) but varies widely, and is influenced by factors such as pubescence, amount of dockage, unfilled and immature kernels, and grain type. Test weights for typical U.S. long-, medium-, and short-grain types are shown in Table 11.9.

Color and Milling Requirements

For grading purposes, the U.S. Standards for whole kernel regular milled white rice specify that "U.S. No. 1 grade shall be white or creamy, and shall be well milled. U.S. No. 2 may be slightly gray, and shall be well milled. U.S. No. 3 may be light gray, and shall be at least reasonably well milled. U.S. No. 4 may be gray or slightly rosy, and shall be at least reasonably well milled. U.S. No. 5 and No. 6 may be dark gray or rosy and shall be at least lightly milled." Colors of raw milled rice range from white to dark gray or rosy whereas, parboiled rice is graded from "parboiled light" to "parboiled dark."

Degree of Milling

The extent to which the bran layers and germ are removed from

TABLE 11.9. Range of Average Test Weight Among Typical U.S. Commercial Long-, Medium-, and Short-Grain Rice Types (13)

Grain Type	Average Lb/Bushel	Average Kilograms/ Hectoliter
Long	42 to 45	54 to 58
Medium	44 to 47	57 to 60
Short	45 to 48	58 to 62

Data based on measurements of fully developed mature rough rice grains of varieties within each grain type.

the endosperm during the milling process is referred to as the degree of milling. For grading purposes, The U.S. Standards recognize four degrees of milling known as well milled, reasonably well milled, lightly milled, and undermilled.

Red Rice

Kernels and pieces of milled rice kernels which are distinctly red in color or which contain an appreciable amount of red bran are classified as red rice by U.S. Standards. Red rice (Oryza sativa L.) is a severe problem for the U.S. rice industry. It is a major weed pest on large acreages of the rice growing states. Red rice is objectionable because the red bran is incompletely removed in regular milling and this detracts from general appearance and ultimately the market value of the rice. Also, when rice containing large amounts of red rice is milled, an undesirable rosy color develops because the bran from the red rice, stains the milled endosperm of the cultivated variety. For grading purposes, a rice kernel is classified as red if it has a streak of red bran which extends 1/2 or more of the kernel length or if it has two or more streaks which total 1/2 or more of the length of the kernel.

Dockage

In rice grading, dockage is defined as "any matter other than rice which can be removed readily from the rough rice by the use of appropriate sieves and cleaning devices, and underdeveloped, shriveled, and small pieces of kernels of rough rice which are removed in properly separating the dockage and which cannot be removed by properly rescreening and recleaning." Impurities such as metal and glass fragments and certain weed seeds which are difficult to remove are classed as "objectionable materials."

Damaged Kernels

Kernels and pieces of kernels of rice which are distinctly discolored or damaged by water, insects, heat, or any other means are classified as damaged kernels. Parboiled rice kernels when found in nonparboiled rice are classed as damaged kernels. Kernels materially discolored and damaged by heating are placed in the special class "heat damaged" because this form of kernel damage is considered to be more serious than other kinds of damage. Distinctly discolored kernels, regardless of the source of discoloration, are considered by parboilers and others in the rice industry as "pecky" rice. In parboiled rice, "peck" is a particularly serious problem because the parboiling process tends to intensify kernel discoloration.

Odors

The grade of rice is severely affected by off odors. Rice which is musty or sour, or which has any commercially objectionable foreign odor, is graded U.S. sample grade. Objectionable foreign odors include odors of fertilizers, hides, oil products, skunk, smoke, and decaying animal and vegetable matter.

Special and Numerical Grades

Parboiled, smutty, weevily, coated, granulated, brewers, and undermilled listed in Table 11.8 as special grades for the various types and classes of rough, brown, and milled rices are additional quality factors that describe the conditions or processing treatments of the sample. Numerical grades of the rice are determined by trained inspectors based on size of whole and broken kernels, uniformity, cleanliness, damage, general appearance, infestation, and odor.

REFERENCES

1. Rice Millers Association. Rice Acreage and Production in the United States 1981-1982. Rice Millers Association, Arlington, VA (1982).
2. Rice Council. Rice Council Workshop Manual, Rice Council for Market Development, Houston, Texas (1983).
3. Rice Council. Rice Council Fact Sheet Brochure, Rice Council for Market Development, Houston, TX (1983).
4. Holder, S. H. and Dorland, D., U.S.D.A. Economic Resarch Service Statistical Bulletin NO. 693 (1982).
5. Adams, C. F., Agriculture Handbook No. 456, pp. 135-136. Agricultural Research Service, United States Department of Agriculture (1975).
6. American Association of Cereal Chemists, Inc., Cereal Laboratory Methods. Method 46-13. St. Paul, MN, (1962).
7. International Rice Research Institute, Proceedings of Workshop in Chemical Aspects of Rice Quality, Los Banos, Laguna, Philippines (1979).
8. Rice: Production and Utilization, (Luh, B. S., ed.), AVI Publishing Co., Inc., Westport, CT (1980).
9. Handbook of Processing and Utilization in Agriculture, (Wolff, I. A., ed.), Vol. II, Part I, CRC Press, Inc., Boca Raton, FL (1982).
10. Adair, C. R., Bollich, C. N., Bowman, D. H., Jodon, N. E., Johnston, T. H., Webb, B. D., and Atkins, J. G., in "Rice in the United States: Varieties and Production," pp. 22-75. U.S. Dept. Agr., Agr. Handbook 289 (1966, rev. 1973).
11. Webb, B. D. and Stermer, R. A., in "Rice Chemistry and Technology" (D. F. Houston, ed.), pp. 102-123. American Association of Cereal Chemistry, St. Paul, MN, (1972).
12. Webb, B. D., Bollich, C. N., Johnston, T. H., and McIlrath, W. O., in "Proceedings of the Workshop on Chemical Aspects of Rice Grain Quality," pp. 191-205. International Rice Research Institute, Los Banos, Laguna, Philippines, (1979).
13. Webb, B. D., in "Rice Production and Utilization" (B. S. Luh, ed.), pp. 543-565. AVI Publishing Co., Inc., Westport, CT, (1980).
14. Buttery, R. G., Ling, L. C., and Juliano, B. O., Chemistry and Industry, 958-959 (1982).
15. Williams, V. R., Wu, W. T., Tsai, H. Y., and Bates, H. G., J. Agric. Food Chem. 6, 47-48 (1958).
16. Juliano, B. O., Cereal Sci. Today 16(10), 334-338, 340, 360 (1971).
17. Webb, B. D., Cereal Sci. Today 17(9), 141 (1972).
18. Little, R. R., Hilder, G. B., and Dawson, E. H., Cereal Chem. 35, 111-126 (1958).

19. Halick, J. V. and Kelly, V. J., Cereal Chem. 36, 91-98 (1959).
20. Halick, J. V., Beachell, H. M., Stansel, J. W., and Kramer, H. H., Cereal Chem. 37, 670-672 (1960).
21. Webb, B. D. and Adair, C. R., Cereal Chem. 47, 708-714 (1970).
22. Sanjiva Rao, B. S., Vasudeva, A. R., and Subrahmanya, R. S., Proc. Indian Acad. Sci., Section B, 36, 70-80 (1952).
23. Halick, J. V. and Keneaster, K. K., Cereal Chem. 33, 315-319 (1956).
24. Juliano, B. O., in "Proceedings of the Workshop on Chemical Aspects of Rice Grain Quality," pp. 251-260. International Rice Research Institute, Los Banos, Laguna, Philippines (1979).
25. Houston, D. F. and Kohler, G. O., "Nutritional Properties of Rice." Natl. Res. Council-Natl. Acad. Sci., Washington, D.C. 65 pages (1970).
26. Eggum, B. O., in "Proceedings of the Workshop on Chemical Aspects of Rice Grain Quality," pp. 91-111. International Rice Research Institute, Los Banos, Laguna, Philippines (1980).
27. Kennedy, B. M., in "Rice Production and Utilization" (B. S. Luh, ed.), pp. 439-469. AVI Publishing Co., Inc., Westport, CT (1980).
28. U.S. Department of Agriculture. United States standards for rice. U.S. Dept. Agr. Federal Grain Inspection Service (Rev.) (1983 rev.).
29. U.S. Department of Agriculture. Rice Inspection Handbook. U.S. Dept. Agr. Federal Grain Inspection Service (1982; rev. 1983).

12

Oats: Factors Contributing to Its Role as a Cereal Product

VERNON L. YOUNGS

INTRODUCTION

Grandpa and Grandma enjoy that bowl of hot oatmeal cereal, and being fixed in their ways, they are dismayed that their teenage grandchildren hate the stuff. The young cereal chemist who reviews the potential nutritional contributions of various cereal grains becomes quite excited when he sees data on oats. They are impressive. The cereal manufacturer accepts oats as a convenient cereal crop that comes gift wrapped, and accepts the fact the old timers, such as rolled oats and Cheerios, bring in a steady profit.

The farmer identifies oats as a "second thought," or fill-in crop, and as a cash crop, not the best. Yet he accepts oats as one of his best animal feeds. He knows that oats can be harvested green and stored as haylage, and if he is a dairy farmer, oat straw is invaluable to his animals as bedding. But the "ho-hum" attitude of the farmer towards oats is kindled to fire with the addition of one word--wild (oats). Now he talks with vengeance about North America's worst weed.

The _Avena_ L. family, perhaps like many families of _Homosapiens_, is relatively low key, has its meritorious and notorious members, and the notorious members get the most attention.

AVENA

Oats _(Avena)_ is a grass with many diverse species ranging from the huge _Avena murphyi_ kernels to the diminutive _A. brevis_. Common oats, _A. sativa_ L., is the most important cultivated oat, and it is adapted to temperate climates. _A. byzantina_ C. Koch is generally cultivated as a winter oat, but is frequently crossed with _A. sativa_. _A. fatua_, the common wild oat, is a very bad weed, especially in the U.S. and Canada. Detailed information on oat species is available from Baum (1), Coffman (2) and Rajhathy and Thomas (3).

*Mention of trademark name, proprietary product, or specific equipment does not constitute a guarantee or warranty by the U.S. Department of Agriculture and does not imply its approval to the exclusion of other products that may be suitable.

Cereals and Legumes in the Food Supply, edited by Jacqueline Dupont and Elizabeth M. Osman 1987 Iowa State University Press, Ames, Iowa 50010

OAT PRODUCTION

Oat production is generally for farm use; only about 8% is purchased by industry for direct use in cereals to supply the annual 3.2 lbs. per capita consumption by residents of the U.S. Minnesota, South Dakota, Iowa and North Dakota are the major producers of oats in the U.S., and collectively they produced over 50% of the estimated U.S. oat crop in 1981 (Table 12.1). Oats is a major cereal crop in the U.S., but along with barley, rice and rye, falls behind the giants wheat and corn (Table 12.2). In the world, the North American continent raised about 24% of the oats in 1981, and nearly all of this was grown in the U.S. and Canada (Table 12.3).

Oat production as a cash crop is always under heavy competition from other crops. For instance, one acre of hard red spring wheat produced $108.50 for the North Dakota farmer while one acre of oats

TABLE 12.1. 1981 Oat Production, Selected States (4)

State	1,000 Metric Tons
Iowa	867
Minnesota	1,308
N. Dakota	641
S. Dakota	1,024
USA	7,375

TABLE 12.2. Comparison of Major U.S. Cereal Crops, 1981 (4)

Cereal Crop	1,000 Metric Crops
Oats	7,375
Corn	208,314
Wheat (Total)	76,025
Barley	10,414
Rice	8,408
Rye	473

TABLE 12.3. World Oat Production (1981 (4)

Continent or Country	1,000 Metric Tons
N. America	11,025
Europe	14,942
USSR	15,200
S. America	769
Africa	165
Asia	1,952
Australia and New Zealand	1,604
World	45,657

was worth only $62.10 (5). However, acreage planted to oats in North Dakota in 1983 was over 60% greater than 1982 because of the government's Payment-in-Kind Program.

THE OAT KERNEL

The oat plant produces spikelets which typically contain two or three kernels: the primary, which is largest; the secondary, somewhat smaller, and sometimes a tertiary which is quite small. These kernels have two-part hulls, the lemma and palea, and when removed, the oat groat is exposed. While the hull is essentially non-digestible by non-ruminants, the groat is a nutritious package. It has a typical cereal seed structure with bran, endosperm, embryonic axis and scutellum.

MAJOR BIOCHEMICAL CONSTITUENTS OF THE OAT KERNEL

Some discussion is warranted here, especially on the protein and lipids of oats, because these two constituents contribute to the unique qualities of oats as compared with other cereals.

Protein

Crude protein concentration varies considerably in oats. Robbins et al. (6) reported a range of 12.4%-24.4% protein (dry basis) in 286 groat samples selected from the world collection. These values put oats at the top of the cereals for protein concentration. Table 12.4 shows a typical distribution of protein within the oat kernel. Although protein percentages are high in the embryo (embryonic axis and scutellum), most of the protein is found in the bran and endosperm, because these are the major parts of the kernel by weight.

The amino acid balance of oat protein is unusually good for cereal proteins, which are generally low in lysine (Table 12.5). This amino acid balance is also reasonably stable at different protein levels (11) and since these levels are quite high, actual intake of essential amino acids per given serving is probably the best of the cereals.

For several years, U.S. oat breeders have considered the increase of oat protein as a major goal in breeding. Some progress has been

TABLE 12.4. Protein concentration in Oat Groats, Hulls, and Groat Fractions[a] (7)

Variety	Groats (%)	Hulls (%)	Embryonic Axis (%)	Scutellum (%)	Bran (%)	Starchy Endosperm (%)
Orbit	13.8	1.7	44.3	32.4	18.8	9.6
Lodi	14.6	1.6	36.5	26.2	19.6	10.7
Garland	14.8	1.4	40.5	28.9	18.5	10.9
Froker	15.5	1.4	26.3	28.0	20.7	9.7
Portal	16.5	1.9	35.3	29.1	23.0	10.3
Dal	20.8	1.5	40.9	24.2	26.5	13.5
Goodland	22.5	1.9	40.7	32.4	32.5	17.0

[a]All results shown are on a dry basis and were determined by multiplying nitrogen values by 6.25.

TABLE 12.5. Essential Amino Acids in Oat Groats

Amino Acid	Oat Flakes[1]	FAO Scoring Pattern[2]
Protein	17.6	
Lysine	4.2	5.5
Methionine Cystine	4.1	3.5
Valine	5.3	5.0
Isoleucine	3.9	4.0
Leucine	7.4	7.0
Phenylalanine	4.2	6.0
Tryptophin	1.7	1.0

[1]Data from Pomeranz et al. (8) except tryptophan, which is from Tkachuk and Irvine (9).
[2]From FAO/WHO (10).

made, although dramatic increases in protein concentration are generally accompanied by lower grain yields.

Lipids

Oat groats have the highest lipid concentration of any of the cereal grains. In a review of 11 papers reporting quantitation of oat lipids, the range among 5,987 samples was 2.0%-11.8% lipid (12), and lipid concentration is a heritable characteristic. Most of the current varieties being grown range between 5% and 9% lipid. About 95% of the fatty acids involved are either palmitic, oleic or linoleic acids. Most of the fatty acids are chemically bound to glycerol as glycerides. Considerable variation in fatty acid concentration occurs among varieties (Table 12.6).

Other major lipid components have been identified and include triglycerides, partial glycerides, sterols, glycolipids and phospholipids. Triglycerides represent the major component in oat lipids as they do in other cereal lipids.

The large variability of total lipids and of fatty acid composition in oats strongly suggests that both factors could be altered through plant breeding. Some selection for high lipid concentration has been done by Canadian oat breeders, but generally much greater effort has

TABLE 12.6. Range of Fatty Acid Concentrations Reported in Oat Lipids (12)

Fatty Acid	Range %
Myristic	0.4- 4.9
Palmitic	15.6-25.8
Stearic	0.8- 3.9
Oleic	25.8-47.5
Linoleic	31.3-46.2
Linolenic	0.9- 3.7

been directed toward breeding for high protein. No direct relationship between protein and lipid percentages appears to exist in oats, hence, it should be possible to develop oats varieties with variable protein-oil combinations (13). There is good agreement that specific relationships occur between oil concentration and concentration of specific fatty acids. Palmitic and linoleic acid concentrations are negatively correlated, and oleic acid is positively correlated with lipid concentration (12). Hence, breeding for different lipid concentrations would also affect the unsaturation of the lipids.

Carbohydrates

Oats contains less carbohydrates than wheat or rice, and the starch granules are small. Amylose-amylopectin ratios are about 25%/75% and no waxy oats (high amylopectin) or high amylose oats have been reported.

Gums

While the oat hulls contain considerable pentosans, the gums of endosperm and bran are largely beta-glucans (14,15). The gums are responsible for increasing viscosity in cooked oat cereals.

Minerals and Vitamins

Ash values of oat groats are consistently higher than wheat, hence total minerals would also be higher. Greatest concentrations appear in the bran. Vitamins in oats are also concentrated in the bran, aleurone and embryo. Like wheat, oats is a good source of B-vitamins. No effort has been made to alter vitamins genetically, although it may be possible (16).

Fiber

Fiber in the oat groat is located primarily in the bran. Rasper (17) analyzed oat bran dietary fiber by the Southgate procedure and reported that it contained 53.4% hemicellulose, 25.8% crude cellulose, and 20.9% crude lignin. Lignin content was nearly three times the concentration reported in wheat bran.

Enzymes

While several enzymes have been studied in oats, lipase is of greatest interest. This enzyme is very active in oats once the groat has been ground. It releases free fatty acids primarily from the glycerides, and these fatty acids are probably acted upon by lipoxygenase to release hydroxy acids and other short chain compounds that contribute to rancidity. Since lipase is heat labile, the activity of lipase can be curtailed by steaming the oats before processing, thus essentially eliminating the problem of rancidity in oats.

QUALITY EVALUATION OF OATS

In the U.S. oats are generally graded and sold according to the U.S. Grain Standards, with 32 lbs. constituting one Winchester bushel. However, much of the oats sold today weighs 36-42 lbs/bu. For food or feed, large, plump oats are desirable because this generally designates large, plump groats. An exception is the occurrence of double, or "bosom" oats, where the groat of the secondary kernel did not develop. The primary and secondary kernels appear together, giving the impression of a very large oat kernel, when actually the hull percentage is very high.

Percent hull is not a grading factor; however, it is an extremely important economic factor for buyers representing food processors because it directly affects milling yields. Milling yields are often determined by how many pounds of oats are required to produce 100 lbs. of final product--generally flaked groats. Most oat breeders seriously make selections based on low percent hull.

Since protein in oats is good, oat breeders usually include protein measurement as part of their quality evaluation. For over 15 years the USDA Oat Quality Laboratory, Madison, WI, has evaluated thousands of samples each year for U.S. oat breeders.

OATS FOR FOOD

Oat Milling

Oat milling generally denotes cleaning, heating the oats to improve flavor, dehulling, and grinding or rolling the oat groat. Oat flour finds use as a raw material for dry cereals. Currently some separation of bran and endosperm has been done because of an interest of oat bran for food use.

Most of the oats destined for food are rolled. Two major types of rolled oats (actually rolled groats) are produced: regular or old fashioned, and quick rolled oats. Regular rolled oats are made from primary groats, and the entire groat is rolled producing large flakes, which take longer to cook, but give a texture desired by many. Quick rolled oats are made from secondary groats, or small groats, which are steel cut into two or three pieces, and these are rolled thinner than regular to produce a product that cooks rapidly. In all cases, the groats are steamed before rolling to stop lipase action and to condition the groats for rolling. Details of dehulling oats and rolling groats have been described by Salisbury and Wichser (18).

Impact of Oats on the Cereal Market

History. The history of rolled oats is interesting. In the U.S., steel cut groats were first packaged for sale in glass jars by Ferdinand Shumacher of Akron, Ohio in 1854. Later, they were packed in barrels and shipped to grocers, who sold them by the pound (19). Rolled oats probably did not appear until several years later. The Quaker Oats Company was established in 1901, and they marketed the first cereal in the world to be sold under a brand name. It was the first cereal to be sold in packages, the first to be nationally advertised, the first to be reproduced in miniature for house-to-house sampling, and the first to carry recipes and offer premiums (20). This company currently is still the leading producer of rolled oats for food; the second is National Oats Company, which began operation as a feed company in 1904 and about 1917 changed its emphasis to food.

Comparison of cereals. In a study of 22 commercial hot cereals sold in six North Dakota grocery stores, oats was the first and only ingredient in 40.9% of all hot cereals. This was exceeded only by wheat (Table 12.7). In 82 other dry cereals, normally not served hot, oats was the first ingredient in 14.6% of the cereals, exceeded by corn and wheat (Table 12.8).

In an attempt to observe consumer preference of cereals, shelf space (volume) allotted to each cereal was measured, and the weight of that cereal which could be stored in this space was calculated. It was hypothesized that grocery managers assign shelf space to cereals relative to the consumer's demand for that cereal. Tabulations for cereals listing corn, oats, wheat, rice, sugar or rye as the first ingredient are shown in Table 12.9. Oat cereals accounted for 25.82% of the total weight of all cereals measured. In these oats cereals,

TABLE 12.7. Distribution of Ingredients in 22 Commercial Hot Breakfast Cereals[a] (12)

	Order of Occurrence (percent of cereals listing ingredient)				
Ingredient	1st	2nd	3rd	4th	5th
Oats	40.9	-	-	-	-
Corn	9.1	-	-	-	-
Wheat	45.5	4.5	9.1	-	-
Rice	4.5	-	-	-	-
Sugar	-	13.6	4.5	13.6	-

[a]Survey conducted in September 1980. These data are not related to consumer acceptance. Relative grocery shelf space devoted to rolled oats exceeds that of the other hot cereals.

TABLE 12.8. Distribution of Ingredients in 82 Breakfast Cereals, Excluding Those Normally Served Hot[a] (12)

	Order of Occurrence (percent of cereals listing ingredient)				
Ingredient	1st	2nd	3rd	4th	5th
Oats	14.6	6.1	9.8	2.4	2.4
Corn	26.8	7.3	3.7	2.4	1.2
Wheat[b]	36.6	8.5	9.8	9.8	7.3
Rice	13.4	-	2.4	-	2.4
Barley	-	2.4	1.2	-	-
Rye	-	2.4	1.2	-	-
Soy Flour	-	-	2.4	-	-
Raisins	-	6.1	1.2	-	-
Sugar[c]	8.5	57.3	15.9	17.1	14.6

[a]Survey conducted in September 1980. These data are not related to consumer acceptance.

[b]Wheat occurred more than once in the ingredient distribution in seven cereals as flour, bran or germ, gluten or starch.

[c]Sugar occurred more than once in the ingredient distribution in 28 cereals as sugar, brown sugar, honey, or corn syrup.

TABLE 12.9. Shelf Space Assigned to First-Ingredient Cereal Crops in Six ND Grocery Stores[1]

Cereal-- 1st Ingredient	Maximum weight of cereal stocked[2] (lbs)	% of total
All oat cereals	4,833.20	25.82
100% oats; no other ingredients	2,209.09	11.80
Corn	3,233.23	17.28
Wheat	7,879.56	42.10
Rice	1,662.88	8.89
Sugar	1,039.57	5.55
Rye	66.81	0.36

[1]Three stores were in Fargo, two in Carrington and one in Zeeland, ND, with populations of 60,000, 2,500 and 300 respectively.

[2] $\frac{\text{Shelf volume}}{\text{Volume of pkg.}}$ x wt. of pkg. = max. wt. of cereal stocked.

45.71% contained only oats, with no other crops (or sugar) listed. Hence, rolled oats comprised 11.80% of the total cereals studied. No single cereal brand or combined related brands, such as Corn Flakes, Post Toasties and Country Corn Flakes, exceeded rolled oats in weight of product on shelves. On a national scale this is also true. Rolled oats (instant, quick, regular) manufactured by the leading producer alone is reported to command 8% of the total U.S. cereal market (21).

SUMMARY

This discussion has not mentioned uses of oats other than for food, such as furfural from hulls, oat flour as an antioxidant, medicinal and cariostatic properties, and oats as an animal feed. As a food or feed, oats is quite good nutritionally, and genetic variation of the nutritional constituents is possible. It is a cheap source of food. Consumer acceptance has not changed for many years, although percent of total production used as a food has increased because the population has increased. Greater acceptance probably will require new and unusual products--not just other reshaped dry cereals, but products utilizing certain parts of the oat kernel, with special physical and nutritional properties, and much advertising.

LITERATURE CITED

1. Baum, B. R., Oats: Wild and cultivated. Monograph No. 14, Printing and Publishing Supplies and Services Canada, Ottawa (1977).
2. Coffman, F. A., "Oat history, identification and classification." Tech. Bull. No. 1516. U.S. Dept. Agric., ARS, Washington, D.C. (1977).
3. Rajhathy, T., and Thomas, H., "The Oat Species: Morphology and Distribution," pp. 5-19, in Cytogenetics of oats (*Avena* L.). Genet. Soc. of Canada, Ottawa (1974).

4. U.S. Department of Agriculture. Agricultural statistics. U.S. Govt. Printing Office, Washington, D.C. (1982).
5. North Dakota Crop and Livestock Reporting Service. North Dakota crop values drop, Jan. 28, p. 1 (1983).
6. Robbins, G. S., Pomeranz, Y., and Briggle, L. W., J. Agric. Food Chem. 19:536-539 (1971).
7. Youngs, V. L., Cereal Chem. 49:407-411 (1972).
8. Pomeranz, Y., Youngs, V. L., and Robbins, G. S., Cereal Chem. 50:702-707 (1973).
9. Tkachuk, R., and Irvine, G. N., Cereal Chem. 46:206-218 (1969).
10. FAO/WHO. Tech. Report Series No. 522. Expert committee on energy and protein requirements. World Health Org.: Geneva, Switzerland (1973).
11. Peterson, D. M., Crop Sci. 16:663-666 (1976).
12. Youngs, V. L., Peterson, D. M., and Brown, C. M., "Oats." pp. 49-105, in "Advances in Cereal Science and Technology", Vol. V. (Y. Pomeranz, ed.). Am. Assoc. of Cereal Chem., St. Paul, MN (1982).
13. Youngs, V. L., and Forsberg, R. A., Crop Sci. 19:798-802 (1979).
14. Prentice, N., Babler, S., and Faber, S., Cereal Chem. 57:198-202 (1980).
15. Wood, P. J., Siddiqui, I. R., and Paton, D., Cereal Chem. 55:1038-1049 (1978).
16. Frey, K. J., Hall, H. H., and Shekleton, M. C., Agric. Food Chem. 3:946-948 (1955).
17. Rasper, V. F., Chemical and physical characteristics of dietary cereal fiber, pp. 93-115, in "Dietary Fibers: Chemistry and Nutrition". (G. E. Inglett and S. I. Falkehag, eds.). Academic Press, Inc.: New York (1979).
18. Salisbury, D. M., and Wichser, W. R., Oat Milling Systems and products. Bull. Assoc. Oper. Millers. May:3242-3247 (1971).
19. Stanton, T. R., "New Products from an Old Crop," pp. 341-344, in "USDA Yearbook of Agriculture", 1950-51. U.S. Govt. Printing Office, Washington, D.C. (1951).
20. Anonymous. The story of oats. Brochure OB6-80 Quaker Oats Co., Chicago (1980a).
21. Anonymous. Food in the A.M. Time 115:53 (March 31, 1980b).

13

Food Constituents from the Wet Milling of Corn

THOMAS J. AURAND

THE CORN WET MILLING INDUSTRY

History

The corn wet milling industry began in the early 1840s with the building of a small corn starch factory by Colgate & Company where Thomas Kingsford was plant supervisor. By 1848, Kingsford had started his own company and built a major plant in Oswego, New York. Thomas Kingsford went on to become known as the father of the corn wet milling industry.

By the mid 1860s, wet milling plants had begun to appear in the midwest in addition to several that were built on the east coast. By the 1880s, due to increased competition and development of new efficient processes, consolidation of small millers had begun. By the early 1900s the formation of most of the present wet millers had occurred. From the small labor intensive plants, the corn wet milling industry has progressed into a highly automated, capital and technology intensive industry. Due to the advances in starch technology, wet milled corn starch now dominates the food starch markets.

What Is Corn Wet Milling?

The U.S. farmer produced approximately 8.2 billion bushels of corn in 1981. While the majority of this corn is used as animal feed, the corn refining industry uses approximately a billion bushels of corn/year. The corn refining industry can be divided into two segments, dry milling and wet milling. In short, these industries differ greatly in methods of refining and the basic properties of the final products. The dry milling operation basically grinds the corn to produce an assortment of flours, meals and corn oil. The wet milling process actually refines the corn into its constituent parts, producing relatively pure fractions.

The Raw Commodity

The milling of corn, either wet or dry milling, is centered on the transformation of a basic agricultural product into more basic fractions for consumption by man or animals. The corn plant has the unique ability to utilize large amounts of sunlight to produce a concentrated, stable package of matter. This matter is in the form of carbohydrate, oil and protein located in the corn kernel; each kernel a source of energy.

The corn kernel consists of three main parts, the seed coat or pericarp, the endosperm or starch storage area, and the embryo, or germ. On a dry basis, the kernel is about 73% starch, 19% protein, 4% fiber and 4% oil. The starch fraction along with the protein or gluten fraction, is located mainly in the endosperm, while the fiber is located in the pericarp. The oil fraction is located in the embryo. From this complex mixture, the wet miller produces relatively pure ingredients with improved functionality over the raw corn kernel.

The Milling Process

The wet milling process begins with the conveying of the shelled corn from grain elevators into large tanks called steep tanks. The corn is soaked in these tanks from 20 to 58 hours at 115°F to 130°F in a dilute sulfurous acid solution. This steeping softens the kernel, extracts soluble components and helps in the separation of the starch, protein and corn germ. From the steep tanks, the corn is separated from the steepwater and the germ is removed in what are called degerminating mills. Degerminating the corn kernels is an important step, since the germ contains essentially all the corn oil in the kernel. On a dry basis, the oil content is roughly 50% by weight of the germ. This high level of oil permits efficient separation of the germ and kernel by centrifugation.

After degerminating, the kernels are transported to the grinding mills. The grinding mills break the kernels down into a slurry mixture of starch, gluten (or protein) and hull. The hulls are removed by specially designed screens, while producing a slurry of gluten and starch. This slurry is then sent through a hydrocyclone that separates out a purified corn starch stream and a gluten feed stream. This entire process lends itself to continuous operations in the wet milling plants. The entire operation is shown in Figure 13.1.

PROCESSING OF THE CORN COMPONENTS

Corn Oil

Production. The germ is the first major fraction isolated in the corn wet milling process. The germ is removed by centrifugation and sent to a continuous screw press where the oil content of the germ is reduced from roughly 50% to 20%. In large wet milling operations, the remaining 20% residual oil is reduced further by solvent extraction. After solvent extraction, the germ meal contains from 1% to 3% oil. This germ meal is then sold as animal feed.

The crude oil from the extraction process must be refined further before it is acceptable for use in food products. The composition of crude corn oil is shown in Table 13.1.

The triglycerides are the most desirable component of the crude corn oil. The other components must be removed in order to make a high quality refined corn oil. The reasons to remove these minor components are listed in Table 13.2.

The undesirable components of the corn oil are removed during the oil refining process outlined in Figure 13.2. With removal of

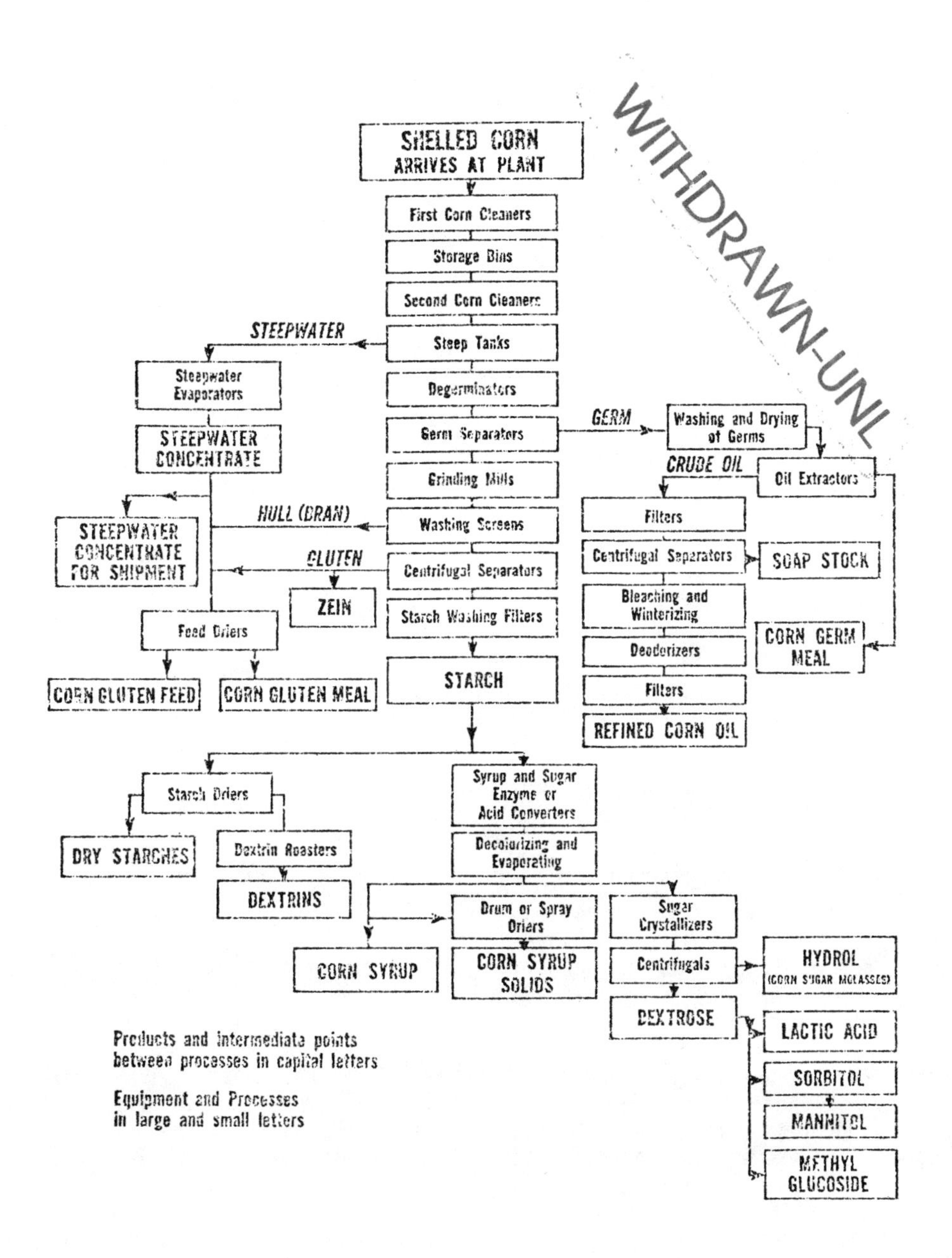

FIG. 13.1. The corn refining process.

TABLE 13.1. Typical Composition of Crude Corn Oil

Component	Typical Percent
Triglycerides	96 or above
Free Fatty Acids	Above 1.5%
Phospholipids	1-3%
Phytosterols	1% or more
Tocopherols	Trace
Waxes	.05%
Carotenoids	Trace
Odors, Flavors	Trace

TABLE 13.2. Reasons for Trace Component Removal

Component	Reason
Free Fatty Acids	Causes oil to smoke at frying temperatures
Phospholipids	Generates color upon heating; Precipitates with trace amounts of water
Waxes	Causes clouding of oil upon cooling
Carotenoids	Undesirable color in final oil
Odors, Flavors	Crude oil flavors and odors are too high

these components, the corn oil will consist of approximately 99% triglycerides, phytosterols and tocopherols. This oil is very bland in flavor and odor and can be used in an assortment of food applications.

Food uses for corn oil. Refined corn oil is generally considered a premium oil due to its clean pleasant taste, high polyunsaturate level and resistance to off-flavor development. The high polyunsaturate level provides a positive image to the consumer due to implications in regard to reducing the risk for cardiovascular disease. This positive image has helped to increase the use of corn oil in a large number of uses.

Refined corn oil can be hydrogenated in order to produce desired functionality for special food applications. The hydrogenation process would involve changing the liquid oil into a solid by the addition of hydrogen to the unsaturated portion of the corn oil. This hydrogenation process can be controlled so various properties can be produced in the final oil. These properties can range from oils that are semi-solid at room temperature to fats that are very "hard" at room temperature. Application for lightly hydrogenated oils would include corn oil margarines where melting of the fat in the mouth is

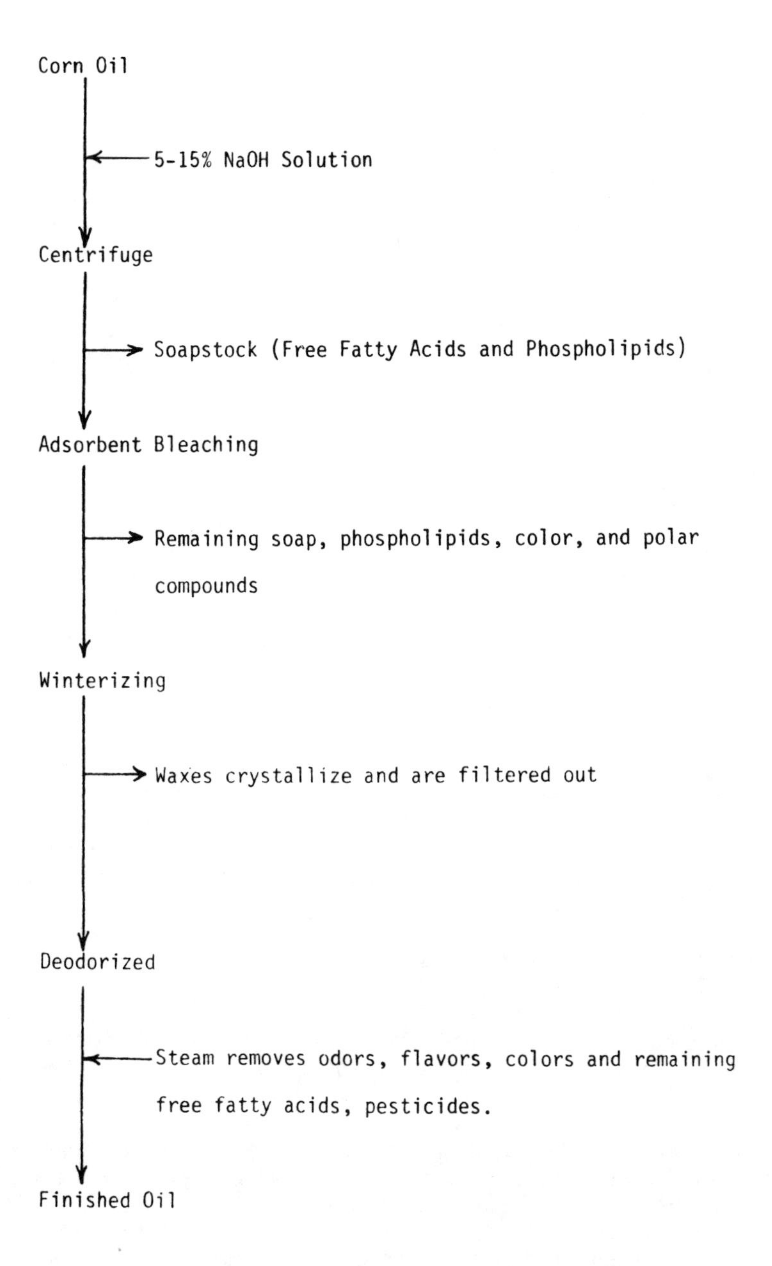

FIG. 13.2. Scheme for corn oil refinement.

a necessary quality. Highly hydrogenated oil is used in cooking or frying oils where stability at high temperatures for a long period of time is necessary.

A major use for refined oil is in salad dressing or mayonnaise productions. In salad dressing, the final product can be either pourable or spoonable and in either application, the oil will act as a lubricant by forming a thin film over the salad greens. If a hydrogenated oil or non-winterized oil is used, a greasy, undesirable mouthfeel will be produced and will not be acceptable to the consumer.

These examples demonstrate the range of uses for refined corn oil in the food industry. By proper selection of the oil, various textures, stabilities and functionalities can be formulated into a food product.

Corn Derived Sweeteners

Production. Corn derived sweeteners are produced by hydrolyzing the starch that has been refined during wet milling. The family of corn sweeteners produced by the wet milling industry is very diverse and can range from little or no sweetness to very high levels of sweetness. The production of corn sweeteners involves the hydrolysis of the raw starch into its basic component, dextrose (D-glucose). The most simple means of identifying a syrup is by referring to its DE value. The abbreviation DE stands for dextrose equivalent and represents a measurement of the total reducing sugars on a dry solids basis in the syrup. The DE value of a syrup can indicate the level of starch hydrolysis used to produce that product. For example, dextrose has a DE of 100 and is the final product of starch hydrolysis. A syrup with a DE of 24 would be considered a low conversion syrup.

The commercial production of corn syrup is shown in Figure 13.3.

Basically, there are three commonly used methods for syrup production with the final type of syrup indicating which method is used. These three processes are:

1) Acid catalyzed hydrolysis.
2) Enzyme conversion.
3) Acid-enzyme conversion.

Each method of syrup production will produce a certain type of syrup with differing saccharide distributions which gives them specific qualities.

Acid catalyzed hydrolysis is a random breaking of the bonds of the starch polymer. Although a random process, time, temperature and starch concentration can be controlled to produce syrups with a predicted DE and distribution of dextrose polymers.

Enzyme conversion utilizes the specificity of amylase to convert the starch to syrups containing large amounts of dextrose and maltose. The three enzymes of commercial importance are alpha-amylase, beta-amylase and glucoamylase. Alpha-amylase produces saccharides of intermediate to high molecular weight. Beta-amylase produces a more specific saccharide distribution due to it hydrolyzing the starch polymer at every other unit bond. This produces syrup with high maltose levels. Glucoamylase is used to produce high levels of dextrose in the syrup. Glucoamylase can be used in conjunction with the other two enzymes, or, it can be used by itself in a less efficient process.

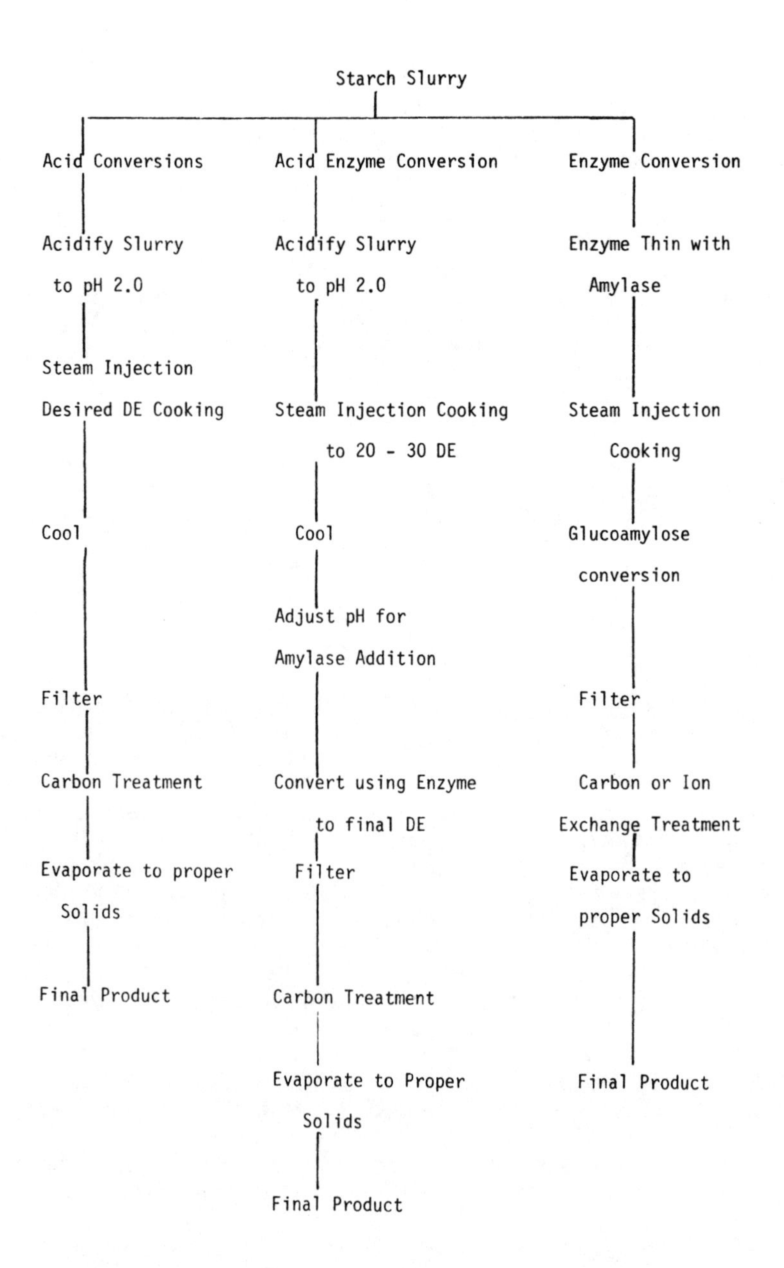

FIG. 13.3. Scheme for corn syrup production.

Acid-enzyme conversion utilizes properties of the two previously described processes. Generally, acid is used to randomly hydrolyze the starch polymer into soluble saccharides. After acid hydrolysis, enzymes are added to break the saccharide chains down into syrups composed typically of dextrose and/or maltose.

High fructose corn syrup (HFCS) is produced by isomerization of dextrose into fructose. Either enzyme converted or acid-enzyme converted corn syrup can be used as the dextrose source. Depending on the extent of isomerization, syrups can be produced that have a large range of fructose content and, correspondingly, sweetness.

High dextrose content syrups are used in HFCS production where an isomerase is used to change the dextrose into fructose. While the equilibrium point is roughly 51% dextrose:49% fructose, in commercial practice it is simpler to end isomerization at 42% fructose. Syrup produced with a 42% concentration of fructose is termed "first generation" HFCS.

Higher levels of fructose may be obtained by chromatographic fractionation of the first generation syrups. Using this process, it is commercially feasible to obtain fructose levels as high as 90-95%.

To produce the popular 55% fructose syrups, the highly concentrated fructose syrups are blended to the desired level of fructose. HFCS containing 55% fructose is becoming a widely used product due to functionality and sweetness that is very similar to inverted sucrose. The 55% fructose syrup is termed "second generation" HFCS while the 90% fructose syrups are termed "third generation" HFCS.

Functional properties and uses of corn syrups. The functional properties of corn syrup are extremely varied and each class of syrup has its own characteristics. The uses of corn syrup are wide spread and can be found in most types of food applications. For this reason, the functionality of corn syrups has been widely studied and characterized by the food industry. Various properties or functionalities of corn syrup are summarized in Figure 13.4. The most important functional properties are listed in Table 13.3. These properties may be utilized in a product either separately or in combinations to produce the final desired characteristics.

Sweetness. The most common reason for using corn syrup in a food system is for sweetness. Sweetness and the sweetness response is a complex phenomenon and an extensive thesis has been written by Shallenberger (1). Sweetness of the various sugars is usually judged on a relative sweetness scale, where sucrose is assigned a rating of 100.

When corn syrups are ranked for relative sweetness, the 90% HFCS exhibits the greatest sweetness, while the low converted syrups exhibit

TABLE 13.3. Major Functional Properties of Corn Syrups

1) Sweetness
2) Viscosity
3) Osmotic Pressure
4) Humectancy/hygroscopicity
5) Browning
6) Flavor enhancement

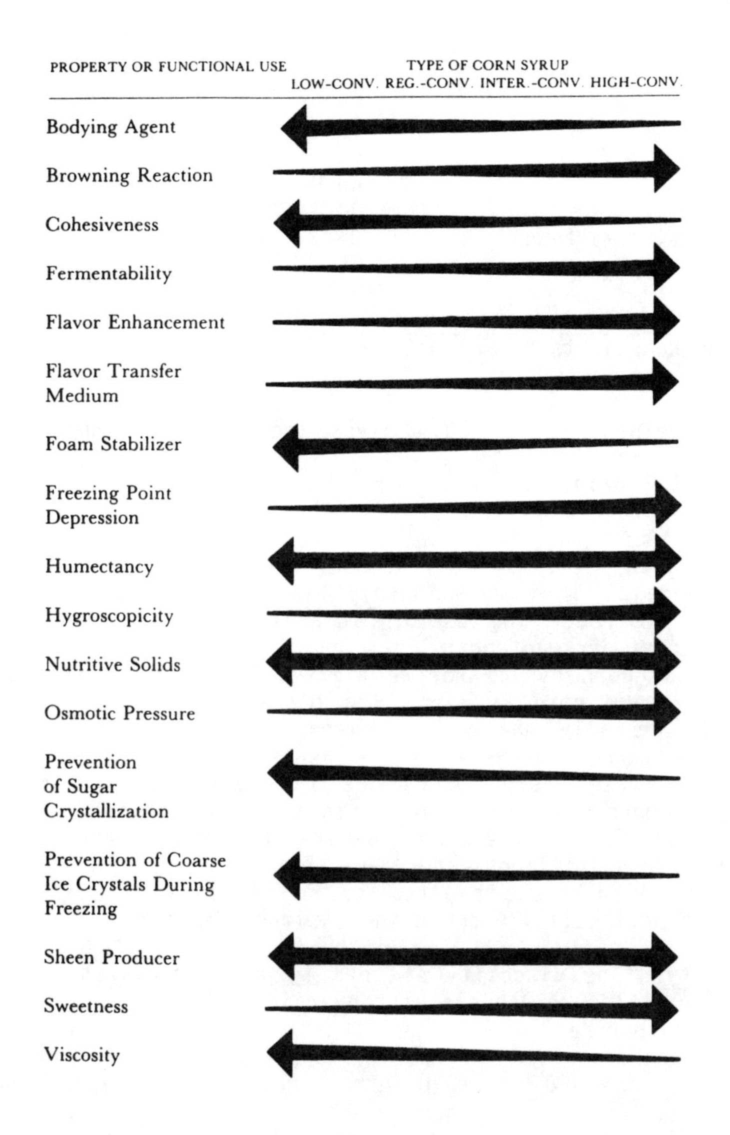

FIG. 13.4. Properties and functional uses of corn syrups.

TABLE 13.4. Relative Sweetness of Corn Derived Sweeteners

Syrup	Relative Sweetness[a]
HFCS - 90*	106
HFCS - 55*	99
HFCS - 42*	91
63 DE	45
42 DE	30
High Maltose (44% Maltose)	35
36 DE	25
Dextrose	67

*Number represents % of fructose.
[a]Sucrose equals 100.

the lowest sweetness, Table 13.4 gives the relative sweetness of corn derived sweeteners.

Table 13.4 indicates that the degree of conversion found in a corn syrup will have a bearing on its relative sweetness. Higher degrees of hydrolysis of the starch will give higher levels of sweetness. By isomerizing the dextrose to fructose, even higher levels of sweetness can be obtained. This relationship can be applied to foods where sweetness is desired. The use of relative sweetness can be used for reducing the cost of sweeteners.

The use of corn syrups for adding sweetness is due to the economic advantage of using corn syrup in the place of sucrose. This economic advantage is probably the major reason why HFCS has been increasing its sweetener market share at the expense of sucrose. Another major factor for increased corn syrups use is their ease of handling, i.e. they can be pumped. Prominent products that contain HFCS would include carbonated and still beverages, bakery fillings, canned and frozen fruit, fruit pie fillings and jams or jellies. The use of corn sweeteners in foods is projected to be increasing, as the ratio of cost versus functionality becomes increasingly important.

Viscosity. The viscosity of corn syrups is related to the degree of conversion. At lower levels of conversion, there are higher molecular weight polysaccharides present and therefore, viscosity will be higher than in syrups composed of mainly dextrose and/or fructose. This difference can be explained by the fact that more water will be associated by the higher saccharides, thereby increasing viscosity. Viscosity differences between corn syrups are shown in Table 13.5.

The formulating of various food products can take advantage of differences in corn syrup viscosities. The viscosity of a product can be increased with low conversion corn syrups which will impart a "fuller or richer" tasting character. Food products that take advantage of syrup viscosities would include fruit fillings, chocolate syrup, toppings, cordials, fruit drinks and candies.

Osmotic pressure. In two solutions of equal solids, there may be fewer high molecular weight molecules than lower molecular weight molecules. This greater number of molecules of lower molecular weight

TABLE 13.5. Viscosity of Various Corn Syrups

Corn Syrup	Viscosity (cps) at 100°F Solids
26 DE	22000
35 DE	20000
43 DE	12500
54 DE	7500
64 DE	5500
42% HFCS	75
55% HFCS	270

will produce a greater osmotic pressure than the smaller number of higher molecular weight molecules. Corn syrups will vary greatly in the molecular weight of the saccharides present, therefore will impart a wide range of osmotic pressures in solutions of equal solids. As a general rule, the higher the degree of conversion, the higher the osmotic pressure exerted by that syrup. Because of this large variety of saccharides, a large number of applications can be found that can use corn syrups for the control of water migration in a product. For example, HFCS is composed mainly of monosaccharides and will exert more osmotic pressure than an equal weight of sucrose. This fact must be taken into account when using HFCS in curing maraschino cherries or pickles, since simple pound replacement of sucrose with HFCS-55 would draw water out of the product causing the product to shrivel. While the migration of water into the syrup portion may seem undesirable in that case, some applications at high levels of solids can benefit from this phenomenon. Products such as jams and jellies, pancake syrups and candies will exhibit better microbial stability due to the higher osmotic pressure exerted by the HFCS.

Hygroscopicity--Humectancy. In a manner closely related to osmotic pressure, corn sweeteners will exhibit differing degrees of water holding ability depending on the saccharide distribution. Syrups with relatively high levels of fructose tend to be very hygroscopic. This can be beneficial if the product must remain moist until time of consumption. HFCS can be used as a means of providing moisture control for textural improvement in such items as cookies. This hygroscopicity does present problems for candymakers where moisture uptake from the atmosphere would produce a sticky mass. To overcome this, candymakers use low converted syrups which also helps in controlling graining or sucrose crystallization.

Browning. Browning, as used in this text, is the development of color in food products. This browning is due to non-enzymatic or Maillard browning reaction as well as caramelization reactions which can occur at high temperatures. The Maillard browning reaction involves reducing sugars and protein and can be considered detrimental or useful in food products, depending on the application. Corn derived sweeteners, especially HFCS, are a ready source of reducing sugars. Therefore, the browning reaction becomes an important consideration when formulating with corn sweeteners. Bakery items depend on the reaction for production of acceptable products, while the browning reaction would not be desired in such items as asceptic vanilla pudding.

Summary. Corn syrups represent an important family of food ingredients that possess many characteristics that are beneficial to the food formulator. Corn sweetener usage has been increasing in the past few years due to a large extent to the development and use of high fructose corn syrup. The functional and general characteristics of corn syrups lend themselves to a wide range of applications. Corn syrups have become an economical source of functional food ingredients that the food industry can rely on.

Corn Starch

The last major food ingredient derived from corn wet milling is corn starch. Corn starch is a widely used stabilizer in food systems where it is used to impart desired attributes. The world of corn starch is as varied as its uses. In order to comprehend the properties imparted to food systems by corn starch, it is first important to understand the basic structure of corn starch.

Unmodified corn starch. Corn starch is a polymer comprised of anhydroglucose residues linked together to form either linear or branched chains. The amylose fraction is the linear chain polymer, while amylopectin is the highly branched polymer. The amount of amylose or amylopectin in corn starch imparts unique properties to the starch. Amylose chains, being linear, will align themselves in solution with the resulting matrix entrapping water and forming a gel. This gel will retrograde with time and temperature, causing the gel to exude water. The amylopectin fraction, being highly branched is unable to form a gel matrix but will form a soft salve-like paste. These properties are exhibited in native starches and may or may not be a benefit in a food system.

The amount of amylose and amylopectin can vary in a starch depending on its genetic source. The corn wet milling industry today has three genetic variations of dent corn it can process. These genetic variations contain differing amounts of amylose and amylopectin. These variations are:

1) Common dent corn, with roughly 27% amylose and 73% amylopectin.
2) High amylose dent corn, with 50 or 70% amylose, the remainder being amylopectin.
3) Waxy maize, with essential 100% amylopectin.

Each variety has its own unique properties and uses in the food industry. The corn wet miller will use these varieties to produce starches that will suit a particular application.

Another property of corn starch that is important to understand is the basic package in which the polymers are found. This package is the starch granule. The corn starch granule is a minute discrete particle that ranges in size from 5 to 25 microns and contains the starch polymers arranged in a crystal type structure as shown in Figure 13.5.

These granules are insoluble in cold water due to the hydrogen bonding that is found between the starch polymers. Upon heating, this hydrogen bonding is broken and the starch granules begin to imbibe

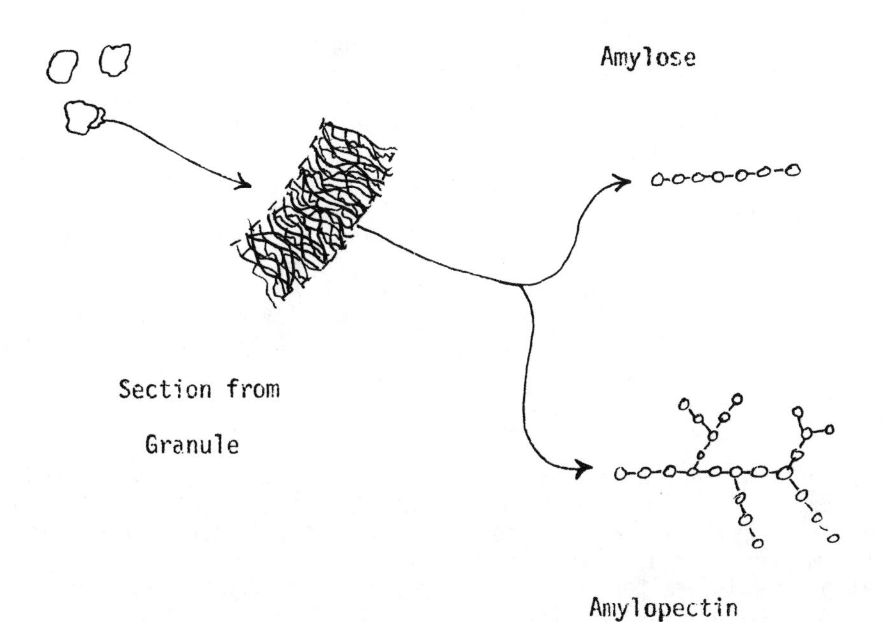

FIG. 13.5. Structure of the corn granule.

water, causing them to swell and the solution to thicken. Starch granules are able to imbibe large amounts of water, but will reach a maximum, after which they will burst. Upon bursting, there will be a reduction in viscosity. The loss of viscosity of native starch granules is dependent on such conditions as heating time, temperature, concentration of starch, shear and pH. These conditions can have dramatic effects on unmodified starches and are the reasons why modified corn starches are used in a greater variety of products.

Modified food starches. Modified food starches were developed to improve the characteristics of native starch that can be undesirable in some applications. These characteristics are:

1) The fragility of the hydrated granules.
2) Tendency of starch solutions to syneresis at room or refrigerator temperatures.
3) Gel forming properties of amylose.
4) Development of high viscosity during heating.
5) Removal of color components from the starch.
6) Provide better flow characteristics to the dry starch.

The major modification procedures approved by the FDA and regulated in 21 CFR 172.892 are shown in Table 13.6.

Each of these methods provides unique properties over native starches. Modification methods are used individually or may be used in conjunction with each other if approved in 21 CFR 172.892. With modification, the food starch will function better than an unmodified starch. Modified starches can be designed to withstand extremes in pH, temperature, shear, holding times at elevated temperatures and interaction with other food ingredients. This changing of the chemical and physical properties of starches permits them to perform more efficiently and produce food products not possible with unmodified starches.

Oxidation. Oxidation of corn starch is accomplished by the addition of an oxidizing substance to the starch. These agents are listed in 21 CRF 172.892(b) and may include hydrogen peroxide, sodium hypochlorite, potassium permanganate and sodium chlorite. The main reason for oxidation is to whiten the color of the raw starch which typically has a yellowish tint to it. This tint is due to the presence of carotene, xanthophyll and other pigments. In addition to whitening, oxidation provides other attributes, such as lowering the microbial levels, improving the flow properties of the dry starch, decreasing cooked viscosity, increasing adhesive properties and lowering the gelatization temperature of the starch. Oxidized starches find use in pharmaceutical applications, breading and battering systems, fluidizing agents in sugar and in canning operations.

TABLE 13.6. Major Modification Processes for Food Starch

Oxidation
Acid-Modifying
Crosslinking
Substitution

Acid modified. Acid modification may involve one of two different processes which result in two entirely different products. The first process is a wet or slurry process in which acid, usually hydrochloric or sulfuric acid, is added to the starch. The acid is used to partially hydrolyze the starch by breaking the glucosidic linkage. This hydrolysis produces lower molecular weight polysaccharides than what are found in the native starch. After reacting for a prescribed length of time, the slurry is neutralized, washed and dried. The resulting starch is termed "thin boiling." Thin boiling starches (when compared to unmodified starches) exhibit low hot viscosities when cooked in high soluble solids systems. The major use of thin-boiling starch is in the candy industry where they provide the set and texture to starch jelly candies, such as gum drops and jelly beans.

The other acid modifying process is a dry process where acid, usually hydrochloric acid, is added to granular starch and roasted. This process of pyrolysis produces dextrins and can produce a wide range of products. Dextrins are usually rated as a basis of cold water solubility from 0 to 100%. Dextrins can also be classified in three categories on a basis of color. The first is termed white dextrins, which have an appearance of the original starch but have reduced viscosities and increased cold water solubilities. The second category is the yellow or canary dextrins. Canary dextrins are produced with high temperatures and will produce very tacky pastes at high concentrations. The final category is the British gums, which are heated the longest time and have colors ranging from tan to brown. Due to flavor characteristics, British gums are not used in food applications.

Crosslinking. The modification of starch by crosslinking the polymers represents a major classification of modified food starches. The native corn starch is very susceptable to "over cooking" where the starch granule will swell to its limit and then begin to rupture. Also, well hydrated starch granules are very sensitive to shear and other food processing conditions which tend to rupture the starch granule. Crosslinking is used to enhance the strength of the swollen granule in order to make it more resistant to rupture.

Crosslinking is accomplished by adding very small amounts of bifunctional reagents, which react with the hydroxyl groups on two different glucose molecules. This reaction takes place in an aqueous slurry of starch granules and would typically involve less than one crosslinking bond per 1000 glucose units. After crosslinking, the hydrogen bonding within the starch granule is reinforced by chemical bonds that are better able to maintain the integrity of a swollen granule. Crosslinked starches are better able to resist extremes in pH, temperature and shear when compared to native starches. The viscosity of crosslinked starches is generally lower than unmodified, although the modified starch will maintain its viscosity much better than unmodified starches. This is shown in the Brabender curves in Figure 13.6.

A practical application for crosslinked starches is in retorted fruit pie fillings. In these systems, the acid found in the fruit will increase the hydration rate of starch. Without crosslinking, an unmodified starch granule would rupture before adequate preservation

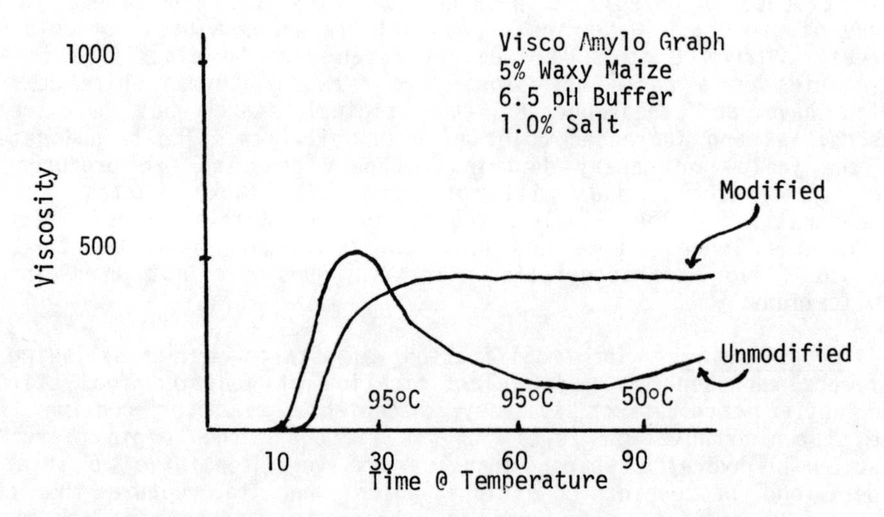

FIG. 13.6. Brabender curves of unmodified versus crosslinked waxy corn starch.

temperatures are reached. A crosslinked starch would be able to withstand the typical temperature of 190°F used in canning of fruit products and would not break down. In this example, very high levels of unmodified starches would have to be used to obtain an adequate viscosity in fruit pie fillings and this high level would produce an unacceptable product. By using a crosslinked starch, granule integrity would be maintained, thereby reducing the levels of starch used. Other applications for crosslinked starches would be in retorting operations where increased heat penetration of the product can be obtained by retarding the hydration of the starch until later in the retort cycle.

In summary, the crosslinking reaction is a very beneficial reaction for increasing the stability of the starch granule. By adding less than one chemical bond per 1000 residues, the starch granule becomes much more resistant to pH, temperature and shear. Crosslinking also gives control of viscosity development during processing, inhibits gel formation and increases the gelatinization temperature of the starch.

Substitution. Substitution is the addition of monofunctional agents to a starch slurry in order to add the substituent group to a hydroxyl group of the starch polymer. The only monofunctional agents used are those approved in 21 CFR 172.892 (d), (e) and (f). These substitution groups may be any one of the following: acetate, succinate, phosphate, hydroxypropyl or octenyl succinate. Each of these groups have unique functionality which can be used in formulating food products.

Substitution serves several functions in food starches. The most prominent is their ability to increase the binding or holding of water by the starch polymer. This additional water holding capacity of the starch is why substituted starches exhibit more viscosity than the native starch. The presence of a substitution group on the starch also changes the granule structure to a more open configuration. This opening is due to steric hindrance introduced by the substituted group. This steric hindrance does not permit close association between starch polymers. The substitution group destroys many of the hydrogen bonds and permits water to enter the granule much easier. This explains the reduction of the gelatinization temperature and decreased gel forming properties in substituted starches. This reduction in gelatinization temperature is shown in the Brabender curve in Figure 13.7. Substitution also increases the water holding ability of starches at low or freezing temperature. This gives the starch freeze-thaw properties that the native starch does not possess. Also, by increasing the amount of water that the granule can hold, the clarity of the starch paste is increased. Drawbacks to substitution would essentially be the opposite of its attributes. Because substitution permits water to enter the starch granule easier, the granule becomes very susceptible to destruction and loss of viscosity. Substitution, by itself, does not produce a starch very stable to such conditions as high acid, shear or temperature. This attribute can be utilized in the food canning industry by using substituted starches as filling aids in retort food system. By exhibiting high viscosities, the starch solution can suspend particulate particles such as vegetables and meat. When the product reaches retort temperatures, the starch will break down to a water-like consistency, producing a desirable product with the optimum amount of particulate matter present.

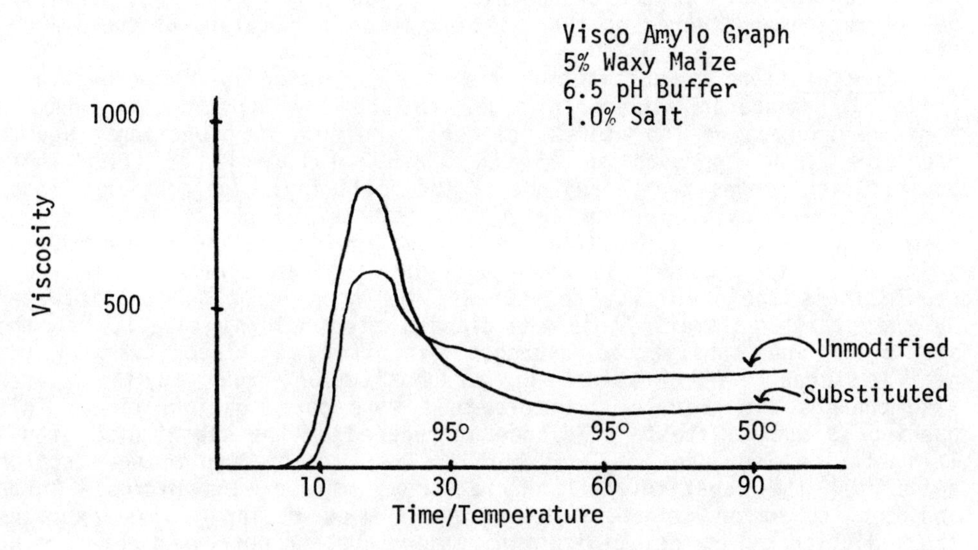

FIG. 13.7. Brabender curves of unmodified versus substituted waxy corn starch.

Substitution is usually not used by itself, but is used in conjunction with the crosslinking reaction. By combining both types of modification on a starch, the corn wet miller is able to combine the best of two worlds and produce starches with very specific functionality. By using this philosophy, the production of a wide variety of starches is possible. Starches with both modifications can vary from high to low viscosity and from low resistance to breakdown to very high resistance to breakdown. This wide range is dependent on the amount and type of modification found on the starch.

Summary. From the above discussion, it becomes evident that the world of corn starch is very comprehensive. Specific properties of the starch can be traced to the genetic variety of corn, types of modification present or to characteristics found in the food system. The development of modified starches has enabled the food processor to develop new products or improve old products. The use of modified food starches has helped improve quality of food textural and functional characteristics.

CONCLUSION

From its humble beginning in the early 1800s, the corn wet milling industry has become a cornerstone in serving the food industry. The corn wet miller supplies a wide variety of ingredients that can be found in a majority of food products. These ingredients range from the oil found in margarine, to the sweetener in soft drinks, to corn starch used as stabilizers in many foods. The proper selection of the food ingredients produced by the corn wet miller can help the food processor produce high quality food products while maintaining cost control.

REFERENCES

1. Shallenberger, R. S. and Birch, G. G. "Sugar Chemistry". Avi Publishing Co., Inc. Westport, Conn. (1975).
2. Corn Starch 4th Edition, Corn Refiners Assoc., Inc., Washington, DC. (1979).
3. Corn Sweeteners 2nd Edition, Corn Refiners Assoc., Inc., Washington, DC. (1979).
4. Hanover, L. M., in "Chemistry of Foods and Beverages; Recent Developments," p. 211. Academic Press, New York. (1982).
5. Peckham, B. W. "Economics and Invention: A Technological History of the Corn Refining Industry of the United States." PhD Thesis. University Microfilms International, Ann Arbor, Michigan, (1979).
6. Products of the Corn Refining Industry in Food. Seminar Proceedings, Published by Corn Refiners Assoc., Inc., Washington, DC. (1978).

14

Use of Cereal Products in Beverages

AUBREY J. STRICKLER

INTRODUCTION

The growth of the use of cereal products in beverages has increased dramatically within the past 10 years. This can be attributed to the introduction of High Fructose Corn Syrup (HFLS) in the late 1960s. Corn was the best selection of raw material because of its high percentage of starch available in addition to its relative ease of separation by the wet-milling process. The corn abundance and economic value as a low cost raw material has not been challenged by other cereals. However, any edible starch could be used such as wheat, tapioca, oats, barley and beans.

The high fructose corn syrup usage has increased to the point where now approximately 6.3 billion pounds (dry weight) is used in our everyday food supplies. Over 60% of these products are used in the manufacture of beverages. To put this amount in total perspective, if you were to line up approximately 115,000 tank trucks end to end, assuming tractor and trailer length were approximately 50 feet, they would cover a distance of slightly over 1,000 miles. Over 100 million bushels of corn is required to produce the amount of HFCS used in beverages, however, when compared to an 8.2 billion bushel corn crop, it is equivalent to only 1.2% of total corn grown.

HIGH FRUCTOSE CORN SYRUP

Carbonated Beverages

The first use of corn sweeteners in carbonated beverages occurred to a very limited amount during World War II. The products of that time were not designed for the use and the few independent bottlers who used some were very sorry they did. Products produced were not sweet enough and most of them encountered spoilage. I recall an elderly owner of a bottling company, in the early 1970s, telling me his problems with corn syrup and didn't even want to hear corn syrup mentioned for his use.

The first use of a corn sweetener of any significance in the carbonated beverage industry was dextrose. Dextrose was used in a blend with sucrose at blend levels of 60-80% sucrose and 20-40% dextrose.

Cereals and Legumes in the Food Supply, edited by Jacqueline Dupont and Elizabeth M. Osman © 1987 Iowa State University Press, Ames, Iowa 50010

Dextrose is 75-80% as sweet as sucrose on a dry basis and there have been reported synergistic effects in sweetness. One franchise had formulas for an 80/20 blend of sucrose/dextrose they approved for use in their products. Due to higher sweetness levels of both the 42% and 55% high fructose corn syrups, easier handling, and greater economy, dextrose usage declined. Today, I know of no major carbonated beverage bottler which is using dextrose.

The first product offered was 42% fructose HFCS and it was first accepted by independent bottlers based on a considerable amount of of testing. Taste panel studies, accelerated storage evaluations, quality checks, and replacement levels of 25-100% of sucrose and/or medium invert were tested in many different flavors. Sucrose replacement overages were evaluated in the range of 3-5%. These overages were indicated due to inversion of sucrose in the carbonated beverages. The highest overage used in commercial production is 10%. One company indicated an improved fruit flavored carbonated beverage when using this overage. It provided for more than adequate sweetness and enhanced the fruit flavor which was preferred in their taste panel studies.

Every flavor had to be evaluated separately. The acceptance of 42% HFCS in fruit flavors was excellent probably due to monosaccharides in high fructose and dextrose being more similar to the carbohydrates found in the fruits and berries. In many soft drinks it was necessary to make minor formulation adjustments to produce finished products indistinguishable from sucrose based beverages. In addition to solids adjustments, flavor bases and acidity levels are minor changes which produce beverages equivalent in sweetness, acidity, and flavor. Acidity notes especially in root beers can be noted more readily when comparing a sucrose based root beer to a 42% fructose corn syrup root beer and using the same quantity of acid to assure proper benzoate activity for prevention of microbiological spoilage. Some major cola franchises were against additional solids because of increased caloric content in their colas and, therefore, it was evident that a second generation product was necessary. As early as 1972, samples were submitted for further study and evaluation of higher fructose containing products. The final outcome was a 55% fructose corn syrup.

The 55% fructose corn syrup gained acceptance rapidly at higher replacement levels of sucrose in most all carbonated beverages. In addition it has become a sweetener now approved in all major franchise colas of at least a 50% level of sweetener in them. Some users require a 5% overage so that beverages will have the same caloric content as those made with sucrose. The sucrose inverts to fructose and dextrose and in so doing causes an increase in solids level by approximately 5%.

The 90% fructose corn syrup has not been used to any extent in beverages in the United States. However, after the ban of saccharin in Canada during 1979, it was used to produce a diet soft drink with 50% fewer calories. However, when aspartame was approved in Canada, bottlers immediately switched to the di-peptide sweetener to produce the 1-2 calorie beverages. The cost of 90% fructose corn syrup which is considered to be 15-20% sweeter than sucrose in beverage applications is not an economical saccharin or sucrose replacement in the United States.

The high fructose corn syrups are high quality products and have many attributes other than a clean, non-masking sweetness. They are highly refined so that they have low ash, excellent color and color stability especially when they are handled properly. The carbohydrate compositions are nearly the same among the suppliers and show less variation than many medium invert syrups. The fructose and dexrose content are determined by high pressure liquid chromatography and a standard method has been established. Due to high osmotic pressure created by the high monosaccharide levels the microbiological stability is excellent. In fact, studies have been made in which known cultures have been inoculated in the high fructose corn syrups. In all the cases studied, initial counts declined over a few days until they were nearly sterile. Those stored at recommended storage temperatures declined the fastest.

The HFCS are compatible with the other ingredients used in carbonated beverages. In addition, they can generally be handled in existing equipment, pumps, meters and pipes and storage tanks. Some minor modifications may have to be added, such as heat tracing and insulation on use lines and heat at storage tank in form of an enclosed area or heat pads and insulation on storage tanks.

The high fructose corn syrups are delivered to customers in either tank trucks or tank cars. Tank trucks are delivered warm so there is no fear of dextrose crystallization. However, in cold winter months, tank cars may arrive in a crystallized condition especially the 42% fructose corn syrup which contains approximately 52% dextrose. Special instructions are available to user on steaming the tank car to redissolve crystals so that color will not develop in the process. The rail cars are insulated and contain heating coils in the outer shell. There are only a few cars left in service which contain heating coils in the vessel and to the best of my knowledge these cars are not used to transport HFCS.

The 55% fructose corn syrup does not crystallize as readily as 42% HFCS. However, it may be necessary to warm the car after taking the temperature so that viscosity wil be lessened for easier handling through pumps and piping. Below 70°F there is a possibility of dextrose crystallization.

Slight agitation or recirculation in addition to heat will hasten the redissolving procedure. The product, even though it has crystallized and then redissolved, will have the same carbohydrate profile as the original syrup before crystallization. Recommended storage temperature for 42% fructose corn syrup is 80°-95°F and 55% fructose corn syrup is 70°-85°F. Recommended storage temperature for 90% fructose corn syrup is 70°-80°F.

The storage tanks in end users plant may be epoxy lined mild steel tanks or stainless steel. The stainless steel tanks (Schedule 316) are the ultimate material. The only problem encountered is corrosion at the welds which is related to technique and craftmanship. Corrosion at the welds is usually caused by tramp-carbon, and/or the weld not being ground and polished well.

The use of high fructose is well established in carbonated beverages in the continental United States and Canada. Many other countries have high fructose refineries some of which are not based on corn as the raw material.

<u>Fountain Syrups</u>

The fountain syrups are very closely related to the syrups initially prepared in producing carbonated beverages. Some may be served in a carbonated form or in a non-carbonated form. Some are used as flavoring in ices or slushes. They are most generally high solids in the range of 45-60% sweetener solids depending on flavor and dilution ratio supplied by the manufacture. The sweeteners used are nearly identical to those used in carbonated beverages.

Still Drinks

These are the non-carbonated beverages which may be found in cans, plastic, glass, or pouches that have been pasteurized or produced under aseptic conditions. Many of the canned products are pasteurized and in addition to high fructose corn syrups may also contain conventional corn syrup such as 62 D.E. corn syrup which provides more mouthfeel and is usually used at 15-35% of total sweetener solids. In addition to these, syrups are available to the consumer in addition to institutions in the form of concentrates to be diluted 1-4 or 1-5 as per manufacturers recommendation.

Dry Mix Drinks

These drinks can be in the form of either sweetened, non-sweetened or artificially sweetened. However, in most cases, there is generally crystalline dextrose, corn syrup solids or malto-dextrins used in their formulation. The artificially sweetened use corn syrup solids such as 24 D.E. or the malto-dextrins as the flavor and/or the carrier for artificial sweetener. They are less hygroscopic and the products will stay free-flowing. Flavors used may also be spray dried using special starches and/or low D.E. products as carriers encapsulating the flavor oil. This may also be the technique used in forming a cloud or opaque solution.

BEVERAGE INGREDIENTS

Most all of the aforementioned drinks contain citric acid. Much of the citric acid that is produced may make use of cereal products. Citric acid may be produced by fermentation. In addition to molasses, 95 D.E. corn syrup, corn starch and corn flour has been the starting material for the fermentation.

In those drinks which require color, especially the colas, there are significant quantities of caramel color used. The caramel color produced uses as a major ingredient the "mother liquor" from dextrose crystallization which is evaporated to at least 71% solids and sold to caramel color manufacturers.

FERMENTED BEVERAGES

Beer

The largest user of cereal products in the fermented area is beer, both full calorie and "light". The major ingredient of all beers is barley but not really employed in its original form. It must be converted to barley malt. The conversion of barley to barley malt is very important to producing consistent quality beer. Major breweries produce their own malt from high quality barley which they select for their type of beer. During the malting process, diastase and peptase activity is formed which are important in conversion of starch and proteins available in the malt in addition to adjuncts used in the U.S. Some of the adjuncts used are corn grits, corn flakes, rice grits, rice meal, refined corn starch and corn syrups. The advantage of refined corn starch is the high yield and low residual oil, protein, and fiber. The corn syrup used as an adjunct is usually a 62 D.E. which is an acid-enzyme type.

As beer usually contains a significant amount of residual carbohydrate, one major brewer uses liquid dextrose as an adjunct which is totally fermentable. The use of dextrose reduces the amount of residual sugars and therefore is preferred for "light" beers. Other processes can be used.

Some brewers in the U.S. use rice as an adjunct. In Japan, rice is the basis for "sake."

Most all of the cereals, including oats, rye and wheat have been evaluated and used to a small extent. Rye contains protein substances which influence and rye produces beer with less brilliancy. Oats have been shown to be advantageous to wort filtering but cause turbidity because of undesirable proteins. Primary reason for corn and rice is that they do not cause haze in wort after boiling and cooling as do wheat, rye and oats.

Dry corn milling produces brewer's corn grits. The corn grits can be gelatinized to form corn flakes. At one time, they were used by many smaller brewers who did not want to cook the adjunct to add to the mash. Most of these smaller brewers are not in existence today.

The primary adjuncts used are brewer's corn grits, rice, refined corn starch, and 62 D.E. corn syrup and dextrose in "light" beer.

Wine

Rules and regulations in different states vary as to using sucrose, dextrose or high fructose corn syrup for ameloriation. New York state grapes are not high enough in sugar levels and generally high fructose corn syrup is used to increase sweetener solids to arrive at an alcohol level typical after fermentation. California grapes are high enough in sweetener solids so they do not allow additional sugars before fermentation.

High fructose corn syrup replaced liquid dextrose, once used as an ameloriating agent. It is also used as an added sweetener for sweetened and "pop" wines.

High fructose corn syrup is an excellent fermentable in wine as yeasts used in wine production are well suited for fructose fermentation. However, in "beer" production, it has not been accepted because of longer fermentation times as the "brewer's" yeast is not nearly as effective on fructose.

Distilled Products

In addition to the beverage for the automobile, gasohol, there are many distilled products which use cereal based products. Corn mash bourbons, blended whiskeys, vodka, gin and many others have their roots based on corn, rye, etc. Again corn seems to be the raw material of choice probably because of high starch and easy accessability to starch and germ removal.

Cordials and Liqueurs

In addition to source of the alcohol being cereals, many are sweetened with high fructose corn syrup. One area of difficulty is in "white" goods such as schnapps. Research has been done to overcome the color reaction between mint oils and high fructose corn syrup, when compared to a highly refined sucrose in "white" goods.

OTHER BEVERAGES

Instant breakfast drinks generally have corn syrup solids and some contain isolated soy protein. These are the types that are mixed with milk to provide a nutritious breakfast. In this class, we also find dietetic beverages which contain corn syrup or corn syrup solids.

There are baby's formula which contain soy protein isolate and also corn syrup usually a 25 to 30 D.E. variety, primarily for those infants who may not tolerate milk. Most all baby formulas do contain corn syrup.

Then there are the instant grain beverages which may contain barley malt syrup and/or wheat. Pushing this area just a little further, there are also malted milk powders which change a milk shake into a malt. These contain usually wheat or malted barley.

In the area of coffee, we find some of the international coffees contain corn syrup solids. In addition, the dairy and non-dairy creamers sold in the dry form most generally contain corn syrup or corn syrup solids. This may be stretching the beverage area but there is a significant amount of product sold to this industry. There are still a significant number in the population who like their coffee to be creamed.

The last area is the frozen ade and punch concentrates. Lemonade and limeade may be sweetened with high fructose corn syrup in addition to punch concentrates. One advantage of the high fructose corn syrup is that it reduces the freezing point so that the container can be emptied easier in the home because product contains less frozen water.

CONCLUSION

Undoubtedly, some beverages in the world have been missed and I apologize. Some that I have heard of are the sweetened, flavored milk beverages which grew significantly in popularity in Australia at one time. However, by now they may have changed significantly in market share.

One major carbonated beverage franchiser also sold a nutritious drink in South America and the status of that product is unknown to me.

The previous two beverages have not been successful in the United States. We usually think of a beverage as being "thirst quenching" or as a source of enjoyment and not as a nutritional supplement to our bountiful supply of nutrients, even though many do not realize their diets can be low in some nutrients.

15

Quality Aspects of Pasta Products

JOEL W. DICK

PASTA DEFINED

Pasta is the generic word for spaghetti, elbow macaroni, lasagne, egg noodles and other similar products. The U.S. Food and Drug Administration (FDA) in its standard of identity uses the phrase, "macaroni and noodle products," to define these same products (Table 15.1).

Basically, pasta is made from a mixture of wheat and water which is then formed into a desired shape usually by extrusion through a die. The shaped product is then eaten fresh or preserved in some way for later consumption (Table 15.2).

Pasta products can be categorized into four main types (Table 15.3). These include the "long goods" such as spaghetti, "short goods" such as elbow macaroni, "egg noodles" which can be long or short but must contain a minimum of 5.5% egg solids by weight, and "specialty items" such as lasagne. The name of a pasta identifies its shape, and also gives an indication of its relative size and thickness.

TABLE 15.1. Pasta Defined

--the generic word for spaghetti, elbow macaroni, lasagne and egg noodles, etc.
--the official phrase used by the FDA for those same products is "macaroni and noodle products"

TABLE 15.2. Forms of Pasta Products

freshly prepared
canned
refrigerated
frozen
dried
dried instant

Cereals and Legumes in the Food Supply, edited by Jacqueline Dupont and Elizabeth M. Osman © 1987 Iowa State University Press, Ames, Iowa 50010

TABLE 15.3. Pasta Categories and Shapes

Long goods--e.g. spaghetti, linguine
Short goods--e.g. elbow macaroni, rigatoni
Egg noodles--pasta with a mininum of 5.5% egg solids
Specialty items--e.g. lasagne, jumbo shells

TABLE 15.4. Pasta Quality Defined

mechanical strength
microbiological stability
appearance
cooking and eating quality

PASTA QUALITY

Pasta quality can be defined by a few major factors (Table 15.4). The consumer is the ultimate judge of quality and for that reason the processor and merchandiser must also be concerned about it. The discriminating consumer wants a product that is appealing in appearance and upon cooking has a texture and flavor which contributes positively to its eating quality.

To ensure that the product is appealing to the consumer the processor must take steps to produce a pasta that is microbiologically stable. A dried product must have sufficient mechanical strength to resist cracking or breakage during cutting, packaging, handling and shipping. Excessive cracking and breakage will not only detract from the appearance of the product but might also reduce cooking and eating quality. A dried pasta product is generally considered to be very acceptable in appearance if it is translucent and bright-yellow in color, is free from cracking and checking, and has relatively few black, brown or white specks which detract from its appearance. The appearance of canned and frozen products is not quite so critical as for dry pasta since the consumer usually cannot see the canned or frozen product in the package and because the pasta is already covered with sauce and other ingredients.

Cooking and eating quality are of prime importance for all forms of pasta since they are the final measurements of quality by the consumer. Dried pasta when cooked properly should be completely hydrated but should be tender yet firm to the bite and should neither be mushy nor sticky. A good quality pasta weighs approximately three times its original dry weight at optimum cooking time, has the ability to withstand a reasonable amount of overcooking, and will not slough off an excessive amount of solid material during cooking.

FACTORS AFFECTING PASTA QUALITY

Factors affecting the quality of pasta include the quality of the raw material ingredients used, the procedures used to refine the raw materials, the actual manufacturing process of the pasta, and finally the techniques used to prepare the pasta for eating (Table 15.5). The manufacturer of pasta has considerable control over all of these factors except the final preparation for eating.

TABLE 15.5. Factors That Affect Pasta Quality

Raw Material Ingredients--wheat, water, eggs, etc.
Milling--cleaning, tempering, grinding, purification
Processing--mixing, extrusion, drying, freezing, canning, packaging
Preparation--cooking and handling

Raw Material Ingredients

Wheat is the raw material ingredient which makes up the greatest percentage of a pasta formula. Durum wheat, which is one of five major U.S. wheat classes (Table 15.6), is generally considered to be the raw material of choice to manufacture pasta products though other hard wheats are sometimes used alone or in combination with durum wheat. Durum wheat differs from other classes of wheat in that it is a separate species of wheat, has a relatively high test weight and one-thousand kernel weight, has a long kernel in relation to its height and width, has an amber kernel color, has a relatively hard kernel and tough endosperm, and has a yellow-colored endosperm due to the high level of xanthophyll pigments present (Table 15.7).

Durum has similar total protein and gluten content to the bread wheat classes though as a class its gluten protein generally lacks some of the elastic strength present in most strong bread wheats. Greater gluten mixing strength has been associated with pasta cooking quality, therefore the more recently developed U.S. durum cultivars have been bred for stronger gluten properties.

The durum wheat plant breeders have also been successful in enhancing the natural yellow pigments present in the endosperm of the durum wheat kernel which allows pasta processors to produce pasta products with a bright yellow color. Though purely an aesthetic

TABLE 15.6. Major Wheat Classes in the USA

hard red spring
hard red winter
soft red winter
white
durum

TABLE 15.7. Durum Wheat Compared To Other Wheat Classes

separate species of wheat
high test weight and 1000 kernel weight
kernel size longer in relation to height and width
amber kernel color
hard kernel
tough endosperm
yellow endosperm from high level of xanthophyll pigments

consideration, yellowness in pasta is indirectly associated with good quality since yellow color is an inherent attribute of durum wheat.

Because the durum kernel endosperm is tough and hard it is ideal for milling into semolina, the granular middling stock which is preferred to manufacture the best quality pasta with respect to appearance and cooking quality. Semolina has a granular consistency similar to table salt or granular sugar, as opposed to a fine powder such as flour.

Routine laboratory tests have been developed to help predict the potential quality of new durum cultivars as sources for use in commercial pasta production (Table 15.8). Assuming the new cultivar first meets the agronomic and wheat yield requirements of the farmer, it should possess quality that is equal to or better than presently grown cultivars for milling, pasta processing properties and final pasta quality before it is released for general seed increase and production.

The other main raw materials used in pasta are water and eggs. Water used for pasta products should be pure and suitable for drinking, and should be free of off-flavors and minerals which might diminish the quality of the final product. Eggs when used in pasta must meet the FDA requirements for purity and must make up at least 5.5% egg solids by weight of the finished product. Eggs used may be fresh, frozen, or dried.

Milling

The durum wheat miller can have a significant influence on pasta quality by the way in which he selects wheat and how he handles the wheat during the milling process. Wheat for milling should be selected for its inherent quality traits as well as its official quality grade. The process of milling durum wheat consists of four major steps: cleaning, conditioning or tempering, grinding and purifying.

The purpose of cleaning wheat before milling is to remove foreign matter which might be detrimental to the appearance and eating quality of the finished product. Dark objects such as ergot or dirt when ground-up affect pasta appearance, whereas ground-up stones or metal result in having hard gritty particles in the pasta.

Conditioning is usually done by tempering which, more specifically, is the controlled addition of water to the wheat prior to milling. Tempering is done so that an efficient separation of bran and endosperm can take place. Proper conditioning promotes increased yields of

TABLE 15.8. Quality Tests for Predicting Commercial Potential of Durum Wheat Cultivars

Wheat	Semolina	Spaghetti
Test weight	Milling characteristics	Processing properties
Vitreousness	Semolina & Flour Yield Color	
1000 kernel weight	Protein Cooked weight	
Kernel distribution	Ash Cooking loss	
Moisture	Gluten Strength Cooked firmness	
Protein	Specks	
Ash	Color	
Falling number		
Gluten strength		

semolina and better semolina color, and results in a broad separable bran, while it minimizes the production of durum flour and finely ground bran. Too much fine bran in the semolina reduces pasta color quality and increases the number of dark specks in the pasta.

Grinding of the wheat into a granular but narrow particle size range is important to promote uniform hydration of the semolina particles when they are mixed with water to form a dough just prior to extrusion through a die. Semolina which has a wide range in particle size is likely to produce pasta which has numerous white specks because of incompletely hydrated large particles. These specks detract from the appearance of the pasta and create potential weak spots which are more susceptible to fracture.

Because it is impossible to produce only broad bran during grinding of the wheat, the ground wheat after sieving must be purified to remove as many small bran particles as possible. This is accomplished by passing the middling stocks over a series of reciprocating sieves in the presence of an air current which lifts the light bran particles up and separates them from the semolina. Thus, with the proper selection of wheat and the proper use of milling techniques the miller can provide a semolina from which the pasta manufacturer can make a good quality pasta.

Processing

Just as the raw materials and the milling affect pasta quality so, of course, does the actual pasta manufacturing process. Whether using a batch system or a continuous processing system, which is now most common, the proper controlled conditions of dough mixing, dough kneading, extrusion, drying and canning or freezing are necessary to ensure the consistent production of pasta with uniform, acceptable quality.

The purpose of mixing is to incorporate all ingredients into a homogeneous dough mixture. The word homogeneous is the key since a dough mixed evenly throughout will yield a pasta with a uniform structure and texture. Factors such as semolina moisture, semolina particle size distribution, make-up water temperature and hardness characteristics, consistency of other ingredients, the ingredient incorporation system and the type of mixing equipment used all influence dough mixing results. For example, water temperature, semolina particle size and mixer type all affect semolina hydration time. Also, eggs can be added as a powder or a slurry and, therefore, must be incorporated into the blend in different ways because of their different characteristics.

In continuous pasta processing systems the loosely formed dough from the mixer passes to the extrusion barrel where it is kneaded and eventually forced through the die to be shaped into pasta. The quality of the extruded pasta is directly related to temperature and pressure within the extrusion barrel, vacuum at the extrusion head and the type of extrusion die used. Too high an extrusion temperature caused by high pressure and friction can destroy the integrity of the pasta by denaturing the gluten protein which ultimately results in poor cooking quality. Applying a vacuum to the dough just prior to extrusion reduces the occurrence of air bubbles in the pasta which improves its mechanical strength and appearance. Extrusion dies containing teflon inserts give a finished pasta that is smoother and more attractive than similar products extruded through dies with a metal surface. On the other hand, dies with a metal surface yield pasta with a rougher surface and better sauce absorption properties.

Drying is one means of preserving pasta and can also be used to improve pasta quality. Pasta should be dried in a manner which results in a product that is microbiologically stable, mechanically strong, eye appealing and has good cooking and eating quality. Drying too slowly can cause souring of the product while drying too quickly without regard to the relative humidity of the drying air can produce checking and cracking in the product. In recent years the use of high drying temperatures under controlled conditions has been shown to not only reduce drying time, but also to reduce microbial count, improve the mechanical strength and cooking quality, and enhance the color in the final product.

Canned or frozen pasta has been increasingly popular because of its convenience. Processors of these products must be conscious of over-abusing the pasta in the preparation or preservation steps in such a way as to significantly reduce its eating quality.

Instant dried pasta and refrigerated pasta are also unique products. Instant pasta is rendered fast cooking either because of its thickness or because of the nature of the process. Refrigerated pasta is usually relatively high in moisture and packaged in transparent flexible bags.

Preparation

All of the efforts in raw material selection, milling and processing can be for naught if pasta is not prepared properly by the chef or homemaker who does the final cooking of the product. Whereas, the cook has little control over the quality of canned or frozen pasta, how one handles a dried product in its preparation can determine its final eating quality. While a good quality pasta should withstand some overcooking, if cooked too long it will eventually lose its integrity and will fall apart or become mushy. When boiling pasta, cooking time is dependent on the water temperature and the thickness of the dry product. Boiled pasta is completely cooked when each piece is just saturated with water, although people who desire a softer pasta should cook beyond that point.

ATTRIBUTES OF PASTA

Pasta is gradually increasing in popularity as a food source for several reasons. It is a versatile, economical, convenient, good tasting and nutritious food (Table 15.9).

Pasta is versatile because it comes in so many different shapes and sizes and can be served in numerous ways on its own or as a complement to other food types. In addition, it can pass as an

TABLE 15.9. Attributes of Pasta

Versatile
Economical
Convenient
Good Tasting
Nutritious

appetizer, main dish, side dish, salad, soup or dessert, thus adds flexibility to any menu.

Pasta is also inexpensive and convenient relative to many other foods. It is convenient because it can be preserved by canning, refrigerating, freezing, or drying and because it is simple to prepare.

Pasta is good tasting, and recently has been recognized as being a valuable source of nutrition despite having had the past image of being a fattening food. Not only is pasta highly digestible, but it also provides valuable amounts of complex carbohydrates, protein, B-vitamins and iron and it is low in sodium and total fat content.

SUMMARY

Pasta is essentially made from wheat, usually durum wheat, and water, but sometimes contains small amounts of other ingredients. There are many quality aspects of pasta products. Factors that affect its quality are the raw material ingredients used, milling and pasta manufacturing procedures, and the final preparation for consumption. Despite having had a past image of being a fattening food, pasta has recently been recognized as being a valuable source of nutrients.

16

Extrusion Processing as Applied to Snack Foods and Breakfast Cereals

EDMUND W. LUSAS and KHEE CHOON RHEE

GENERAL PRINCIPLES OF EXTRUSION

Introduction

Snack foods and breakfast cereals are examples of "engineered" or "fabricated" foods. They are products compounded with multiple ingredients, processed, formed, and flavored to achieve characteristics that do not occur in nature. Other examples of food fabrication include bread making, known to have been practiced by the Egyptians 5,000 years ago, sausage making, practiced in Europe at least since the Middle Ages, and the more recent Japanese innovation of surimi-type restructured seafood analogs, prepared from proteins extracted from underutilized fish species.

Extrusion is simply the operation of shaping a plastic or dough-like material by forcing it through a restriction or die. The more technical definition by Rossen and Miller (1): "Food extrusion is a process in which a food material is forced to flow, under one or more of a variety of conditions of mixing, heating, and shear, through a die which is designed to form and/or puff-dry the ingredients" is often quoted. However, many of the same types of physical changes can be imparted to products by other operations, like hand working, rolling, sheeting, and flaking, and by various types of mechanical depositors and fillers.

History of Extrusion

The first commercial application of the extrusion principle is believed to have been the manufacture of seamless lead pipe in England in 1797, using a hand-operated batch-type piston press. This principle was later applied to production of clay pipe, tile, soap, and macaroni. Continuous twin-screw extruders were first developed in England in 1869 for sausage manufacture and were adapted to processing plastics in Italy in the late 1930s. The first known single-screw extruder was used in Germany in 1873 for processing rubber (2).

General Mills, Inc. was the first processor to use an extruder in manufacture of ready-to-eat (RTE) cereals in the late 1930s. The initial application was for shaping of precooked cereal dough prior to drying, flaking, or puffing. Expanded corn curls, or "collets",

Cereals and Legumes in the Food Supply, edited by Jacqueline Dupont and Elizabeth M. Osman © 1987 Iowa State University Press, Ames, Iowa 50010

were first extruded in 1939, but the product was not marketed until 1946 when it was introduced by the Adams Corporation. Cooking extruders were pioneered in the animal feeds and pet foods industries in the late 1940s. Continuous cooking and forming of RTE cereals was initiated in the early 1960s, as was extrusion of soybean flour or protein concentrate to produce texturized soy protein (TSP) and Texturized Vegetable Protein (TVP)(R) products with fibrous characteristics of cooked meat (3).

Simple, inexpensive extruders were initially developed in the United States on-the-farm cooking of soybeans and cereal feeds in the late 1960s. These machines were quickly adapted in the mid-1970s for use as low-cost extruders (LCEs) in nutrition intervention and supplementation projects in many developing countries (4).

Capabilities and Types of Extruders

Depending on complexity of the selected process, one or more extruders can be used in a processing line. The following operations can be performed, singularly, simultaneously, or in sequence, depending upon design and selection of components: mixing, conveying, homogenization, deaeration, heating/cooling, cooking, sterilization, forming/shaping, expansion, texturization, flash drying, center-filling, and package and filling. Extruders are especially useful in operations where high productivity per man hour is required, and where avoidance of effluent wastes, as from soaking or steeping processes, is desired. Reviews on design and application of single-screw extruders have been prepared by Johnston (5) and Harper (6), on twin-screw extruders by Janssen (2), and on both types of equipment by Harper (3, 7) and Linko et al. (8).

The principal components of a cooking extruder are shown in Figure 16.1. The most critical element is the screw, which performs the function of conveying the ingredients through the feeding, compression, and metering sections. The last two sections also are called "kneading" and "cooking" zones, respectively. The objective of the feed section is to receive and to sometimes mix the ingredients. In the compression zone, air and gasses are expressed, and the mixed ingredients are compacted into a dense mass that has pseudo-plastic flow properties. This mass also acts as a plug to prevent "blow back" of vapors from the cooking section to the feed section. In the cooking section, proteins are denatured and starch is gelatinized and assumes plastic-like flow characteristics. With application of sufficient pressure, the resulting "melt" will flow for subsequent shaping, even at moisture levels as low as 10%. Pressures of 200PS1 or higher and temperatures of 200°C are sometimes attained in extrusion processing.

Shear and heat are induced as the result of friction between the ingredients and with the screw and barrel surfaces as they are conveyed through the extruder. Shear and friction are further accentuated by a reduction of volume in the compression zone, by tight fit between the screw and barrel, by groving or rifling the barrel, by use of cut flights in the screw and protruding mixing bolts, by placing of tight fitting "locks" between screw sections over which the ingredients must pass, and by backpressure against the melt resulting from constriction at the exit or die face. These effects are achieved by various screw, barrel, and die designs, including changing the depth and pitch of flights in a one piece screw, by use of modular-type screws consisting

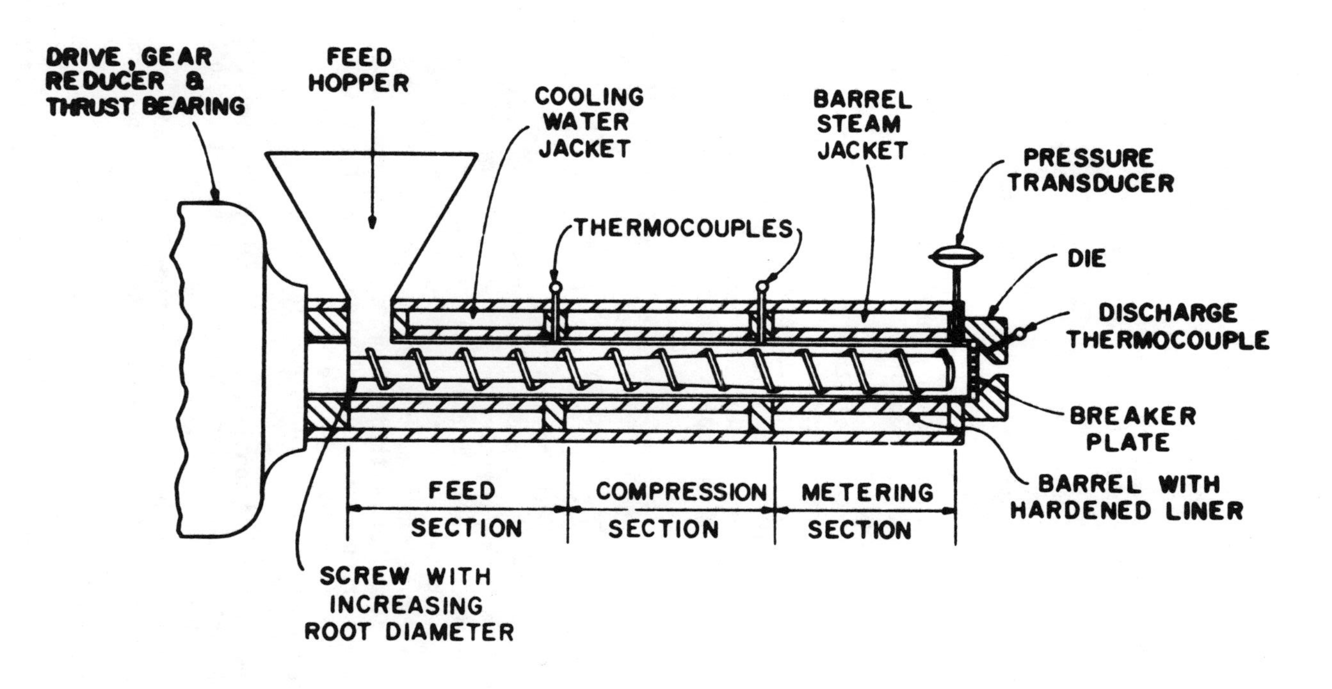

FIG. 16.1. Components of a single-screw cooking extruder (9).

of various worm sections slipped over a central shaft, and by assembling the barrel using a series of differently-designed modular sections.

Sections of the barrel may be jacketed to selectively heat or cool the product. Rossen and Miller (1) have developed the following thermodynamic classifications for single-screw extruders: autogenous--machines where the entire temperature increase is the result of viscosus dissipation of frictional heat induced by mechanical energy inputs; isothermal--machines where constant product temperature is maintained throughout the entire length of barrel, either because of limited friction as in cold forming of previously mixed doughs, or by use of water cooled jackets; and polytropic--machines where heat may be alternately added by friction or steam jacket and removed by cooling jacket.

Figure 16.1 shows a generalized extruder where the raw materials are compacted by reduction in depth of the screw flight. Heating of product is induced by the resulting friction. The barrel is jacketed for steam to allow additional contact heating in the metering section. In certain situations, it is desirable to cool the ingredients in the feed section to increase their friction and pickup by the screw, and to solidify the back section of the plug seal formed by the melt. If puffing is desired, the cooked melt is allowed to pass through the breaker plate and die at a temperature above its boiling point, with the result that internal steam forces expand the product just prior to cutting. If expansion is not desired, the product is either not allowed to reach its boiling point, or its temperature is reduced by an additional cooling jacket prior to exit. The "flash" temperature is the boiling point of water and dissolved product solutes, and is higher than the boiling point of water alone. Cross dimensional shapes of product pieces are controlled by the design of the die, and their length controlled by speed of the rotating cutter knife in relation to linear speed of the exiting product. The extruder in Figure 16.1 is shown equipped with thermocuples positioned after the feed, compression and metering sections, and a pressure transducer placed just before the die. These are minimal sensors for manually-adjusted production-type machines. Research machines are typically equipped with considerably more sensors, recorders, and control instrumentation.

The optimal type of extruder for a specific product application is determined by product moisture content, by the extent of cooking and shear desired, and by the proprietary screw and barrel designs of the extruder manufacturer. A simplified generalization for selection of extruders is shown in Figure 16.2.

Twin screw extruders consist of two parallel screws in a barrel with a figure-eight cross section. Four design variations of co-rotating or counter-rotating, and intermeshing or nonintermeshing screws are possible. Twin screw extruders are generally one and one-half times or more expensive than single screw machines for the same capacity, but possess certain advantages especially in handling viscous or sticky ingredients and products. Ninety percent or more of the world's food extrusion capacity is still of the single-screw type. Generally, the same formulation and processing principles apply to twins screw extruders as to single screw machines.

Product Changes During Processing

Certain principles apply to extrusion processing, regardless of whether the final product is a snack food, RTE cereal, breader, pet

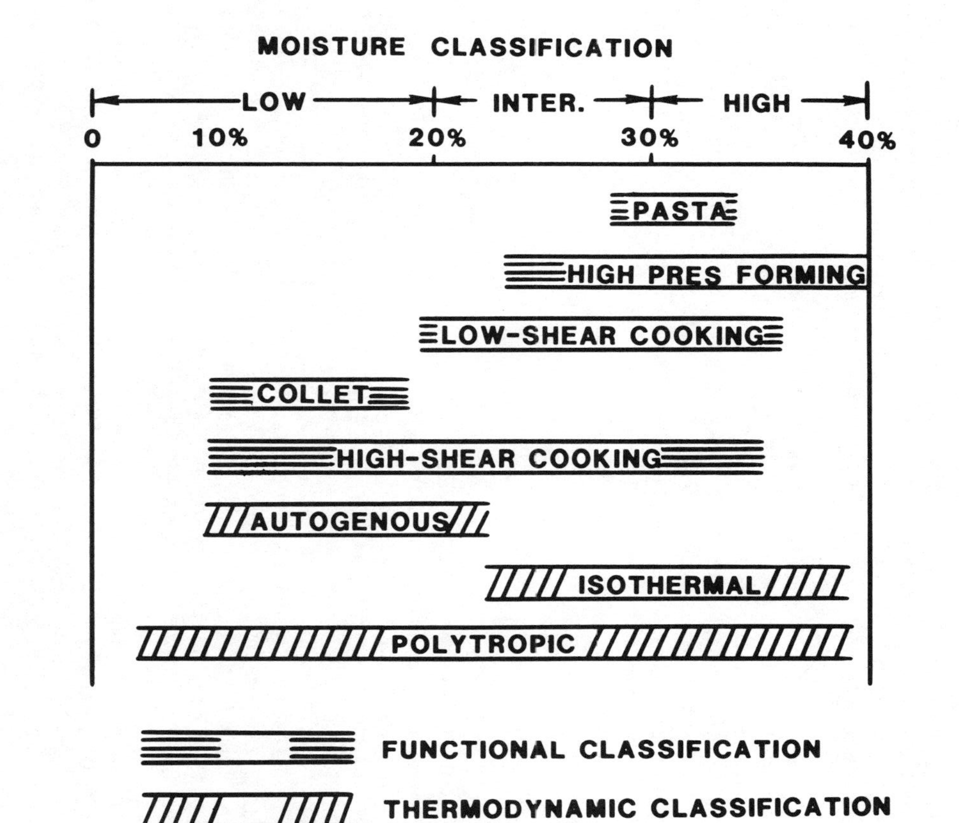

FIG. 16.2. Single-screw extruder classifications based on product moisture ranges (10, with permission).

food, or animal feed. Heat treatment and moisture content during processing must be sufficient to pasteurize undesirable micro-organisms, and to deactivate enzymes that might limit product shelf-life. If a puffed product is desired, the ingredients must be sufficiently mixed and homogenized to achieve a uniform texture. A continuous matrix of the main cohesive material, usually gelatinized starch, must be achieved. Certain materials, like fat and coarse particles, may interrupt the cohesive matrix and appreciably reduce product expansion. For these reasons, ingredients are finely ground, and fat is usually applied topically to product after extrusion. Certain ingredients, like β-carotene (used for coloring), are unstable during extrusion (11) or are lost by volatilization, like the antioxidant butylated hydroxyanisole (12), and are omitted from the extrusion mix and applied topically after extrusion. Dryers, slurry tanks, tumblers, powder dispensers and electrostatic salters, used for post-extrusion applications, are described by Harper (7). Some ingredients, like reducing sugars, vitamins and minerals, may become highly reactive with the product during the high heat treatment of extrusion, or with exposure to light, fat, and air on the surface of extruded products, and may substantially reduce product shelf-life. It sometimes is necessary to precoat such ingredients, or to apply post-extrusion surface coatings as several layers in a specific order to minimize their reactions.

SNACK FOODS

Consumption Patterns

Snacking is increasing in America's life style and results partially from factors such as increases in one-person households, a higher proportion of working mothers and more school-age children obtaining their own meals and refreshments, a highly mobile population, and ready availability of snack foods in vending machines and neighborhood convenience markets. Various products, which were once consumed mainly on impulse, are becoming accepted as side-dish items, for example, corn chips served in place of mashed potatoes. Based on a 1978 survey, at least 60% of the United States population (including 40-64% of adults and 59-70% of children and teenagers) eat at least one snack per day. This survey also estimated that, for this population group, snack foods provided an average of 20% of food energy intake, 12% of protein, 16% of fat, 25% of carbohydrates and 13-21% of various vitamins and minerals (13). The established position of snack foods in the American diet was further demonstrated by their continued growth in sales during the economic recession of the early 1980s (14-16).

Types of Snack Foods

Although this review is concerned with extruded snack foods, they are part of a much larger market consisting of cookies and crackers, pretzels, natural potato chips, fabricated potato chips, corn/tortilla chips, popped corn, meat snacks, frozen hot snacks, snack nut meats, granola products, snack cakes and pies, toaster pastries, and dried fruits. Sales of these products amounted to approximately $10.25 billion in 1981 (15). Total sales of all snack foods, including the previous products plus doughnuts, pastries, candy, ice cream and frozen desserts, and yogurt, were $20.2 billion (14). Extrusion-type techniques are used to varying extents in the production of cookies and crackers, pretzels, fabricated potato chips, corn/tortilla chips, and meat snacks. However, sales of the specific "extruded snack" category, consisting primarily of cheese-flavored corn puffs and other formulated cereal products, amounted to approximately $313 million during that period.

Production of Snack Foods

Processing methods of various types of snack foods, including processes and patents, are discussed in the books of Lachman (17), Gutcho (18), Inglett (19), Matz (20), Duffy (21), and Harper (7). Matson (22) described a three-generation evolution of snack foods: First Generation--processed natural products, like fried potato chips and popped popcorn; Second Generation--single ingredient, simple-shaped products like corn tortilla chips and puffed corn curls; and Third Generation--multi-ingredient formed products, made by extrusion cooking and expansion, redensification into intricate shapes by a second forming extrusion, drying to approximately 7% moisture, and bagging for sale for frying at another location or at the point of consumption.

Extruders are used in four major ways in the production of snack foods: 1) preparation and precooking of ingredients, as in making pellets for puffing or in continuous preparation of alkali-treated corn flours; 2) cold forming of doughs, as in production of pretzels and corn chips; 3) production of expanded products like corn collets, wheat, puffs, and fabricated chips for dipping; and 4) preparation of Generation Three snacks.

Two general types of extruded snacks are produced--products prepared by collet-extruders, and those prepared by cooking-forming extruders. Collet extruders typically have short screws (Length/Diameter ratios of 3:1 to 6:1), and are autogenous machines where all the heat is produced through friction. The usual starting material is defatted corn, with approximate specifications as shown in Table 16.1 (23).

Two types of collets (baked and fried) are produced. In making baked collets, corn of approximately 13% moisture is used and the

TABLE 16.1. Typical Specifications for Corn Grits Used for Extrusion of Collet-type Snack Foods (21, with permission)

Parameter (%)	Typical	Range
Moisture	13.5	12.5 -14.0
Protein (as is)	7.2	6.5 - 8.0
Ash (as is)	0.25	0.20- 0.30
Far (as is)	0.30	0.25- 0.40
Fiber (as is)	0.30	0.25- 0.40
Starch (as is)	78.0	76 -80
Granulation		
On US 16	1.0	0 - 5
On US 20	70.0	60 -80
On US 30	27.0	20 -35
On US 40	1.0	0 - 2
Thru US 40	1.0	0 - 2

extruder die head is designed to achieve maximum puffing. Although considerable moisture is flashed-off, it is necessary to further reduce the content to less than 2% by drying. Vegetable oil then is sprayed on the baked collets and flavors are applied in a slurry with the oil, or in powder form with the salt. In preparation of fried collets, a slightly modified extruder, with an adjustable die that enables control of pressure at the discharge, is used. Wetted corn grits (containing approximately 20% moisture) are gelatinized, but only partially expanded. The resulting collets are then fried in oil under conditions selected to control oil absorption and texture. Flavorings and coloring materials are then applied slurried in oil, or in powdered form. Snack foods also have been made from cracked rice, potato granules, and other high-starch content ingredients by collet extruders (23).

Cooking/forming extruders are of more recent design, have long L/D screw ratios, are polytropic in nature, and have considerable versatility for controlling product temperature and shear and for achieving a variety of product shapes. In the production of snack foods by cooking/forming extruders, premixes of several ingredients are processed at 22 to 30% moisture levels. After puffing, the products are dried to a moisture level of 2-4%, before coating with oil, flavorings, colorings, and salt. A processing line, as would be used for production of snacks with this type of extruder, is shown in Figure 16.3. Performance characteristics of various starch-based ingredients in extruded snack foods are presented in Table 16.2. Typical operating data for production of snacks by collet and cooling/forming extrusion processes is shown in Table 16.3.

"Half-products" or "intermediates" are gelatinized starch doughs that have been processed at 25 to 30% moisture, and then formed into chips and dried to a horny consistency at 8.5 to 10.5% moisture. In this form, they can be stored for many months. In their final preparation, they are fried in oil at 175° to 210°C for 10 to 40 seconds and expanded from 4 to 10 times their volume (20, 23). Half-products differ from Generation Three snacks, mainly in that the latter are produced from more than one ingredient, and a second extruder is used to form and produce the more elaborate textures and shapes.

In recent years, nutritionists and the general public have become concerned about the possible role of salt in hypertension. Although, the sodium content of salted snack foods is often less than many common foods (24), salt is readily noticed in snack foods because of visibility of surface-applied crystals and desiccating effect of the low-moisture background on the perceived taste. The snack foods industry has responded to these concerns by: 1) providing comparative analyses of snack and traditional foods; 2) by adopting voluntary nutritional and sodium labeling, 3) by using different granulations to achieve the same flavor with lower sodium intake; 4) by use of sodium chloride-potassium chloride blends (25); and 5) by introduction of "reduced-salt" and "unsalted" products (16).

Distribution of Snack Foods

Light, moisture, oxygen, heat, time, and various inherent thermodynamic tendencies of the products to become stale and develop off-flavors limit the shelf-life of snack foods. Antiodixants, and sequesting agents to arrest the catalytic effects of copper and iron, are sometimes used to extend shelf life. Cost-benefits trade-off must

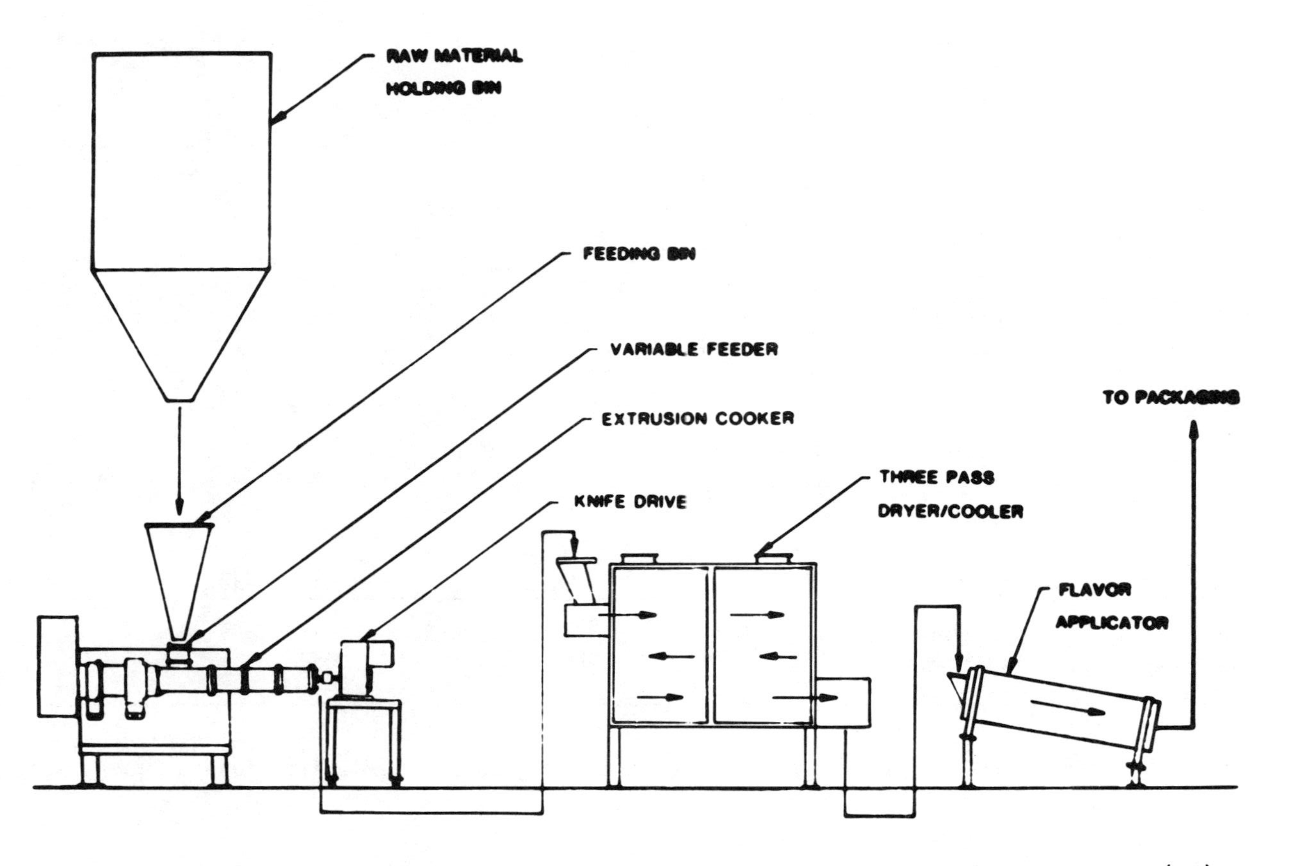

FIG. 16.3. Equipment for production of baked collet and cooked/formed snack foods (22).

TABLE 16.2. Performance Characteristics of Starch Based Ingredients Used in Extruded Snack Foods (22, with permission)

Ingredient	Expansion	Flavor	Color	Moisture required	Die Temperature required (°F)
Cornmeal	Very good	Corn, strong	Yellow	10-15	300-350
Corn flour					
Potato flour	Very good	Potato not strong	Gold to light brown	12-20	250-300
Rice flour	Excellent	Bland	White		
Wheat flour	Good	Breadlike, not strong	White to off white	12-20 18-25	250-300 300-350
Tapioca flour	Good	Bland	White		
Oat flour	Poor	Oat, strong	Off white to light brown	12-20 18-25	300-350 325-375

TABLE 16.3. Operating Data for Three Preparations of Snack Foods by Three Types of Extrusion (23, with permission)

Parameter	Baked collet	Fried collet	Cooking/Forming
Feed moisture, %	11-14	15-18	25-30
Extrudate moisture, %	7-9	11-14	24-28
Auger length/diameter ratio	2-3	2-3	10-20
Auger speed, rpm	300-450	300-450	50-100
Internal pressure, psi	20,000-30,000	5,000-10,000	300-1,000

be made between the price of protective multi-laminated packaging materials and their effectiveness in extending total shelf life. Snack foods manufacturers have generally chosen to ensure product quality by direct-to-store distribution systems with their own route men doing the racking. Direct distribution systems achieve rapid delivery of fresh products, retain control over age and location of products, and ensure placement of unbroken products on store shelves. In 1981, 54.2% of snack foods were sold through producers truck routes, 25.4% through independent distributors, and 16.1% through warehouses (16).

BREAKFAST CEREALS

Description of Market

Although breakfast cereals were originally intended to be eaten at the first meal of the day, in fact they are convenience foods consumed, at any hour, in dry form or with milk. In a national survey, it was found that, although children under 12 made up 20.5% of the domestic population in 1978, they consumed 30.3% of RTE cereals. Also, single-person households spend 16% more on cereals per person than multiperson households, and households headed by persons over 65 years of age spent 27% more on cereal and bakery products than households where the head was younger than 65. Further, nearly half of all school-age children fixed their own breakfast at least twice a week (26).

Although both industries utilize extruders and somewhat similar processing equipment, the breakfast cereals industry is vastly different from the snack foods industry. Breakfast foods have been successfully promoted to the public on the basis of nutrition, while conscientious efforts to introduce nutritional snacks have failed. Snack foods generally have shelf lifes of less than 6 weeks and require deliveries by route men. However, breakfast cereals are stable for 6 to 9 months or more, and can be processed in centralized plants and distributed through the general nonperishable grocery products distribution system. As a result, breakfast cereals usually utilize more costly protective packaging than snack foods. Also, because of reduced distribution problems, and a more mature market, the breakfast cereals industry consists of less than 2 dozen manufacturers, with about 75% of the market held by three companies. In contrast, hundreds of snack foods manufacturers exist, and the largest company producer, with 44% of total industry sales, maintains 39 processing plants (27).

The category of breakfast cereals includes hot cereals (like oatmeal and farina), shredded cereals, puffed cereals, flaked cereals, and granolas and nutritional products, in addition to extrusion-cooked expanded products. Although extruders are sometimes used in precooking of hot cereals (such as baby foods and instantized products) and doughs for shredded RTE and flake-type cereals, this review primarily summarizes preparation of RTE products that are expanded by extruders.

Extruder Processing of RTE Cereals

Harper (7) has summarized the various roles of extruders in cooking, pellet-forming and cooking/forming of cereals, as shown in Figure 16.4. Some types of Generation Three RTE cereals are also processed using multiple extruders, with the mixture of cereal ingredients cooked by

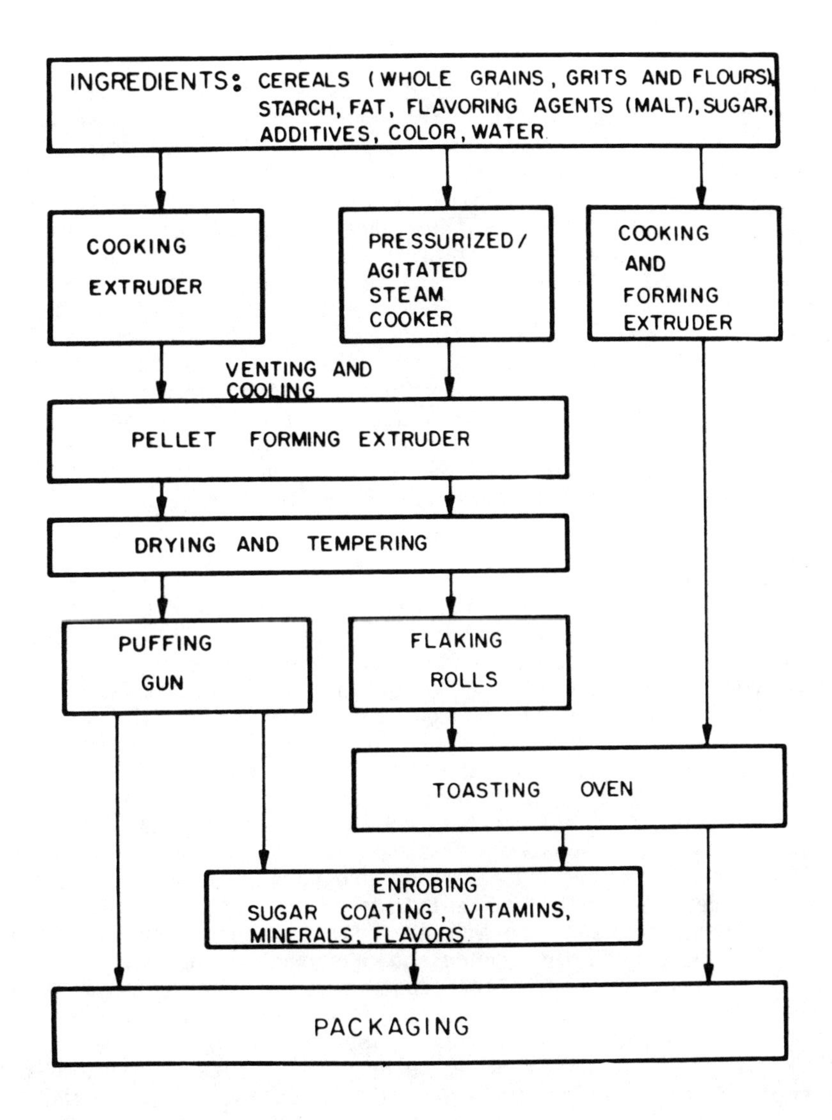

FIG. 16.4. Role of extruders in processing of RTE breakfast cereals (7, with permission).

the first unit, and shaped into intricate shapes in the second extruder. A diagram of a RTE flaked cereal process line in shown in Figure 16.5, and a flow chart of a Generation Three type RTE process is presented in Figure 16.6.

As in preparation of snack foods, components that would interfere with product expansion, or that are unstable at the temperature and moisture conditions of extrusion, are withheld and enrobed later. Post extrusion processing of RTE cereals differs from that of snack foods in that many of the additives are enrobed in an aqueous base which is subsequently dried, rather than being sprayed on as oils or applied by dusting. Products typically are enrobed with sucrose syrups, containing 60 to 85% total sugars, and selected flavorings and colorings. A frosty coating results if sugar is allowed to crystalize on the surface of the product: however, a clear glaze can be obtained if 1 to 8% of other sugars (such as invert sugar, glucose or fructose) is included in the syrup (29). Fats, with high-temperature melting points, and distilled monoglycerides, have been used in enrobing syrups to prolong crispiness of RTE cereals in milk (7). Additional minor ingredients, like minerals and protected forms of vitamins, may be applied by spraying or dusting in the later stages of drying while product surfaces are still sticky.

Nutritional Labeling

Success in marketing nutritional concepts to RTE consumers has been varied. Attempts to emphasize protein quality (as determined by Protein Efficiency Ratio, or "PER", values) have generally failed, although products marketed on the basis of increased total protein content have been successful. Low-sodium extruded RTE cereals have generally not been developed, probably because of the beneficial effects of salt on texture formation of the multi-ingredient fabricated products. However, shredded wheat-type products, with no added salt, are made.

The RTE cereals industry has come under criticism for marketing "empty calorie foods", apparently as the result of high levels of added sugar (50% or more in some presweetened products). Also, public concerns have been expressed about pro-dental caries effects of sugar residues that might lodge between teeth after eating. In response to public concerns, RTEs have been introduced and promoted on the basis of "reduced or low sugar content" and "no added sugar". Products artificially sweetened with sodium cyclamate were introduced in the 1960s but were withdrawn when FDA suspended approval of this additive. Saccharin-sweetened products were sold in the interim, but have been mainly replaced by products sweetened with aspartame. Since the caloric content of sugar and cereal ingredients are similar, formulation changes that increase formula ratios of cereal ingredients to sugar have essentially no effect on caloric density per unit weight of product, except for replacement of the solid fats used to preserve bowl crispiness.

The Food and Drug Administration has not issued a regulation specifying appropriate fortification of RTE cereals. The majority of the industry fortifies dry cereals to contain 25% of U.S. Recommended Daily Allowances (U.S. RDA) of selected vitamins and minerals, although several RTEs are fortified to contain as much as 100% RDAs. Typically, vitamins A, C, thiamin, riboflavin, niacin, E, B_6, folic acid, and B_{12}, and the minerals calcium, iron, phosphorous, magnesium, zinc,

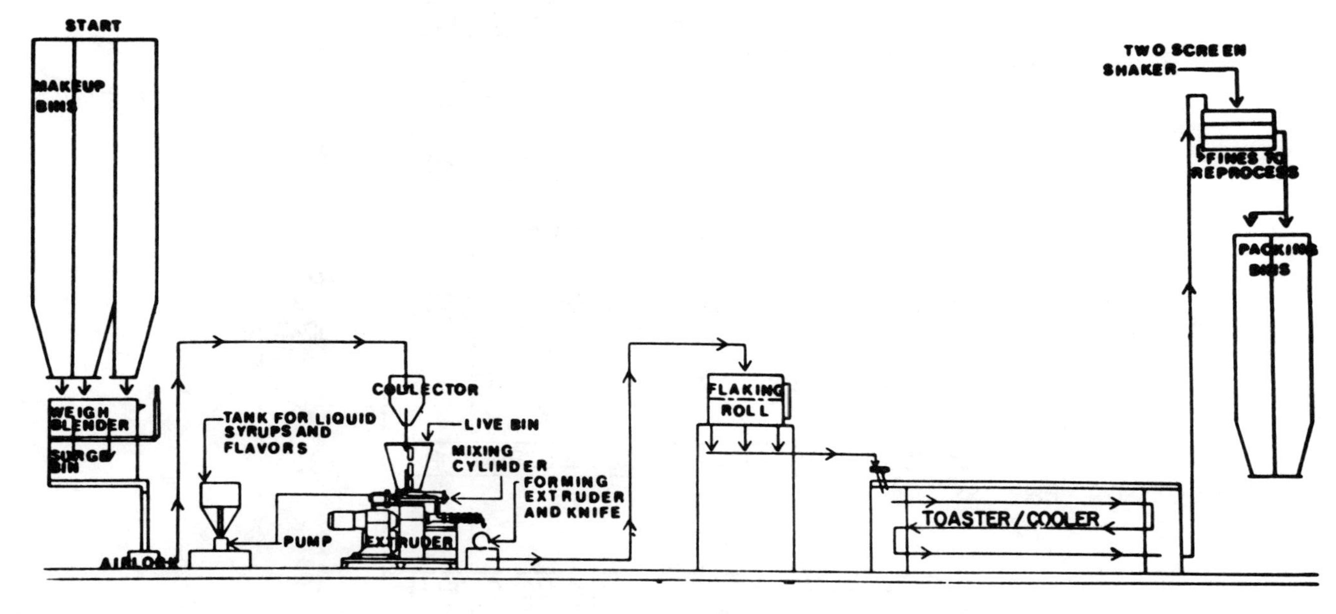

FIG. 16.5. Process line for production of RTE cereals using an extruder to make precooked pellets (28, with permission).

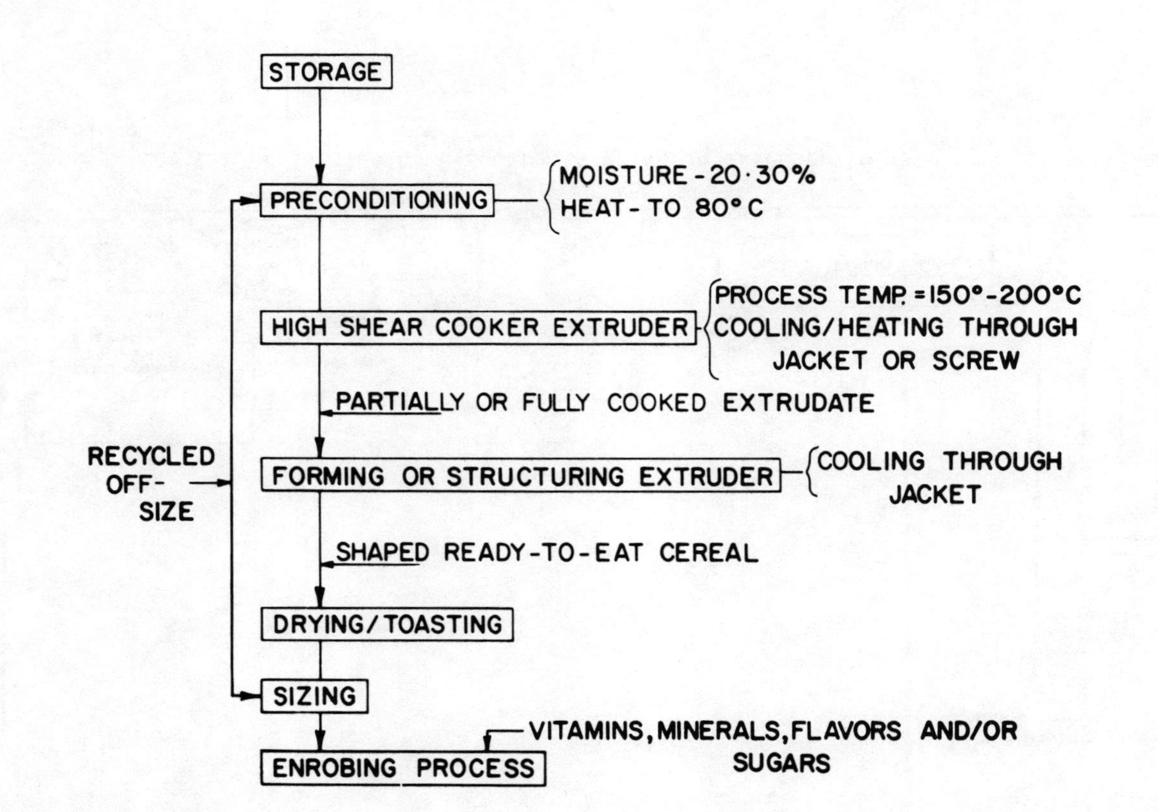

FIG. 16.6. Flow chart of RTE cereal process utilizing sequential extrusion (10, with permission).

and copper are supplemented. However, selection of vitamins and minerals for fortification is not consistent within the industry, or even between products of the same company, probably because of compatability problems with the product or between the forms of fortifying ingredients used.

Nutritional information per serving of RTEs usually is shown on the package as two columns--in terms of the recommended serving size (usually 1 oz. of product), and the recommended serving eaten with one-half cup of whole milk. Claims are made for calories, protein (g), fat (g), and sodium (mg). Packages also typically show carbohydrate information, in terms of starch and related (complex) carbohydrates, and as sucrose and other sugars in grams per serving. Some labels also provide information on dietary fiber content in grams per serving.

REFERENCES

1. Rossen, J. L., and Miller, R. C., Food Technol. 27(8), 46 (1973).
2. Janssen, L. P. B. M., "Twin Screw Extrusion." Elsevier, New York, NY (1978).
3. Harper, J. M., "Extrusion of Foods, Vol. I." CRC Press, Boca Raton, FL (1981a).
4. Crowley, P. R., in "Low-Cost Extrusion Cookers, Second International Workshop Proceedings (Tanzania)" (D. E. Wilson, R. E. Tribelhorn, eds.), p. 11. Dept. Agr. and Chem. Eng., Colorado State University, Ft. Collins, CO (1979).
5. Johnston, G. L., International Seminar: Cooking and Extruding Techniques, Solingen-Grafrath, Germany, November 27-29 (1978).
6. Harper, J. M., Crit. Rev. Food Sci. Nutr. 11(2), 155 (1979).
7. Harper, J. M., "Extrusion of Foods, Vol. II". CRC Press, Boca Raton, FL (1981b).
8. Linko, P., Colonna, P., and Mercier, C., in "Advances in Cereal Science and Technology, Vol. IV". (Y. Pomeranz, ed.), p. 145. American Association of Cereal Chemists, St. Paul, MN (1981).
9. Clark, J. P., J. Texture Studies 9, 109 (1978).
10. Triebelhorn, R. E., and Harper, J. M., Cereal Foods World 25, 154 (1980).
11. Lee, T., Chen, T., Alid, G., and Chicester, C. O., AIChE Symposium Series Food, Pharmaceutical and Bioengineering 1976/77, 74, 192 (1978).
12. Fapojuwo, O. M. O., and Maga, J. A., J. Agric. Food Chem. 27, 822 (1979).
13. Morgan, K. J., Cereal Foods World 28, 305 (1983).
14. Anonymous, Vending Times 22(9), 24 (1982a).
15. Anonymous, 14th annual state of the snack food industries report. Snack Food June, M-M23 (1982b).
16. Morris, J. B., Chipper/Snacks, September 1982, CS1 (1982).
17. Lachman, A., "Snacks and Fried Products." Noyes Development Corporation, Park Ridge, NJ (1969).
18. Gutcho, M., "Prepared Snack Foods." Noyes Data Corporation. Park Ridge, NJ (1973).
19. Inglett, G. E., "Fabricated Foods." AVI Publishing Co., Westport, CT (1975).
20. Matz, S. A., "Snack Food Technology." AVI Publishing Co., Westport, CT (1976).

21. Duffy, J. I., "Snack Food Technology: Recent Developments." Noyes Data Corporation, Park Ridge, NJ (1981).
22. Matson, K., Cereal Foods World 27, 207 (1982).
23. Toft, G., Cereal Foods World 24, 142 (1979).
24. Stauffer, C. E., Cereal Foods World 28, 301 (1983).
25. Bravieri, R. D., Research and Development Associates Semi-annual Meeting, Norfold, VA. (1983).
26. Hayden, E. B., Cereal Foods World 25, 141 (1980).
27. Hollingsworth, P., Prepared Foods 152(3), 60 (1983).
28. Smith, O. B., Division of Agr. and Food Chem. Meetings. American Chemical Society, Las Vegas, NV, (1974).
29. Vollink, W. L., U.S. Patent 2,868,647 (1959).

17

Combinations of Cereals, Legumes, and Meat Products in Extrusion Products

ISAAC O. AKINYELE

INTRODUCTION

Protein-calorie malnutrition continues to be the major public health problem in many developing countries. This problem manifests itself primarily within the vulnerable group comprised of infants, children, the elderly, pregnant and lactating women. To millions of people in developing countries the availability of daily bread to provide the necessary nutrients to maintain good health is often remote where hunger, poverty and starvation reign supreme. Unfortunately, resources abound in many of these countries to provide the needed food and, where not available, food surpluses exist in many developed countries which could provide the needed relief for the malnourished if appropriate policies existed to achieve such an objective. The problems related to providing food are not simple and there are conflicting attempts at resolution. There are those who believe that if a person is dying of hunger you give him food. There are those who believe that if a person is to live tomorrow you should teach him how to produce his own food properly. Finally there are those who believe that opportunities should be provided for all those who have learned the proper production methods to produce enough food and a means of marketing and preserving the surplus so they can eat tomorrow.

The green revolution program in many developing countries came about based on the belief in self reliance. This program has resulted in tremendous increases in yields especially of cereals and legumes due primarily to research efforts by scientists, who have developed new varieties of crops to provide the higher yields, requiring shorter growing seasons and which are to some extent more resistant to pests, drought and diseases. Their acceptances by many farmers in developing countries have been very good and improved cropping practices have also added greatly to the crop yields.

While it is undeniably important to increase yields of major food crops an even greater increase in the amount of food available for human consumption could be realized by reducing post-harvest losses. It has been estimated that wastage runs as high as 30-50% in some countries which means that any increases in production may simply replace the amount of food lost post-harvest. Consequently, the green revolution program has in many areas been deemed a failure because no real growth

Cereals and Legumes in the Food Supply, edited by Jacqueline Dupont and Elizabeth M. Osman

was evident in the amount of food available and the risk of developing malnutrition was still endemic among the vulnerable group.

The enormous amount of post-harvest losses from farmers to the consumer, supports the need for an appropriate technology for the processing and preservation of cereals and legumes, the main food staples in many countries, to reduce post-harvest losses and increase food availability.

Any such technology adopted for processing in these countries must be low-cost to obtain or develop, since large capital ventures which operate in industrialized countries are too expensive and inappropriate for local application in developing countries. Furthermore, primarily processed products would be available for use in individual home recipes to meet local tastes and customs. It has been established (1) that low cost methods must possess the following characteristics:

a. Processing systems must be cheap comparatively.
b. Production rates must be moderate ranging between 250 and 1000 kg/hr.
c. Operation must be simple requiring little sophisticated technical capability to operate and maintain.
d. To the greatest possible extent, the system should not require auxiliary boilers, dryers or pre and post processing or conditioning equipment that can increase cost and lead to product contamination.
e. Process should be versatile; able to handle a wide variety of cereal/legume blends.
f. System should be easily maintained utilizing locally available parts and equipment to the greatest possible extent.
g. Equipment must be cleanable and suitable for producing human food products.

Several types of low-cost processing systems based on extrusion technology are potentially capable of achieving these characteristics and some of them are currently in use. This paper provides a review of the application of some low-cost extrusion cookers for processing cereal/legume blends in many developing countries. The main objective in these countries was to improve the nutritional status of the population by increasing the availability of foods which can provide the necessary nutrients needed for proper growth and development.

CHARACTERIZATION OF LOW-COST EXTRUSION COOKERS

Various authors (2,3,4,5) have discussed the unique characteristics of extrusion technology, which is often defined as the continuous process by which mechanical shear is combined with heat to gelatinize starch and denature protein as they are plasticized and restructured to create new textures and shapes. The advantages of extrusion cooking of cereal/legume blends were clearly stated by Harper and Jansen (1) and are as follows:

a. The blends can be easily fortified with a broad range of vitamins and minerals.
b. By virtue of protein complementation, the extruded blends have been demonstrated to be of high protein quality and are well digested.

c. The use of precooked blends is energy efficient particularly in situations where firewood is scarce and most cooking is done over an open fire.
d. Extrusion increases the calorie and nutritive density of gruels made from the blends.
e. Central processing facilities allow the use of suitable packaging to protect the nutritious products.
f. Extrusion combined with suitable packaging increases shelf life of the product by reducing water activity and inactivating enzymes.
g. Central processing allows consistency of formulation and product quality.

Apart from these advantages, other desirable benefits of low-cost extrusion cookers in developing countries include:

a. Enhancing utilization of locally grown protein and carbohydrate food resources.
b. Serving as a focus for the development of small scale industries in these countries.
c. Reducing transportation costs associated with movement of manufactured goods over long distances.
d. Allowing formulation and processing to meet local tastes and preferences.
e. Minimizing packaging costs required, since products will be consumed locally in a relatively short time.

There are varieties of extruders classified according to the moisture content of the food mixture. Low-moisture or dry extrusion involves the utilization of food mixtures with a moisture content below 20-22% (6,7). The heat for the process is generated by friction from the large drive motor accompanied by pressure and attrition which cook and expand the ingredients to gelatinize the starch and destroy inhibitors. There is usually no pre-processing related to the extrusion function. Product cooling results in 6% moisture loss.

High-moisture extrusion operates with mixtures containing at least 28% moisture. Some of this moisture may be added as steam which condenses and preheats the raw ingredients, minimizing the amount of heat that must be produced by friction. These systems can handle a wider variety of raw ingredients than the low-moisture extruder. High-moisture extruders require a boiler to generate steam and a drier to reduce the products to a safe storage moisture level.

Intermediate-moisture extrusion operates with mixtures containing between 20% and 28% moisture. Some product drying is required to remove moisture in excess of 12% in the finished product.

Low-moisture or dry extrusion systems are also called low-cost extrusion cookers (LEC) because of their relatively low power input requirement and self-contained processing. The development of these types of extruders is fairly recent even though extrusion processing has been used since the 1940s. Until the 1960s, farmers who raised soybeans might sell part of their crop to crushers and then buy back the meal for use as animal feed. Therefore, attempts were made to develop a simple extruder that would dry-cook soybeans and enable the farmer to prepare his own feed supply. These extruders were unique in that they were inexpensive, required few skills to operate, needing no elaborate accessories and had a fairly large output.

The Brady crop type extruder and the Insta-Pro extruders are the two most widely used and studied low cost extrusion cookers. Insta-Pro is manufactured by the Triple F Feeds Inc., Des Moines, IA, while the Brady is distributed by the Brady Farm Division of Koehring Company.

Cereal/Legume Blends with the Brady Crop Extruder

Most of the work reported in the literature has been a direct result of studies initiated at Colorado State University by Dr. Julian M. Harper and his team either alone or in conjunction with the United States Department of Agriculture, Agency for International Development and some international foundations based in the United States. These studies have led to the adoption of low-cost extrusion cookers or research activities on the technology in many developing countries of Central America, The Philippines, Indonesia, Sri Lanka, Tanzainia, Kenya and Thailand.

The first extrusion equipment tested in Central America was the Brady crop extruder loaned to the Institute of Nutrition of Central America and Panama (INCAP) in 1975. Since then Bressani et al. (8) have undertaken considerable research to optimize the running parameters of the extruder so as to obtain food systems with the best nutritional and storage stability properties. Of the many food mixtures tested (Table 17.1) the Maisoy formula (70 parts corn and 30 parts soybean) has been considered the best mixture for Honduras (9) basically because it has a high nutritional value and a flavor similar to cooked corn flour, a staple food in the traditional Honduran diet. Other combinations are still being evaluated in many countries of South and Latin America. The Maisoy has also gone into commercial production in Bolivia in the form of flakes using a combination of 85 parts corn and 15 parts soybean (10). The wide acceptance of these products and the apparent well-being of the consumers is an indication of the success these programs have achieved.

Similar reports from Thailand (11), Indonesia (12) and the Philippines (13) have demonstrated the effectiveness of the Brady crop extruder to fulfill the requirements of the low-cost methods and the needs of the local people from one country to the next. The combinations of cereal, legumes and fish products in Thailand (11) (Tables 17.2 and 17.3) have formed the basis for a baby food formula and various snack foods. Based on research results it was concluded that extrusion cooked grain products appear to have potential as low-cost, nutritionally improved foods for use in Thailand (11).

Muchlis (12) reported experimental extrusion processing of cereal/legume blends in Indonesia in which all ingredients used were derived locally. The Brady extruder used was found to operate well and to give products acceptable to a sensory evaluation panel. The results of the study are presented in Table 17.4. Six formulations were developed in the Philippines (13) and were successfully extruded using the Brady crop extruder. Pablo (13) reported that a cereal concentration of 10 to 75% is suitable for extrusion when processed in combination with legumes or coconut. The different combinations were found acceptable in local recipes. The mixtures and proximate composition are presented in Tables 17.5 and 17.6.

The work at Colorado State University (14) has emphasized corn/soy blends and full-fat soy flour extrusion since these represent the major food products in those area of the world where low cost extrusion cookers

TABLE 17.1. Some Food Mixtures Processed by Extrusion at INCAP

Ingredients	Formula (%)	Formula (%)	Formula (%)
Corn/Soybeans	85/15	82/15	70/30
Rice/Soybeans	82/18	70/30	
Oats/Soybeans	82/18		
Sorghum/Soybeans	70/30		
Cassava/Soybeans	54/30		
Cowpea/Corn + Methionine	65/25	72/28	
Cowpea/Cassava + Methionine	81/19		
Soybean/Sesame	50/50		
Corn/Soybean/Sesame	72/14/14	72/21/7	
Rice/Soybean/Sesame	72/14/14	72/21/7	
Wheat/Soybean/Sesame	72/14/14	72/21/7	
Corn/Pigeon Peas/Soybeans	60/24/6		
Corn/Redbeans/Pigeon Peas/Soybeans	60/10/10/20		

Source: Brusani et al. (8)

TABLE 17.2. Formulation of Mixed Ingredients Passed Through the Brady Crop Extruder (11)

Samples	Ingredients	Formula (%)	Texture
1	Rice:Soy:Fish	75:20:5	Well-Cooked, Crisp
2	Rice:Soy:Sesame	65:25:10	Well-Cooked, Crisp Dark in Color
3	Rice:Peanut:Fish	75:15:10	Well-Cooked, Crisp
4	Corn:Peanut:Fish	75:15:10	Well-Cooked, Crisp
5	Cassava:Peanut:Fish	55:30:15	Well-Cooked, Sticky

TABLE 17.3. Proximate Analysis of the Processed Mixed Ingredients Shown in Table 17.2 (11)

Sample	Moisture	Protein	Fat	Carbohydrate	Energy kcal/100g
	%				
1	9.26	16.3	4.32	66.4	373
2	8.55	15.3	11.35	61.4	344
3	10.04	16.7	7.46	63.0	410
4	11.39	18.4	10.09	57.0	379
5	8.58	23.2	14.55	49.6	409

TABLE 17.4. Proximate Analysis of Indonesian Mixtures Extruded with a Brady Crop Extruder (12)

Mixtures	Water	Ash	Protein	Fats	Carbohydrate	Energy kcal/100g
	%					
Rice 70%, Soybean 30%	6.6	2.4	16.6	7.2	67.2	400
Rice Polish 70%, Soybean 30%	6.8	2.9	16.8	7.2	66.8	398
Corn 70%, Soybean 30%	6.2	2.6	17.0	9.5	64.7	412
Cassava 70%, Soybean 30%	7.4	3.2	13.0	8.1	68.3	408

TABLE 17.5. Mixtures of Cereals, Legumes and Nuts Successfully Extruded in the Philippines Using a Brady Crop Extruder (12)

Formula	Cereal	Legume	Nut	Extrusion Temperature (°C)
I	62% Corn	25% Pigeon Pea	13% Peanut	160
II	70% Corn	20% Mungbean	10% Peanut	160
III	35% Corn, 35% Rice	30% Winged Bean		129
IV	10% Corn, 10% Rice	70% Cowpea	20% Coconut	127
V	70% Rice		30% Dessicated Coconut (Partially Defatted)	185
VI	60% Rice		40% Dessicated Coconut (Partially Defatted)	185

TABLE 17.6. Proximate Analysis of Six Blends Shown in Table 17.5 (13)

	Formula					
	I	II	III	IV	V	VI
	%					
Moisture	8.0	9.0	7.0	7.0	5.0	6.0
Protein	16.0	15.0	17.0	18.0	9.0	12.0
Fat	6.0	8.0	7.0	3.0	4.0	5.0
Ash	2.0	2.0	3.0	2.0	2.0	3.0
Crude Fiber	2.0	3.0	3.0	3.0	2.0	3.0
Carbohydrate	66.0	63.0	62.0	64.0	79.0	72.0

TABLE 17.7. Chemical Analysis of Corn/Soy Blends (70/30) Extruded at 163°C with a Brady Extruder (14)

Components	Degermed Corn Dehulled Soy	Whole Corn Dehulled Soy	Whole Corn Whole Soy
Moisture %	2.8	3.3	3.6
Fat %	6.8	9.2	8.6
Protein (N x 6.5) %	17.0	18.6	17.8
Ash %	4.2	5.1	5.1
Fiber %	0.8	1.9	2.3
Carbohydrate (by difference) %	68.4	61.9	62.6
Nitrogen Solubility %	4.8	6.4	6.7

TABLE 17.8. Effect of Temperature on Extrusion of Soybeans Using a Brady Extruder (14)

Temperature °C	Nitrogen Solubility Index	Urease pH Units	Trypsin Ihnibitors		Corrected Per[2]
			TIU/mg	% Destroyed	
Unextruded	55.6	2.07	64.5	--	1.01 ± .08[3]
121	41.9	1.96	64.8	0.0	1.35 ± .04
127	56.1	1.82	57.2	11.2	1.42 ± .05
132	44.3	1.46	45.5	29.5	1.41 ± .04
138	47.1	0.34	47.2	26.7	1.55 ± .04
143	21.6	0.02	28.0	56.7	1.94 ± .05
149	16.6	0.01	16.8	74.0	1.78 ± .08

[1]Composition of dehulled soybeans: protein 39.2 ± 0.3, ash 5.9 ± 0.1, fiber 2.4 ± 0.1, moisture 5.9 ± 0.5, fat 21.0 ± .02.

[2]Protein efficiency ratios corrected to casein at 2.50.

[3]Means ± standard error.

TABLE 17.9. Nutritional Effects of Extrusion of Soybeans Using Brady and Insta-Pro Extruders (14)

Sample Description	Trypsin Inhibitor		Urease	
	TUI/mg[1]	% Destroyed	pH Units	Corrected[2]
Raw, Whole Soy	74.8	--	1.92	
Whole Soy, Brady, 138°C	22.7	69.7	0.03	1.83 ± .07[3]
Whole Soy, Insta-Pro, 143°C	22.7	69.7	0.03	1.83 ± .07
Whole Soy, Insta-Pro, 143°C	14.0	81.3	0.09	2.24 ± .05
Raw, Dehulled Soy	80.0	--	1.97	1.49 ± .06
Dehulled Soy, Brady, 138°C	36.2	54.7	<0.01	1.78 ± .10
Dehulled Soy, Insta-Pro, 147°C	11.2	86.0	0.02	2.25 ± .08

[1]Trypsin units inhibited.
[2]Protein efficiency ratio corrected relative to casein at 2.50 (4 weeks' growth).
[3]Mean ± standard error.

TABLE 17.10. Trials for Test Production of Proposed Formulas with MFM-Kist Extruder (16)

Formula	Composition (%)	Machine Performance	Product Quality
BSS-4	Barley (65) Defatted Soy (25) Sesame (2) Other (8)	Smooth	Well Expanded Good Flavor
CSS-1	Corn (55) Defatted Soy (29) Sesame (2) Other (14)	Smooth	Well Expanded Good Flavor
CSS-2	Corn (63) Full-fat Soy (27) Sesame (2) Others (8)	Fluctuating Load	Small Flakes Beany Flavor
CSS-3	Corn (65) Defatted Soy (25) Sesame (2) Other (8)	Smooth	Well Expanded Good Flavor
BS-1	Barley (70) Defatted Soy (28) Other (2)	Smooth	Slightly Expanded Good Flavor
BSS-1	Barley (72) Full-Fat Soy (18) Sesame (2) Other (8)	Fluctuating Load	Small Flakes Beany Flavor
BSS-2	Barley (56) Defatted Soy (28) Sesame (2) Other (14)	Not Smooth	Well Expanded Good Flavor
BSS-3	Barley (63) Full-Fat Soy (27) Sesame (2) Other (8)	Not Smooth	Small Flakes Beany Flavor

have been introduced. Nutritional studies have been conducted to determine product quality and standards in terms of chemical composition, effects of process temperature on nitrogen solubility, urease activity, trypsin inhibitor and corrected protein efficiency ratios. Some of the results obtained are presented in Tables 17.7, 17.8, and 17.9. These results demonstrate the effectiveness of process temperature in increasing the nutritional quality of extruded foods. This is more so when the Brady is compared to the Insta-Pro extruder (Table 17.9).

The Korean Institute of Science and Technology (KIST) has fabricated a low cost food extruder which was originally designed by Meals for Millions (MFM) Foundation (15). The MFM-KIST extruder is mechanically simple and easy to operate and maintain (16). Some of the combinations tested are in Table 17.10 while the chemical composition is in Table 17.11. The blends were made on the basis of barley-soybean-sesame seed (BSS) and corn-soybean-sesame (CSS) combinations. Products from formula CSS-1, CSS-3 and BSS-4 were found most acceptable in terms of machine performance and product quality.

Cereal/Legume/Meat Product Blends Made Using Insta-Pro Extruders

The problem of adequate nutrition in Nigeria is not much different from those of other developing countries especially with regard to protein nutrition. Most Nigerian adults and children can adequately meet the recommended daily allowance for calories but total protein intake as well as animal protein intakes are usually just half of the recommended levels. This fact has consistently resulted in the incidence of protein malnutrition especially in infants and children. The climate in Nigeria is suitable for producing a number of cereals and legumes which if blended together could provide adequate intakes of protein supplement in all age groups. This fact could be used to cause the most desirable change in the food consumption pattern of many Nigerians based on an adequate supply of the traditional diet. Crops which show the most promise as blends include corn, cowpeas, groundnuts, beniseed, sorghum, millet, cassava and yams.

Several studies have been conducted to develop cereal/legume blends and evaluate them in traditional Nigerian diets as a measure of their potential for alleviating the protein problem in Nigeria. Marks (17),

TABLE 17.11. Chemical Composition of Proposed Products with MFM-K Grinder (16)

	CSS-3	BSS-4
Moisture (%)	6.18	6.68
Carbohydrate (%)	68.76	67.9
Protein (%)	17.99	18.32
Fat (%)	2.74	2.89
Ash (%)	4.33	4.21
Ca (mg/10 g)	332.5	348.0
Fe (mg/200 g)	4.7	5.75
Vitamin A (IU/g)	53.0	30.0
Thiaman (μg/g)	1.7	1.7
Riboflavin	6.2	5.1
Niacin (μg/g)	85.0	49.0

using the Insta-Pro 500, extruded various blends including corn/peanut, cowpea/fish, cowpea/beef and cowpea alone (Table 17.12). Nutritional evaluation of these products indicated that protein digestibility was improved as a result of the destruction of trypsin inhibitors (Table 17.13). The addition of beef and fish resulted in slightly greater reductions of trypsin inhibitors in the extrusion of cowpeas during run II (Table 17.12). Sensory evaluation of extruded cowpeas incorporated into some Nigerian foods were carried out. The results obtained indicate that acceptable dishes comparable to traditional ones can be prepared using low cost extruded products. These products have the advantages of being low-cost, time-saving, shelf-stable and nutritious (17).

We have used the Insta-Pro 2000R extruder at the Triple F plant in Des Moines, Iowa. Like most low cost extrusion cookers, the Insta-Pro 2000R uses a dry extrusion process which creates heat through pressure and friction. The heat cooks and expands ingredients to gelatinize the starch and destroy inhibitors.

The studies were initiated basically to process cereal/legume blends which are based on indigenous Nigerian commodities for use as a weaning food supplement for infants and children between the ages of 6 months and 4 years. This period is considered most critical in that many children are underfed due to a number of reasons. Furthermore, it was believed that the use of this type of technology would go a long way toward providing a boost to both the development of cottage-type industries and the utilization of a variety of by-products for animal feeding. The availability of extruded products could lead to expansion in the use of composite flours for breadmaking considered an appropriate vehicle for increasing the protein intake of many adult Nigerians. If successful the process would provide the additional advantage of decreasing Nigeria's dependence on imported protein foods while conserving the country's foreign exchange since entirely local food resources would be used in extrusion.

Three varieties of cowpeas--Ife brown, Vita 5 (Vigna unguiculata) and California blackeye pea (Vigna sinensis)--were blended with either rice, corn, rice and corn or rice and banana puree before extrusion[1].

The cowpeas were used whole and dehulled in an effort to test the effect of testa and hypocotyl removal on product quality. The Ife brown and Vita 5 varieties of cowpeas were from Nigeria while other ingredients were obtained locally in Iowa. The legumes and cereals were ground in a Fitzmill separately prior to extrusion using a .093 screen. Combinations were cowpea:corn or cowpea:rice mixture 70:30, cowpea:corn:rice 40:30:30 and cowpea:rice:banana puree 50:40:10. The mixtures contained about 10% moisture prior to extrusion with water being injected in to the barrel to raise the moisture content of the mixtures to 18%. Double flight screws were used in the extruder with a setting of 11R, 11 and 6. Each mixture was blended to achieve uniformity and force-fed into the extruder using an auger. Samples of extruded products were collected after a steady-state operation

1. Akinyele, I. O., Love, M. H., Ringe, M., and Dupont, J., Production and evaluation of weaning foods using low-cost extrusion technology. Personal communications (1983).

TABLE 17.12. Protein Digestibility of Raw Products and Those Using the Insta-Pro 2000R (17)

Product	% Protein Digestibility	Changes in % Protein Digestibility
Extruded Cowpea (Run I)	83.80	+3.04
Extruded Cowpea (Run II)	84.06	+3.30
Extruded Cowpea/Fish	85.04	+4.28
Extruded Cowpea/Beef	85.31	+4.55
Extruded Cowpea/Lard	84.01	+.25
Raw Cowpea	80.76	---
Extruded Corn/Peanut	85.49	+1.65
Raw Corn/Peanut	83.84	---

TABLE 17.13. Trypsin Inhibitor Assays on Cowpea Products (17)

Product	TUI/mg	% Reduction
Raw Cowpea	7.92	----
Extuded Cowpea (Run I)	1.50	81.06
Extruded Cowpea (Run II)	0.71	91.04
Extruded Cowpea/Beef	0.12	98.48
Extruded Cowpea/Fish	0.67	91.54

TUI = Tripsin Units Inhibited

had been attained. Two percent corn oil was added to each product to facilitate extrusion. Extruded products were milled, sampled, dried and analyzed for nitrogen, protein, trypsin inhibitor, and in-vitro digestibility using approved chemical methods. The color of the products was also measured.

Twelve products were extruded and summaries of the process temperature, energy, nitrogen and protein content of each product are shown in Table 17.14. Apart from the triple mixes, the products contained on the average, about 20% protein and 440 kcal per 100g (Table 17.14). Protein digestibility improved with extrusion (Table 17.15) in relation to trypsin inhibitor destruction (Table 17.16).

Color measurements on the D25D2 Hunter Laboratory instrument indicated the extruded products had chromaticity coordinates (x, y) which plotted as light yellow to greenish-yellow depending upon the combination. They were measured as yellow to orange when light colored legumes and corn composed the blend. The legume/cereal combinations of Ife brown with rice or corn was greenish yellow whether they were extruded from dehulled or whole beans. Hunter "L" values ranged from 65 to 72, depending upon the legume color and cereal. However, all the extruded products could be classified as light products with little browning resulting from the process. It is expected that the products would be flavored, reconstituted using hot water to achieve desired consistency, and served hot to infants and preschool children in Nigeria.

Composite flours were made by substituting an all purpose wheat flour with 25% of some extruded products. One pound loaves were made with the composite flours and the resulting bread products were found comparable to 100% wheat flour in acceptability both in appearance and taste. More studies on product development and evaluation are planned for the target population in Nigeria.

Implications of Extruded Cereal/Legume Blends for Weaning

Before using extruded cereal/legume blends in weaning foods, plans must be made to first assess the nutritional status of weaning age children, and current practices of weaning in the particular locality with emphasis being placed on the types and combinations of foods used. Combinations which are not native to the locality should be avoided as much as possible since they may not be compatible with local tastes and preferences.

There are many advantages in the use of extruded cereal/legume blends to make weaning foods. Such products are energy efficient in that the powder is simply reconstituted with boiling water and is ready for consumption. This would save a tremendous amount of time for the mother since she would spend less time preparing the weaning food and time saved could be used to work in other areas. Use of firewood, which is becoming scarce, would be reduced also due to the reduced cooking time.

Extrusion-processed weaning foods must be highly nutritious and less expensive than imported formulas for them to have any meaningful impact. Local production of the food resources used in the extrusion processing would be stimulated by demand for the weaning food. Since the target population is both rural and urban, belonging to the low and middle class groups one can expect such additional benefits as employment income from the plant and increased buying power in the area. Since the Government is usually an active participant in the

TABLE 17.14. Energy and Protein Content of Extruded Cereal/Legume Combinations

Blend	Process Temperature °C	Energy Kcal/100 g	% N	% Protein
70:30 California Cowpea:Corn	168	447	3.05	19.06
70:30 California Cowpea:Rice	161	542	3.06	19.12
40:30 Ife Cowpea:Corn:Rice	168	434	2.47	15.44
70:30 Dehulled Ife Cowpea:Corn	172	439	3.33	20.81
70:30 Vita Cowpea:Corn	170	430	3.41	21.31
70:30 Dehulled California Cowpea:Corn	170	436	3.11	19.42
70:30 Dehulled Ife Cowpea:Rice	172	466	3.32	20.74
70:30 Dehulled California Cowpea:Rice	170	445	3.05	19.06
70:30 Dehulled vita Cowpea:Rice	170	444	3.35	20.92
50:40:10 California Cowpea:Rice:Banana	168	437	2.85	17.81
100% Ife brown cowpea	168	448	4.16	26.00
100% California Cowpea	168	440	3.70	23.13

TABLE 17.15. Protein Digestibility of Raw and Extruded Products

Product	% Protein Digestibility (Raw)	% Protein Digestibility Extruded	Changes in % Protein Digestibility
100% Ife Cowpeas	77.97	83.84	+5.87
100% CAP	78.50	83.16	+4.66
70:30 California Cowpea:Corn	78.58	81.13	+2.55
70:30 Vita Cowpea:Corn	75.64	83.16	+7.52
70:30 California Cowpea:Rice	78.87	82.86	+3.99
40:30:30 Ife Cowpea:Corn:Rice	77.97	82.49	+4.52
50:40:10 California Cowpea:Rice:Banana	77.30	82.10	+4.80
70:30 Dehulled Ife Cowpea:Corn	77.00	83.09	+6.09
70:30 Dehulled California Cowpea:Corn	77.98	82.49	+4.51
70:30 Dehulled Vita Cowpea:Rice	74.14	83.84	+9.70
70:30 Dehulled Ife Cowpea:Rice	78.05	83.61	+5.56
70:30 Dehulled California Cowpea:Rice	75.27	82.78	+7.51

TABLE 17.16. Trypsin Inhibitor Destruction of Cowpea Products by Extrusion

Product		TUI/mg		
	Temperature °C	Raw	Extruded	% Reduction
100% California Cowpeas	168	10.12	1.67	83.5
100% Ife Brown Cowpeas	168	6.12	0.98	84.0
Dehulled California Cowpeas:Rice	150	5.21	10.00	81.0
Dehulled Ife Cowpea:Rice	172	5.20	1.71	67.1
Dehulled Vita Cowpea:Rice	170	6.29	1.97	68.7
Vita Cowpea:Corn	170	5.80	2.14	63.0
Dehulled California Cowpea:Corn	170	6.46	2.35	64.0
Dehulled Ife Cowpea:Corn	172	6.60	1.39	70.0
Dehulled California Cowpea:Rice:Banana	168	5.65	0.90	84.1

economy, one would expect that there would be an initial price subsidy for the producer to ensure that the final price is kept stable and low.

Most weaning foods used in supplementary feeding programs provide 17-20% of their total calories from their protein content. This percentage of protein is high and has a built in safety allowance for children with increased protein requirements due to malnutrition and/or disease, and the low digestibility of plant protein when compared with animal proteins.

Addition of sugar to extruded mixtures has been suggested as a way of increasing caloric density and improving the taste and texture of the product. This practice has, however, been found to cause the extruded product to become sticky and hard to handle. The occurrence of the Maillard reaction between glucose and the amino acid lysine causes a reduction in protein quality thus making the addition of sugar a negative proposition. Care must always be taken when using sugar as an additive to food products because sugar creates a taste preference easily acquired and required by children in their other diets. The use of natural fruit flavors may alleviate this problem.

Finally, with any technological change, there are ramifications beyond those factors most obviously affected by the technology. The effects of extrusion processing of weaning foods in many developing countries, including Nigeria, can best be considered based on the assessment of need and available complementary services such as finance, raw material supply, quality control, packaging, storage, distribution, evaluation and implementation of feeding programs. The extruded cereal/legume blends would be successful if they provide the daily allowance of both protein and calories reducing the number of infants and children suffering from malnutrition and those at risk. Evidence provided by the review on the use of low cost extrusion cookers to make these blends demonstrates the desirability of this type of technology in developing countries and it is hoped that, if it is effected in the proper fashion, a great battle in the war against hunger and malnutrition will have been won giving greater hope that the supplication for daily bread to the Almighty by the poor and needy, the hungry and the starving, will come close to being answered.

CONCLUSION

Low-cost extrusion technology has been shown to be an appropriate technology for the production of a highly safe and nutritious product which could be used either as a weaning food or incorporated into many local diets. Though more studies are required with the targeted population in many instances, indications are that current products would be easily acceptable to the people and effective in the war against hunger and malnutrition.

ACKNOWLEDGEMENT

I would like to acknowledge the cooperation of Dr. Jacqueline Dupont, Dr. Mark H. Love and Mitchell Ringe of Iowa State University in making possible the project I was involved with during the 1982/83 year. The financial support of the Triple F plant in Des Moines, the

World Food Institute, Iowa State University, the various church organizations and the Senate of the University of Ibadan, is gratefully acknowledged.

REFERENCES

1. Harper, J. M. and Jansen, R. G., LEC 10 report, Colorado State University, Fort Collins, (1981).
2. Harper, J. M., CRC Critical Reviews in Food Science and Nutrition 11:155 (1978).
3. Sahagum, J. F. and Harper, J. M., Journal of Food Process Engineering 3:199 (1980).
4. Insta Pro Division of Triple "F" Inc. Insta Pro: the dry extrusion process--informational paper. Des Moines, Iowa.
5. Smith, O. B., in "New Protein Foods" (A. M. Altshal, ed.), p. 86. Academic Press, New York, (1976).
6. Jansen, G. R. and Harper, J. M., Food and Nutrition 6, 2 (1980).
7. Fox, W., Proc. of the Second International Workshop on Low-Cost Extrusion Cookers, p.159. Colorado State University, Ft. Collins, CO (1979).
8. Bressani, R., Proceedings of the First International Workshop on Low-Cost Extrusion Cookers, p. 75. Colorado State University, Fort Collins, CO (1976).
9. Alvarado, R., Prodeedings of the Second International Workshop on Low-Cost Extrusion Cookers, p. 85. Colorado State University, Ft. Collins, CO (1979).
10. Bleyer, P., Proceedings of the Second International Workshop on Low-Cost Extrusion Cookers, p. 29. Colorado State University, Fort Collins, CO (1979).
11. Bhumiratana, A., Proceedings of the Second International Workshop on Low-Cost Extrusion Cookers, p. 225. Colorado State University, Fort Collins, CO (1979).
12. Muchlis, A., Proceedings of the Second International Workshop on Low-Cost Extrusion Cookers, p. 101. Colorado State University, Fort Collins, CO (1979).
13. Pablo, I., Proceedings of the Second International Workshop on Low-Cost Extrusion Cookers, p. 247. Colorado State University, Fort Collins, CO (1979).
14. Jansen, G. R., Proceedings of the Second International Workshop on Low-Cost Extrusion Cookers, p. 121. Colorado State University, Fort Collins, CO (1979).
15. Anon., Food Engineering Int'l. 42 (1977).
16. Cheigh, H.S., Proceedings of the Second International Workshop on Low-Cost Extrusion Cookers, p. 115. Colorado State University, Ft. Collins, CO (1979).
17. Marks, M., M.S. Thesis, Iowa State University, Ames, Iowa (1982).

18

Roles and Status of Composite Flours

DAVID A. FELLERS

INTRODUCTION

The Ascent of Wheat

At some point or points in history, man learned to collect grains, grind them to flour and produce batters or doughs which were baked on a hot stone. The product became known as bread. It was not necessarily made from wheat. There must have been a great amount of experimentation in the selection of grains or blends of grains and other seeds for attainment of the most savory and beneficial products. Egyptians are credited with discovering leavened breads and were enjoying no fewer than 30 varieties while barbaric tribes of Northern Europe were still subsisting on raw meat and wild plants. In medieval Europe, "The peasants' bread was coarse-grained and dark, made commonly from barley, rye or bean flour. The wealthy townsman or squire bought fine wheaten flour for making white loaves--" (1). In other parts of the world, millet and corn were being used to make bread. In the developed countries of the world today, wheat products have come to dominate the bread market. Wheat breads are highly prized for their coherence, delicate flavor, lightness, airiness, whiteness, lack of grittiness, compatibility with other foods, low cost, nutritional quality and convenience. The motto of the Food and Agricultural Organization (FAO) of the United Nations is "FIAT PANIS"--"Let there be bread".

The ascent of wheat and its role in the human food supply, however, is not complete. Several factors are leading to a sharply expanded role for wheat in many developing countries, most of which are in the tropics where wheat is not a very suitable crop. Dr. Byrd C. Curtis, Director of the Wheat Improvement Program at the International Maize and Wheat Improvement Center in Mexico, recently reported that wheat consumption in developing countries increased 73% over the last decade (2); a 5.4% annual growth rate. Most of the increase is being met by imports. The driving forces for this trend are: the high palatability and nutritional quality of wheat foods; a reliable wheat supply of low cost and prompt delivery to any place in the world; the high population growth rates in developing countries; and, the

Cereals and Legumes in the Food Supply, edited by Jacqueline Dupont and Elizabeth M. Osman

underdeveloped agriculture and agribusiness in the developing countries that cannot consistently meet local demands.

A Problem of the Developing Countries

These increasing wheat imports are a serious problem for the developing countries because they use up scarce foreign exchange needed for development. The imported wheat suppresses and displaces indigenous foods and this has a negative effect on rural employment and income. In addition, a "wheat habit" is being formed which depends on imports and which will be very difficult or impossible to break in the future.

FAO Composite Flour Program

In recognition of this economic drain on the developing countries, the Food and Agriculture Organization (FAO) initiated its "Composite Flour Program" in 1964 (3). The objective was to seek methods for substituting flours, starches, or protein concentrates from indigenous crops for wheat flour in the production of Western-type baked goods and pastas. It is the FAO, then, that has popularized the term "composite flour" and has given it its orientation as a potential contributor to the solution of world hunger. The FAO composite flour program is one of the five major FAO activities in the cereals area (4), but one which is currently undergoing evaluation in the light of 20 years of experience with only limited applications.

In undertaking the composite flour program in 1964, the FAO recognized that there had been substantial advancements in baking technology and equipment, that new and improved processes had evolved for processing nonwheat grains, legumes and root crops, and that there had been a proliferation of new food additives, many potentially useful for composite flours. They were encouraged and optimistic that composite flours could be evolved for most developing countries.

COMPOSITE FLOURS IN THE DEVELOPED WORLD

Food Variety, Nutrition and Surplus Utilization

The idea of composite flours is not new. There are biblical references to mixed grain breads. More recently, adverse political conditions and scarcity of wheat during and after the World Wars led to significant studies and use of composite flours (5). At the end of World War I, as much as 20% barley flour was being added to 90% extraction wheat flour for bread production.

Composite flours are also part of our current every day life. In West Germany, over 60% of the bread is rye bread or blends of rye and wheat (6). In the United States, composite flours find their greatest use or role in providing variety to the diet, e.g., multigrain breads, rye and triticale breads, potato bread, oatmeal cookies, corn bread and buckwheat pancakes. Nutritional enhancement is another role, e.g., soy breads and fiber breads. Variety pan breads are estimated to have increased in the United States from 1.5 billion lbs in 1972 to 2.6 billion lbs in 1982 while white pan bread decreased from 8.6 billion lbs to 6.1 billion lbs (7). In Japan, the government has encouraged millers and bakers to add rice flour to bread in order to utilize surplus rice created by high farm subsidies. At its peak in 1977, 480 metric tons (MT) of rice was utilized by adding 10% to 15%

rice flour to the bread. No problems were encountered with quality but the increased price for the rice bread was a market depressant (8). Nishita and Bean (9) have reported a preference for soft, sticky cooking rices with low amylose content, analogous to short and medium grain rices grown in the U.S., for producing flours for rice breads. Long grain, high amylose rice flours give a gritty texture to rice bread which is quite detectable at 30% substitution or higher.

Case Study--Traditional Macaroni Challenged in the United States

Many wheat products in the United States are manufactured under Standards of Identity that specify the types and amounts of ingredients that can be used. Macaroni and its derivatives are such products. In 1971, the Food and Drug Administration (FDA) granted General Foods Company a temporary permit to market a high protein, "engineered" food, consisting of corn flour, defatted soy flour and durum wheat, called "Golden Elbow" macaroni. The FDA's thinking at the time was to improve nutrition by encouraging the marketing of inexpensive cereal-based foods with increased protein. An early formulation was 60% corn, 30% soy and 10% wheat. The use of corn flour provided an economic edge over an all wheat product fortified with soy. Clausi (10) noted that this Golden Elbow macaroni required only 5 to 6 minutes for cooking as compared to 15 to 20 minutes for traditional macaroni. Unfortunately, it was subject to rather severe disintegration during prolonged holding on a steam table. The product was later reformulated to 40% corn, 30% soy and 30% wheat. This had a more normal cooking time and held up better on a steam table. Seyem et al. (11), however, have criticized the Golden Elbow macaroni as having a corn taste, high cooking loss and poor texture when cooked. The traditional macaroni makers were being challenged. They objected to the product being called macaroni and being produced in traditional elbow shape. The macaroni producers, durum millers and durum growers were particularly concerned about the use of corn in place of durum wheat.

The final FDA compromise was issued in 1972 with the publishing of a Standard of Identity for these new products, referring to them as "Enriched macaroni products with fortified protein" (12). The Standard requires that the portion of the milled wheat ingredient be larger than the portion of any other ingredient, that the protein content be not less than 20% and that protein quality (PER) be not less than 95% of casein. Nonwheat cereals and oilseeds were allowed but the Standard requires their identification in the name of the product.

Golden Elbow macaroni was reformulated to 35% durum semolina, 34% corn flour and 31% defatted soy flour to meet the new Standard. It was licensed for production to a small West Coast company which sought to penetrate the institutional market. The 40% corn, 30% soy and 30% wheat product was licensed in Brazil. In both cases, sales were unsatisfactory; Golden Elbow macaroni did not survive. One scientist at General Foods felt the corn flavor was the main problem in the United States while an inconsistent corn flour supply and inadequate marketing efforts were the major problems in Brazil.

In addition to "Enriched macaroni products with fortified protein", there is also a Standard of Identity for "Wheat soy macaroni products". This Standard requires the cereal to be wheat and the product to be fortified with a minimum of 12.5% soy flour. Various protein fortified pastas are currently available in the United States, especially for use in school lunch programs where they are allowed as a substitute for part of the meat requirement.

WHEAT SITUATION IN DEVELOPING COUNTRIES

Wheat Production and Use

There is a great diversity of wheat production and use in the developing countries. Wheat availability in 1974 varied from 306 kg/capita/year to essentially none among 92 free world developing countries. Wheat imports accounted for from zero percent to 100% of the available wheat supply. In 1980, the developing countries had a net wheat import of 51 million MT or a little more than half of all wheat entering international trade. In addition, they had net imports of 17 million MT of corn and 3 million MT of rice (13). These imports are equivalent to 22 kg of grain for every man, woman and child in the developing countries and represents about 7% of their dietary calories.

Table 18.1 divides the developing countries into four groups according to wheat production, importation, and total available supply. As would be expected, most of the major wheat producers are located in the temperate zone while non-producers are in the tropics. In Brazil, wheat is grown in the more temperate south. Wheat can be grown to

TABLE 18.1. Four Classes of Developing Countries Based on Wheat Production and Consumption. Figures are Availability of Wheat (Production + Imports Exports) in Kg/Capita/Year. Figures in Parenthesis are the Percent of Wheat Supply Imported*

MAJOR WHEAT PRODUCERS					
Normally Self Sufficient			Significant Importers		
	Kg/Cap/Year	% Imported		Kg/Cap/Year	% Imported
Turkey	357	(0)	Iraq	222	(57)
Syria	301	(21)	Chile	180	(53)
Uruguay	104	(0)	Morocco	167	(48)
India	46	(0)	Brazil	60	(63)
			Sudan	30	(59)
			Ethiopia	27	(46)
MINOR OR NON-WHEAT PRODUCERS					
Significant Importers			Minor Importers		
Cuba	121	(100)	Tanzania	7	(49)
Bolivia	75	(86)	Zaire	6	(96)
Peru	50	(91)	Mali	5	(94)
Sri Lanka	47	(100)	Niger	5	(93)
Malaysia	32	(100)	Chad	4	(69)
Philippines	16	(100)	Thailand	4	(100)
Nigeria	15	(98)	Burma	2	(16)
Senegal	14	(100)	Papua N.G.	0	(-)

*Sources: 1981 FAO Production Yearbook; 1980 FAO Trade Yearbook.

some extent in the temperate highlands of the tropics but the amount of suitable land is quite limited.

Significant efforts have been made to improve wheat varieties for use in the sub-tropics and tropics. Spring type wheats are the most suitable. The semi-dwarf Mexican types or green revolution wheats, developed in the 1950s and 1960s, are generally insensitive to day length, highly resistant to disease and produce a short, stiff straw resistant to lodging. These new varieties have expanded the ecological adaptation of wheat but unfortunately, they generally require irrigation, fertilizer and modern pest control methods to achieve their yield potential. Mexico, including its high altitude tropical areas, was able to treble wheat yields and approach selfsufficiency in wheat. India, Pakistan and Turkey also had spectacular gains which resulted in freeing surplus wheat from North America and other wheat surplus countries for other markets.

Overall, however, wheat is not well adapted to the tropics and there are more suitable crops that can be grown. In addition, the use of expensive seed, irrigation, fertilizers and pesticides are costly procedures for developing countries. Dramatic increases in the production of wheat in the tropics seems unlikely.

Getting Started on the Wheat Import Habit

The evolution of food habits in any country is the result of a complex mix of factors. It is a process that is very much in evidence today.

In colonial times, the Europeans introduced wheat around the world both as a crop and in trade. In the Caribbean colonies, bread was the food of the ruling European colonists. It had social status and became sought after by the native populations. Caribbean agriculture developed with a trade orientation; great plantations were developed for crops like sugar cane while agriculture for local food needs was neglected. Consequently, today the Caribbean countries have a strong dependence on imported wheat. It might be interesting to speculate on what food habits would have evolved in the Caribbean without the colonial experience. At the present time, there are still many developing countries that consume very limited amounts of wheat. They are at a crossroad similar to the early experience of the Caribbean. Will they too adopt wheat as a staple requiring importation, or will they develop their own indigenous staples with suitable products and appropriate technology?

Urbanization has been and continues to be a strong driving force for the spread of wheat. When people move to the city and become detached from the rural, agricultural community, they have taken a large step toward the international community. Their interests become much broader and multifaceted and their interests in food tend to focus on low cost, continuous and reliable availability, quality and convenience. City dwellers generally hold the political power and do not feel compelled to buy local products. Entrepreneurs quickly discover the reliable availability of low cost, high quality international wheat and begin to import wheat and produce high quality wheat products that meet consumer demands. As a contrast, local crops are often volatile in price, perishable and inconsistent in quality. They are, also, highly seasonal because of inadequate postharvest processing, storage and marketing facilities. Imported wheat is a

TABLE 18.2. Per Capita Consumption of Wheat Foods, Costa Rica*

Wheat foods	Grams per person per week		
	Urban	Rural	Dispersed
Bread	493	235	59
Pastas (noodles)	47	41	4
Crackers and cookies	6	19	0
Total	546	295	63

*INCAP; 1966 Nutrition Survey.

tough competitor in such environments. Without government limitations on imports, wheat seems destined to obtain a significant market share. As one would expect, wheat consumption is generally higher in urban than in rural areas of the tropics. Table 18.2 provides an example of this in Costa Rica where all wheat is imported.

Periodic or chronic food scarcity in developing countries is another factor leading to increased markets for wheat. When a country is short of food and foreign exchange, it must import the most inexpensive yet acceptable and useable food. In the grains, this means essentially wheat, rice or corn. For over a hundred years now, the price of wheat has been declining relative to the price of rice. Today, wheat flour is about 60% of the price of milled rice on world markets. Both products are highly useable, but for the same amount of money, more wheat can be purchased. Corn has been historically cheaper than wheat in international markets and does represent an interesting alternative for those countries where corn is traditionally used as a food. Unfortunately, the availability of technology and knowledge on how to use corn for food is more limited and not as broadly disseminated as for wheat and rice. Also, corn is often rejected as being an animal feed.

COMPOSITE FLOURS IN DEVELOPING COUNTRIES

Incentives

Table 18.3 gives a list of the incentives or reasons why composite flours are of interest. As already discussed, the major reasons for

TABLE 18.3. Reasons for Considering Composite Flours

* Add variety to the diet
* Enhance nutrition
* Extend a limited wheat supply
* Import substitution to save foreign exchange
* Utilize nonwheat surplus
* Reduce ingredient costs
* Stimulate nonwheat agriculture to improve rural employment and income
* Improve utilization of nonwheat flour production plants

them in the developed countries are related to the consumers' desires for variety and nutrition in their baked foods. While these reasons for composite flours are of interest in developing countries, wheat import substitution and stimulation of domestic agriculture are usually of much greater interest. Import substitution means savings of foreign exchange and increased agriculture means more jobs and income in the rural areas.

Another important factor can be the desire of a nonwheat food processor to utilize a surplus product or by-product by incorporating a small percentage into the wheat flour. Where the nonwheat flour is cheaper than the wheat flour, this has been observed to happen rather spontaneously without government mandate. During a survey in Paraguay in 1977 to assess interest in composite flours, it was found that rice millers were selling low priced rice brokens to wheat millers who presumably were making composite flour. Because the price of wheat flour was set by the government, the low cost brokens allowed the wheat millers an improved profit. In Colombia, in the 1970s, when the subsidy on imported wheat was removed, domestic rice suddenly became an economic adulterant and substantial quantities were diverted to the manufacture of composite flours.

In the industrialized, developed countries, the driving force for composite flours is largely consumer demand. The actual preparation of composite flours is usually done by the baker or other end-use manufacturer. In the developing countries, however, the government is most often the driving force because of its interest to conserve foreign exchange and assist domestic agriculture. To achieve the maximum utilization of indigenous ingredients, national programs are sought which involve the preparation of composite flours at the local wheat flour mills and their subsequent shipment to all the traditional wheat flour users: bakers, pasta manufacturers, biscuit makers and home makers. The government will typically receive support for such a program from the groups involved in growing and processing the favorably affected, indigenous commodity and receive opposition from the wheat importers, wheat millers and bakers. Consumers will defer their judgments until effects on quality and price can be determined in the market place, but generally, they are positive on the idea of composite flours as a way of helping their national economy.

Research

Among the grains, wheat is unique because of its gluten protein. No other grain, legume or oilseed has the properties of gluten that allow the formation of a cohesive, elastic dough. These viscoelastic properties are the basis for many of the desirable qualities of wheat products, especially bread. When the gluten is damaged or diluted, the result is almost invariably poorer baking performance. Composite flour research in the last 20 years, however, has been quite successful in finding ways to minimize this problem. Composite flours are technically feasible.

The Tropical Products Institute (TPI) of the United Kingdom has published three composite flour technology bibliographies (14-16) that reference some 952 papers. In the preface of the most recent, 1979 volume, Dendy comments on the vast amount of scientific and technical research but the lack of published information on the economics and implementation of composite flours. He concludes, "---there must be few technologies that have been so thoroughly researched and so little applied".

An important impetus to composite flour research has been the formation, with FAO encouragement, of the International Association of Cereal Chemistry (ICC) Study Group 32 on Sorghum, Millets, Pulses and Composite Flours. The ICC has sponsored important symposia (17-19) that have helped to focus world attention on composite flour opportunities and problems.

De Ruiter (20), who has reviewed composite flour research up to 1978, has taken a very broad view of the definition of composite flours. He recognizes two types, those where wheat is partially replaced and those where no wheat at all is used and gluten substitutes are required such as various gums and surfactants. This paper is only concerned with composite flours where part of the wheat is replaced. The majority of research carried out has covered the area of partial wheat replacement since this approach seemed the most practical and most likely to lead to commercial applications in the near term. The question was not, "Can it be done?", but rather, "How much wheat can be replaced without significantly changing the traditional wheat processes and the acceptability of the products?" The answer to the question, "How much?", depends on several factors: the specific wheat product to be produced, the process chosen for its manufacture, the quality of the wheat, the type and quality of nonwheat flour to be utilized, the use of special improving additives and the perception of acceptability by the consuming population.

Of the wheat products, leavened breads have the greatest sensitivity to nonwheat flours. Composite flour doughs have reduced cohesiveness and thus their ability to capture and hold the leavening gases is impaired. The result is a less appetizing, dense, heavy loaf. The bread crumb also loses cohesiveness and tends to be crumbly. Gumminess during mastication is another common problem. These physical effects are often apparent before color and flavor changes become objectionable. The physical effects may be reduced somewhat by the use of additives such as oxidants (bromate, ascorbic acid), surfactants (calcium or sodium stearoyl lactylate) or fats, and by modification in the dough production process such as increased absorption and yeast level and reduced mixing and fermentation times. High speed mechanical dough development, a process widely used in the United Kingdom, has been very successful in the making of composite flour breads under research conditions (21).

Table 18.4 suggests practical levels of various types of nonwheat flours that might be used in bread production based on research

TABLE 18.4. Practical Levels of Nonwheat Flours That Can Be Substituted for Wheat Flour in Bread Production as Suggested by Research Results

Type of Nonwheat Flour	Range of Wheat Substitution, %
Purified Starches	20-40
Rice	10-30
Cereal and Root Flours	5-20
Proteinaceous Flour	3-15

experience. The table is obviously a generalization; each individual nonwheat flour or combination of nonwheat flours will have its own specific constraints as to processability, color, flavor and palatability. Purified starches allow the greatest substitution but are disadvantageous because they dilute the nutritional quality by reducing protein content. Gelatinized nonwheat flours and proteinaceous flours require an increased use of water in the bread formulation and, because the extra water further dilutes the wheat gluten, the addition of these types of nonwheat flours is limited. The extra water and the nature of these materials also increase the stickiness of doughs which creates handling problems during dough processing.

The processes for production of pastas and biscuits are not as adversely affected by nonwheat flours as those for leavened bread, thus the levels of nonwheat flours can often be higher. Texture, flavor, color, and stability in the finished products, rather than factors during processing become important in determining the amount of substitution. Long goods such as spaghetti tolerate less wheat replacement than short goods because of the tendency for long goods to stretch and break during drying. Other major quality problems with composite flour pastas are their tendency toward increased cooking losses and disintegration on over-cooking. Sorghum has been reported to cause grittiness (22). Use of pregelatinized corn flour has been quite successful in Latin America (23), and when yellow corn is used, it adds a very attractive amber color. Many wheat pastas in Latin America are made with farina milled from bread wheats, rather than semolina milled from durum wheat, and as a result, the pastas are very dull and grayish.

De Ruiter (20) has noted the high suitability of the molding and depositing type processes for manufacture of composite flour biscuits. The doughs for use in these processes should be short and easily deformed with little or no gluten development, a condition readily attainable with composite flour doughs. The sheeting and cutting process, on the other hand, requires sufficient gluten strength and extensibility to allow the dough to be rolled to a smooth sheet before cutting into the desired shapes. Since nonwheat flours dilute the wheat gluten and weaken the dough, less of the nonwheat flours can be used in the sheeting and cutting process.

Applied Results

Success in the research laboratory has not been mirrored in the market place. There is only one country with a functioning national composite flour program and an additional two countries that have legislated or decreed programs which are not yet implemented at the practicing level. There are also a handful of start-up operations in specific locations that might be described as semi-commercial or perhaps even as demonstrations.

The only long term national composite flour program began in Brazil in the late 1960s (24). This program, assisted by the FAO, considered several aspects: different nonwheat flours such as cassava, corn and soy; various baking methods including the Chorleywood mechanical dough development process; and, the use of modern dough and bread improving additives. The Government of Brazil, however, chose a low technology approach to implementation. Regulations call for 2% wheat flour replacement with cassava flour, corn flour or starch. Optionally, an additional 3% of these ingredients could be added for a total wheat flour replacement of 5%. The program involved low technology from the standpoint of composite flours, in that no special baking methods or additives are required. Considering Brazil's utilization of about 7 million MT of wheat flour in 1980, a 5% replacement in bread flour would have required 210,000 MT of cassava and corn flours and starches, a very significant quantity.

Compliance in this program, a program that is still in effect, is probably not fully known since it is very difficult to detect and quantify such small amounts of nonwheat flours in wheat products. A strongly favorable factor for compliance has been the significantly lower cost of cassava flour, at least in 1973 when the cost was about half that of wheat flour. While the program has had some positive effect on foreign exchange savings, its greater effect has been the stimulus for cassava and corn production and the industrialization and marketing of cassava starch and corn flours. Bakers have experienced increased problems, however, because predictability of the baking quality of flour has declined. Accordingly, bakers have pursued various means to obtain wheat flour without cassava or corn (25).

The two countries that have legislated national composite flour programs are Senegal, in 1979, and Bolivia, in 1982. The composite flour program in Senegal began as a FAO collaboration with the Government of that country. The work has been carried out mostly at the Institute of Food Technology (ITA) at Dakar, starting in 1967 (26). Composite flours containing millet, sorghum, cassava starch, peanut and soy were investigated using both modern and traditional bread making processes and dough improving additives. Starting in 1974, composite flour breads made at ITA and at a private bakery were sold locally, and thus, valuable experience on production, marketing and acceptability was obtained.

The program has also been concerned with methods for preparing millet and sorghum flours, including the use of wheat milling equipment which proved less satisfactory than decortication by pearling followed by hammer milling. Various plans were prepared for the expansion of millet milling capacity in Senegal, though apparently these have not been executed.

The 1979 Senegalese composite flour law requires all bakeries to add 15% millet flour in the production of bread. Most breads in Senegal are of the French, hearth type. The law was suspended for over a year, however, due to a poor millet harvest resulting from drought. Bread with 15% millet flour is definitely darker and changed in flavor. Perten (1969) has described the flavor as similar to whole wheat bread. The texture is heavier and some people have described it as somewhat gummy. While these attributes represent a change from white bread, they do not constitute rejection as proven by the continuing ability to sell the ITA bread. However, bakers have resisted the program. Some have used the excuse of continuing millet and millet flour shortages as a reason for noncompliance. Beyond the continuing production at ITA, there are few if any bakers currently making the 15% bread. Government deliberations have considered requiring the addition of millet flour at centralized wheat mills, perhaps at less than 15%, in order to attain better compliance.

In Bolivia, government workers have long been interested in composite flours. A Government Decree in 1975 required addition of 5% quinoa flour to all wheat flour at the wheat mills. Only a small amount of the blend was ever produced because of the lack of an industrial process to make acceptable quinoa flour, an insufficient supply of quinoa (only 10,000 to 15,000 MT are produced each year in Bolivia and about 70% of this does not reach the market place) and price changes that made quinoa flour substantially more expensive than wheat flour.

Subsequently, a program of collaboration between the Agency for International Development (AID) and the Bolivian Ministry of Industry, Commerce and Tourism (MICT) was initiated in 1977 (27). Composite flours with rice, defatted soy flour, corn and quinoa were studied in the production of typical hearth bread rolls using traditional Bolivian "artisan" and semi-mechanized straight dough bread processes. Dough improvers such as bromate, ascorbic acid and sodium stearoyl lactylate were also evaluated.

In the area of infrastructure, the Department (State) of Santa Cruz, located in the tropical lowlands where corn is an excellent crop, constructed a 78 MT per day processing plant in 1979 to produce pregelatinized corn flour. In cooperation with a private oilseeds processor, plant modifications necessary to provide a capacity of 10,000 MT per year of food grade, defatted soy flour were determined. The company received financing approvals for the conversion in mid-1982. The AID-Bolivian project also cooperated with the owners of a small wheat mill, of 15 MT per day capacity, to install feeders and blenders and to demonstrate commercial production of composite flours. Products from the demonstration were used in bakery and pasta production tests and for consumer testing. In March 1982, the Government issued Supreme Decree No. 18883 which requires the addition of soy and corn to wheat flour but leaves many of the implementing details to the MICT. At the time of the Decree, the favored plan was to add 5% defatted soy flour and 100 ppm ascorbic acid to all wheat flour at the wheat mills and 10 to 25% pregelatinized corn flour to all pastas at the pasta plants. Pasta manufacturers would be required to use the soy fortified wheat flour but would have latitude in the use of corn flour within the 10 to 25% range.

Because the Government owns the wheat and simply pays the miller a fee for his services, and because the Government would also purchase soy and corn flours for blending, the success of the program weighs heavily on the Government. It must provide sufficient incentives, coordination and internal programs to: 1) convince wheat millers to purchase and install storage facilities, feeders and blenders to handle the soy flour and ascorbic acid, 2) develop appropriate pricing and accounting systems, 3) develop a regulatory system 4) develop a program of technical assistance to suppliers, bakers and pasta producers, 5) integrate the program with the national agricultural production plan, and 6) develop a program to inform and attain the support of the people. These activities will require additional financial resources but even more critically, superior management expertise. At the current time, food grade defatted soy flour is not available. In the case of corn, some pregelatinized flour is being used in pastas when the price is favorable. Unfortunately, low plant utilization has tended to make the corn flour higher in price than it might otherwise be.

Colombia initiated a major composite flour study at its Institute for Technological Investigations (IIT) in Bogota, with technical and financial assistance from the Netherlands, during the period 1971-1972 (28). The project was to explore the feasibility of the introduction of bread prepared with composite flours from domestic, tropical crops like cassava, soy, corn, rice and sorghum. Results of technological, marketing and economic studies were positive for a 30% substitution with rice or corn. It was noted that a large-scale introduction of composite flours is a complicated affair, affecting national agriculture, milling and baking industries, foreign trade and consumers' attitudes. Accordingly, it was recommended that the introduction of composite flours be undertaken gradually, one city or region at a time.

IIT eventually chose to produce a blend of 80% wheat flour, 17% rice flour and 3% defatted soy flour. Calcium stearoyl lactylate and potassium bromate, sufficient to give 0.25% and 50 ppm respectively, were first incorporated into the soy flour. IIT made arrangements for the production of the composite flour at the Concepcion mill in Cajica under careful controls. Rice brokens, supplied by IIT, were tempered separately but milled with the wheat to give a 78% flour yield. It was found necessary to set the mill rolls a little closer than is usual for 100% wheat. The soy flour with additives was then blended into the wheat-rice flour mix to complete the composite flour.

In 1978, IIT was acting as the sales agent for the composite flour selling it only to those bakeries whose bakers were trained at IIT. The composite flour was subsidized to a price about 20% below that of wheat flour in order to provide a sufficiently strong incentive for the bakers to try it. In order to involve smaller bakeries, it was necessary for an IIT professional baker to provide training at the bakers' own shops because they could not afford time off for training at IIT. The bread thus produced is often of better volume and quality than that produced at the bakery before the training and introduction of the composite flour. But this appears to be as far as the program has progressed. It has not been convincing enough for the private sector to undertake strictly commercial ventures, nor has it appeared sufficiently advantageous for the Government to tackle the tremendous organizational task and provide the managerial and financial resources required to pursue a national program.

IIT has also developed composite flour pastas. The Colombian Plan for Food and Nutrition utilized 750 MT of them in food programs for the poor in 1978 (29). Two formulations were produced and found very successful; 50:25:25 pregelatinized white corn flour:defatted soy flour:high quality wheat flour, and, 85:15 wheat flour from "second grade" wheat:defatted soy flour.

In the Sudan the UNDP/FAO has been assisting the Food Research Center at Khartoum since 1974 on a project entitled, "Research and development of wheat and sorghum products for industrial application" (30). Two pilot plants, one for pearling and milling sorghum with a capacity of 500 kg per hour and one for baking at 300 kg per 8 hours have been constructed. Badi et al. (31) have discussed some of the sorghum varieties available in the Sudan and the desirability of hard, white types for decortication by pearling. Using sorghum flour from the mill, the bakery is very successful and in 1982 produced 2000 loaves per day of 15% sorghum composite flour bread for sale in local retail stores. In 1978, sorghum was more expensive than subsidized wheat making commerical use of sorghums in bread economically nonviable.

Sorghum is a traditional crop in the Sudan with production on the order of 2 million MT. There is almost no industrial milling of sorghum; instead, it is ground on small stone mills at 100% extraction or pounded in mortars and is mainly used to make a fermented, unleavened, sour pancake called "kisra". Wheat has been making some inroads in the home preparation of kisra and, in some cases, kisra is being replaced altogether by western style wheat breads. It is interesting to note that wheat production has increased in the Sudan from 25,000 MT in 1960 to an estimated 250,000 MT in 1976, mostly on irrigated land. At the same time, modern wheat milling capacity increased sharply to 450,000 MT, a sufficient capacity to handle domestic and imported wheat supplies. Industrial food use of sorghum, however, is still very much in the research and development stage. The use of sorghum in composite flours in the Sudan can only proceed as fast as its industrialization. To be economic, the industrialization process must find suitable markets for the approximately 20% of bran produced when sorghum is pearled and milled (32).

In the United States, a composite bread flour has been commercially produced since 1972 for use in the Food for Peace Program (33). The blend contains 12% defatted soy flour, 88% bread wheat flour, 0.5% sodium stearoyl lactylate and 10-40 ppm potassium bromate. A loaf volume purchase specification is an important aspect to insure good baking quality. The cost of this blend has averaged about 10% above a typical export bread flour. About half the increased cost can be attributed to increased ingredient costs and the other half to increased handling costs. During the period 1972-1982, over a billion pounds were distributed throughout the world with excellent reception. However, this product was born in the era of the "Protein Crisis", a philosophy that has waned considerably, so presently, current purchases are decreasing while purchases of regular flour are increasing.

There are indeed many countries where various composite flours have been tested in diverse and numerous products. Significant composite flour programs or studies have been undertaken in Niger, Ghana, Egypt, Nigeria, Ecuador, Venezuela, Peru, Chile, Dominican Republic, Guyana, Philippines, Sri Lanka, India, Pakistan, Korea and others. None have led to a comprehensive and continuing composite flour program in commercial markets.

Problems

The experience of the past 20 years has revealed and sharpened the understanding of several problems encountered in the attempts to commercialize composite flours in the developing countries. Reaching the goal of a national composite flour is much more difficult and complex than originally believed. Table 18.5 points out that the problems are not just technical but political, economic, and structural as well. Discussion on each of the problems follows.

Lack of realism. There appears to be a lack of appreciation on the part of policy makers for the complexity of a national composite flour program. As a result, there has been an inadequate allocation of dedicated, high level, managerial expertise. This is needed to attain coordination and integration with basic policies in agricultural production, agribusiness, imports, prices, incentives, regulations, technical assistance, and public communications and relations. Technical programs have been approved and supported time and again only to be technically successful but politically impotent. Sometimes, a composite flour program initiated within a certain agency or ministry has evoked

TABLE 18.5. Problem Areas Encountered in Attempts to Commercialize Composite Flours

Problem	Type of Problem			
	Political	Economic	Structural	Technical
Lack of realism in government	X			
Lack of supply of nonwheat flours		X	X	X
Lack of regulatory methodology and systems			X	X
Lack of incentives for certain participants	X	X		
Subsidies	X	X		
Baking quality impaired				X

jealousies resulting in the loss of needed cooperation from other agencies and groups. Instability and frequent changes in Government have also prevented the extended concentration of effort needed to carry a composite flour project to fruition.

Lack of supply of nonwheat flours. The most significant problem is the lack of adequate and reliable supplies of uniform, high quality, nonwheat flours at low and relatively stable prices. Speaking before a 1970 FAO Sub Group on Food Industries and Marketing, Hoover (34) concluded that nonwheat flours of suitable quality are generally not available and that this lack of basic raw material will prove to be the biggest deterrent to the development of a wider use of composite flours. The reasons for this are many; a few of them are discussed here.

Developing countries, by definition, lack infrastructure. Inadequate grain drying, storage, processing and marketing facilities lead to great volatility in supplies and prices. Gluts at harvest cause very low prices and scarcity at other times leads to very high prices. In many cases, for example with sorghum and millet, appropriate industrial technology and equipment does not exist for converting the nonwheat crops, at low cost, into high quality flours and saleable by-products. Instead, makeshift systems are used resulting in low quality nonwheat flours of higher cost because of poor conversion efficiency. This type of system is in sharp contrast to the international wheat system where supplies are reliable, costs relatively stable, processing efficiency high, and markets well established for all products.

Subsistence and low technology crop production methods are still very prevalent in developing countries. Under such systems, many different varieties are grown. On the other hand, developed countries tend toward monoculture in a crop which creates great uniformity. The few varieties or hybrids cultivated have been selected for their processability in industrial systems as well as their agronomic characteristics. Just the opposite situation tends to exist in developing countries. The hodgepodge of varieties presents a profusion of color, size, shape and hardness which leads to processing inefficiencies and products with significant variability. Variability in an ingredient is a trait that bakers scorn. Small farms and low technology crop production methods also result in higher commodity costs. Table 18.6 shows that the majority of land for hard corn production in Ecuador is farmed by low technology procedures resulting

TABLE 18.6. Production Cost and Yield for Hard Corn in Ecuador for 1979. Exchange Rate: 25 Sucres Equals U.S. $1.00

Degree of Technology	% of Land	Production Costs/Ha	Yield Kg/Ha	Cost/100 lbs Sucres
High	3	S/ 9,153	2,273	183
Medium	27	S/ 6,901	1,591	197
Low	70	S/ 4,341	773	255
Ave	--	S/ 5,177	1,039	277

*Source: Commission MAG; Government of Ecuador; April 1979.

in production costs nearly 40% higher than corn produced under high technology procedures. Official government support prices are inflated with the result that few commodities cost less than international wheat. Composite flours that cost more than wheat flour receive little support.

Lack of geographical diversity with high year to year variability in weather patterns can lead to severe fluctuations in crop production. For example, the countries of the African Sahal have suffered periodic droughts causing significant losses in the sorghum and millet production. Under such conditions, the domestic component of a composite flour might have to be dropped because of scarcity and higher costs.

Lack of regulatory methodology and systems. Simple, rapid, quantitative methods are not available to measure the levels of nonwheat flours in blends. Even complex methods are not available. Because nonwheat flours may be more or less expensive than the wheat flour component, there could be a strong temptation to use more or less of the nonwheat flours than legally allowed. Also, regulatory agencies in many developing countries are poorly equipped and inadequately funded making the control of a composite flour program a significant additional burden.

Lack of incentives for certain participants. The adoption of a national composite flour program has both its proponents and opponents. Wheat importers and millers are likely to lose a percentage of their business and millers may be forced to make capital investments in storage and handling facilities, and metering and blending equipment. Unless the government allows millers an increased return, they will suffer reduced profits. The job of both the miller and the baker becomes technically more difficult and exacting, and it is unlikely that they will be rewarded for accepting these higher risks. It is not surprising, then, to find that bakers and millers resist the imposition of a national composite flour program. Consumers can also be opponents if the quality of traditional bakery products is reduced or prices are increased. The proponents are the governments (savings in foreign exchange), the growers, processors, marketers and equipment suppliers of the nonwheat commodity to be used.

Subsidies. Subsidies are financial aids, through government programs, to a particular group. That group might be farmers, farm suppliers, transporters, processors, marketers or others. The subsidy is generally reflected in either higher or lower prices of the materials or services provided by that group.

In the food chain of a developing country, the government may subsidize farmers in one of two ways. It may provide high crop support prices which raise the price of the commodity. The consumer basically pays the subsidy to the farmer through the higher price. In a second method, the government may provide inputs and services to the farmer, such as supplying fertilizer or credit, below cost. This results in lower commodity prices and the subsidy flows through to the consumer. Consumers are also subsidized through government programs where it purchases the large majority of a particular commodity and then resells it to food processors or consumers at a lower price.

Subsidies are extremely important tools for use in stimulating agricultural production and insuring that all citizens can afford to purchase at least the basic food needs. Almost all, if not all, developing countries use subsidies in managing their food systems. Because subsidies favorably impact certain groups, they have strong political connotations. The proposal to change or reduce a subsidy can lead to civil strife or even the collapse of a government. Bread riots in Egypt and baker strikes in Bolivia were the result of efforts to remove consumer subsidies on wheat products.

In Bolivia and Ecuador, the domestic farm price for wheat is set higher than imported wheat to encourage production. Wheat millers are required to purchase all domestic wheat offered to them by the farmers at the higher price. On the other hand, wheat imported by the government is sold to the millers at a price well below the government's cost. Because the amount of imported wheat is much greater, the net result in recent years, has been the availability of wheat products to consumers at prices well below total costs. Consumer subsidies for wheat are provided in a number of developing countries. Urban consumers are the major beneficiaries because their consumption of wheat is often much greater than rural consumers. The urban consumers strongly resist any erosion of this benefit by exercising their potent and centralized political clout.

Composite flours are embroiled in the midst of these pricing and subsidy systems. Composite flours are most likely to thrive where the price of wheat flour is higher than the price of domestic nonwheat flours on a long term, continuing basis; the reality has most often been otherwise. Governments are not willing or have been unable to adjust the pricing subsidy systems to favor composite flours. Too many other important factors are involved and must be considered with the result that requirements for composite flours have received only secondary consideration.

Baking quality impaired. The reality for bakers and other wheat product manufacturers is that composite flours increase their risk of product failure without enhanced profitability. Regardless of additives, process modifications or other measures, the baker can do a better job with pure wheat flour. Until consumers demand composite flour breads and products on their own merits and are freely willing to pay for them, as is the case in developed countries, most bakers will resist them.

Most bakeries in developing countries are quite small. Bakers are often unable to take advantage of new composite flour technology because they lack modern equipment and access to specialized additives. Because additives must be imported, governments have discouraged their use by imposing heavy duties.

CONCLUSIONS

The utilization of composite flours in the developed world is based on an increasing consumer demand for variety breads and nutritionally improved wheat products. In the developing countries, wheat importation and consumption is increasing. Composite flours have not been successful in slowing the wheat import trend. Much research has been conducted that shows composite flours are technically feasible and that many different nonwheat cereals, legumes, oilseeds, and root crops can be used. At the national implementation level, however, the lack of adequate supplies of nonwheat flours at low cost and high quality has impeded progress. There is a need for the industrialization of nonwheat flours. In addition, it would appear that governments of developing countries have not been realistic in their consideration of composite flours and thus have not allocated sufficient high level managerial expertise or adequate financial support to succeed with such a complex project as a national composite flour program.

The outlook is for continued slow progress on the implementation of composite flours in the developing countries, and a shift in research emphasis away from composite flours and the production of Western style baked goods toward utilization research on indigenous crops in support of industrial development of traditional foods or new foods.

REFERENCES

1. Hale, W. H., in "The Horizon Cookbook and Illustrated History of Eating and Drinking through the Ages." Vol. 1, p. 28, 83, and 84. American Heritage Publishing Co. (1968).
2. Anon., Milling and Baking News 61(50), 7 (1983).
3. Asselbergs, E. A., in "Composite Flour Programme." Documentation Package, Vol. 1 pp. 5-6. FAO, Rome (1973).
4. Asselbergs, E. A., in "Reports of the Int'l. Assoc. for Cereal Chemistry." Vol. 10, p. 12. Vienna (1980)
5. Dendy, D. A. V., Clark, P. A. and James, A. W., Tropical Sci. 12(2), 131 (1970).
6. Siebel, W. and Drews, E., in "Encyclopedia of Food Technology." p. 763. Avi Publishing Co., Westport, CN (1974).
7. Gorman, W., in "Proceedings of the 57th Annual Meeting." American Society of Bakery Engineers, p. 40. Chicago, IL (1981).
8. Iwasaki, T., Personal communication. Nat'l Food Res. Inst., Ibaraki, Japan (1983).
9. Nishita, K. D. and Bean, M. M., Cereal Chem. 56(3), 185 (1979).
10. Clausi, A. S., Food Technol. 25:821 (1971).
11. Seyem, A. A., Breen, M. D. and Banasik, O. J., Bulletin No. 504; Agric. Exp. Sta., North Dakota State Univ., Fargo, ND (1976).
12. Anon., Part 139--Macaroni and Noodle Products. Chapter 1, Title 21: Foods and Drugs; Code of Federal Regulation U.S. Govt. Printing Office (1981).
13. FAO., "1980 FAO Trade Yearbook." Vol. 34. Rome (1981).
14. Dendy, D. A. V., Kasasian, R., Bent, A. Clarke, P. A. and James, A. W., in "Composite Flour Technology Bibliography." G89 Second Ed. Tropical Products Institute, London (1975).
15. Kasasian, R. and Dendy, D. A. V., in "Composite Flour Technology Bibliography." Supplement 1 to G89. Tropical Products Institute, London (1977).
16. Dendy, D. A. V. and Kasasian, R., in "Composite Flour Technology Bibliography." Supplement 2 to G89. Tropical Products Institute, London (1979).
17. ICC., "Symposium on the Production and Marketing of Composite Flour Baking Products and Macaroni Goods." Oct. 23-27, 1972, Bogota, Colombia (1972).
18. ICC., Symposium on Sorghum and Millet (D. A. V. Dendy, Chairman), in "Reports of the Int'l. Assoc. of Cereal Chemistry." Vol. 10. Vienna (1980).

19. Dendy, D. A. V., "Sorghum and Millets for Human Foods." Proceedings of Meeting May 11-12, 1976; Vienna. Published by the Tropical Products Institute, London (1977).
20. de Ruiter, D., in "Advances in Cereal Science and Technology." Vol. 2, Am. Assoc. Cereal Chemists, St. Paul, MN (1978).
21. Pringle, W., Williams, A. and Hulse, J. H., Cereal Science Today 14(3), 114 (1969).
22. Miche, J. C., Alary, R., Jeanjean, M. F. and Abecassis, J., in "Sorghum and Millets for Human Food." May 11-12, 1976; Vienna. Published by the Tropical Products Institute, London (1977).
23. Salazar de Buckle, T., Cabrere, J. A., Pardo, C. A. and de Sandoval, A. M., IIT Technologia 17(98), 32 (1975). (Bogota, Colombia).
24. van der Made, C., in "Composite Flour Programme." Documentation Package, Vol. 1. p. 95. FAO, Rome (1973).
25. Hoover, W. J., "Improving the Nutritive Value of Cereal Based Foods." Final Report, Contract No. AID/csd-1586. Kansas State University, Manhattan, Kansas (1975).
26. Perten, H., in "Composite Flour Programme." Documentation Package, Vol. 1. p. 43. FAO, Rome (1969).
27. Fellers, D. A., Cereal Foods World, 28(7), 401 (1983).
28. Anon., "Interpan". Joint Report of the Colombian-Netherlands Composite Flour Project. Report issued by: Instituto Investigaciones Tecnologicas, Bogota; Institute for Cereals, Flour and Bread, Wageningen; FAO, Rome (1972).
29. Miller, J. M., in "Industrial Research Institutes. Their Role in the Application of Appropriate Technology and Development." A series prepared by the University of Denver, Denver Research Institute, Office of International Programs, Denver, Colorado (1979).
30. Perten, H., in "Sorghum and Millets for Human Food." May 11-12, 1977, Vienna. Published by the Tropical Products Institute, London (1977).
31. Badi, S. M., Perten, H. and Abert, P., in "Reports of the Int'l. Assoc. of Cereal Chemistry." Vol. 10. Vienna (1980).
32. Dendy, D. A. V., in "Reports of the Int'l. Assoc. for Cereal Chemistry." Vol. 10. p. 99 (1980).
33. Fellers, D. A., Mecham, D. K., Bean, M. M. and Hanamoto, M. M., Cereal Foods World, 21(1), 75 (1976).
34. Hoover, W. J., in "Composite Flour Programme." Documentation Package, Vol. 1. p. 88. FAO, Rome (1973).

19

Nutritional Implications of Cereals, Legumes, and Their Products

C. E. BODWELL

INTRODUCTION

The broad topic, nutritional implications of cereals, legumes and their products, cannot be comprehensively discussed in this paper. Accordingly, most aspects of the topic are considered in a general way, and detailed data are given for examples and for a few specific areas that are now the focus of considerable attention.

Estimates of the amount of protein available for human consumption on a daily per capita basis are given in Table 19.1. On average, in the developed countries (Europe, Russia, Australia, Japan, etc.), about 46% of the protein available is from plant sources (mostly cereals and legumes). In a few countries, including the United States, New Zealand and Australia, only 20-30% of the available protein is provided by plant sources. In the developing countries, which account for more than half of the world's population, about 80% of the available protein is from plant sources.

The global significance of plant protein sources is obvious and, even in the United States, the contributions of plant protein sources to the national diet are important. In addition, the national dietary goals and guidelines (U.S. Senate, 1977a,b; USDA/HEW, 1980), despite some criticisms (FNB, 1980), will motivate an increased consumption of legumes and cereals.

TABLE 19.1. Estimated Protein Available for Human Consumption (g per day per Capita, estimates for 1977) (1)

	Plant Protein	Animal Protein	Total
Developed countries	41.3	48.8	90.1
Developing countries	43.5	11.9	55.4
All countries	44.8	24.6	69.4
United States	33.5	72.7	106.2

Cereals and Legumes in the Food Supply, edited by Jacqueline Dupont and Elizabeth M. Osman 1987 Iowa State University Press, Ames, Iowa 50010

NUTRITIONAL ADVANTAGES

The national dietary goals or guidelines recommend increased consumption of complex carbohydrates and fiber and decreased consumption of saturated fats. In practice this means an increased consumption of cereals and legumes.

Nutrient Content vs. Calories

The levels of nutrients provided per 1000 kilocalories of some cereal and legume products are given in Table 19.2 together with values for foods, representative of various other classes of foods, that provide high amounts of protein and energy in the national diet. Except for a product such as peanut butter, cereals and legumes provide much higher levels of carbohydrate (for the most part, complex), in relation to both protein and fat, than do foods of animal origin. In addition, cereals and legumes can be generally excellent sources of copper, magnesium, manganese, phosphorous and potassium. Levels of iron and zinc are often high although, nutritionally, these may not be completely available. Cereals and legumes also are reasonably good sources of many vitamins.

Compared to foods of animal origin, the lipids of cereals and legumes have high ratios of polyunsaturated to saturated fatty acids. Most foods that are primarily of cereal or legume origin have no cholesterol or quite low levels per 1,000 kilocalories compared to foods of animal origin (Table 19.2).

Protein Quality

In comparison with an ideal protein (12), most cereals or legumes have low levels of one or more essential amino acids (Table 19.3). The pattern for an ideal protein, however, is based on the requirements of young children. The requirements of adults for essential amino acids are much lower. Recent reports have emphasized that protein from plant sources meets the protein and amino acid requirements of adults and, that protein from some plant sources meets those of young children (13-20).

As indicated in Table 19.3, even the poorer plant proteins, (with respect to their amino acid composition), such as rye and peanut flour, provide an excess of essential amino acids if they are consumed at levels that provide the RDA (Recommended Dietary Allowances) for protein (e.g., for males, 0.8 g protein/kg body weight/day). The RDA values for protein are based on the assumption that protein from our usual diets, which contain various sources of protein, is utilized at an efficiency of only 75% compared to 100% for protein from animal sources such as eggs, milk or meat.

With very few exceptions, most cereals and legumes are utilized with an efficiency which is greater than 75%. Also, even if U.S. citizens should greatly increase their consumption of protein from cereals and legumes, they probably would continue to consume some protein of animal origin. Thus, an increased intake of plant protein probably would not affect the protein nutriture of the majority of the population in the United States.

Source of Complex Carbohydrates, Dietary Fiber, and Polyunsaturated Fats

As noted in the section entitled "NUTRITIONAL ADVANTAGES, Nutrient Content vs. Calories", cereals and legumes are excellent potential sources of polyunsaturated fats and, together with fruits and vegetables, of complex carbohydrates (i.e., starch and other nonfiber polysacharides). Legumes and whole grain cereals are also excellent sources of dietary fiber. Although there are numerous claims for the potential health benefits of the consumption of moderate to high levels of dietary fiber, few claims have been substantiated (23-26). It is generally agreed, however, that adequate dietary fiber prevents constipation, increases stool weights, and can modulate or prevent diverticulosis and hemorrhoidal conditions. (Possible deleterious effects of fiber in relation to the utilization of dietary minerals is considered in the section entitled, "POTENTIAL DISADVANTAGES, Effects on Mineral Utilization".

Possible Effects of Plant Protein on Blood Lipid Status

The influence of dietary soy protein on blood lipids (primarily cholesterol) has been examined in several recent studies (Table 19.4). Sirtori et al. (27) found an average decrease of 20% in total cholesterol levels after two weeks in subjects fed diets high in soy protein. Likewise, Carroll et al. (28) reported a slight, but statistically significant, reduction in cholesterol levels in college-age girls. No effects on cholesterol levels in over 60 subjects were observed by Van Raaij et al. (29). In a study with men of various ages (30), dietary intakes of cholesterol were quite low (200-250) mg/day) and cholesterol levels thus decreased when subjects were fed either an animal protein diet or a soy protein diet. In studies in which 16 or 17 men were fed diets based on either of two different soy preparations or on animal protein and also were given "normal" levels of saturated animal fats, total dietary lipids, and cholesterol (31), no significant decreases in blood lipids were observed.

The differences among the findings in these studies can be explained, in part, by differences in experimental conditions and design. However, one feature of the Sirtori studies (27) is significant; the subjects were all markedly hypercholesterolemic, Type II (heterozygote) individuals. In such individuals, soy protein *per se*, through an undefined interaction with other components of the diet, probably has an important beneficial effect in altering blood lipid levels (35).

For plant proteins other than soy, consistently beneficial effects on blood lipids have not been clearly demonstrated in specific types of individuals (Table 19.4). Soy protein, however, has been studied more extensively than other plant proteins. Extensive study of other plant proteins might demonstrate similarly beneficial effects.

POTENTIAL DISADVANTAGES

Almost every food has some nutritional disadvantages or potential disadvantages. Cereals and legumes and products made from them are not exceptions.

Digestibility

Foods must be digested before their nutrients can be utilized. Few data, however, document the effects of digestibility on the utilization by humans of such nutrients as vitamins and minerals. Some data are available on energy utilization, but the most extensive body of data is on protein (nitrogen) utilization.

TABLE 19.2. Nutritive Value of Some Common Foods per 1,000 Kilocalories[a]

Food	Beef, round, roasted		Pork, fresh ham, roasted	Chicken Fried (floured), with skin		Cod fish, broiled, with margarine	Whole milk
	Lean	Lean and fat	Lean	Light Meat	Dark Meat		
Weight (g)	502.0	397.0	455.0	407.0	351.0	708.0	1,627.0
Proximate:							
Protein (g)	151.1	110.7	128.7	123.8	95.64	143.3	53.5
Total lipids (g)	39.3	58.6	50.2	49.2	59.47	46.0	54.3
Carbohydrate (g)	---	---	---	7.4	14.3	2.83	75.8
Minerals:							
Calcium (mg)	47.0	33.0	32.0	67.0	58.0	100.0	1,940.0
Copper (mcg)	663.0	434.0	492.0	234.0	310.0	---	---
Iron (mg)	13.0	9.8	5.1	4.9	5.3	3.33	.80
Magnesium (mg)	142.0	103.0	107.0	110.0	83.0	225.0	220.0
Manganese (mcg)	183.0	121.0	166.0	105.0	136.0	---	---
Phosphorus (mg)	1,260.0	925.0	1,278.0	866.0	620.0	1,583.0	1,520.0
Potassium (mg)	2,183.0	1,579.0	1,695.0	971.0	810.0	3,142.0	2,467.0
Sodium (mg)	314.0	238.0	289.0	311.0	314.0	1,042.0	800.0
Zinc (mg)	28.6	20.3	14.8	5.1	9.13	3.3	6.20
Vitamins:							
Ascorbic acid (mg)	0.0	0.0	1.60	0.0	0.0	25.0	13.0
Thiamin (mg)	0.47	0.33	3.16	0.33	0.33	0.33	0.62
Riboflavin (mg)	1.07	0.75	1.60	0.53	0.83	0.50	2.27
Niacin (mg)	21.12	14.81	22.46	49.00	24.00	15.16	1.41
Pantothenic acid (mg)	3.91	2.52	4.12	3.92	21.39	---	5.13
Vitamin B_6 (mg)	1.18	0.79	1.44	2.20	4.05	1.50	0.67
Folacin (mcg)	53.0	33.0	53.0	14.0	29.0	---	80.0
Vitamin B_{12} (mcg)	11.89	8.36	3.26	1.34	1.07	4.92	5.81
Vitamin A (IU)	0.0	0.0	37.0	278.0	364.0	2,083.0	2,047.0
Lipids							
Fatty Acids:							
Saturated, total (g)	16.21	24.25	17.27	13.50	16.07	8.50	33.82
Monounsaturated, total (g)	17.93	28.08	22.57	19.57	23.38	19.16	15.67
Polyunsaturated, total (g)	2.19	2.19	6.10	10.96	13.71	14.83	2.00
Cholesterol (mg)	426.0	336.0	428.0	354.0	322.0	400.0	887.0

[a]Adapted from Bodwell and Anderson (6); values for beef (choice grade), cod, peanut butter and fried beans calculated from USDA, HNIS, Nutrient Data Research Group (unpublished data); for pork from USDA (7); for chicken from USDA (8); for dairy and egg products from USDA (9); for oatmeal from USDA (10).

[b]Made with vegetable shortening and whole milk; from McQuilkin and Matthews (11).

Cottage cheese	Cheddar cheese	Peanut butter	Dried beans	Cereal, oatmeal	Bread[b]		Egg
			Navy, cooked	Cooked	White	Whole wheat	Cooked
968.0	246.0	169.0	658.0	1,614.0	347.0	379.0	633.0
120.9	61.9	42.6	56.6	41.4	26.25	34.15	76.8
43.64	82.46	86.35	3.94	16.56	23.06	24.46	70.6
25.94	61.93	32.94	186.9	173.80	167.36	175.16	7.6
581.0	1,789.0	58.0	446.0	174.0	223.0	303.0	354.0
---	---	778.0	2,114.0	890.0	260.0	970.0	---
1.3	1.7	3.07	16.58	11.0	7.5	10.15	13.17
51.0	70.18	296.0	399.0	386.0	76.0	352.0	76.0
---	---	3,032.0	3,316.0	9,441.0	---	---	---
1,276.0	1.272.0	630.0	1,067.0	1,228.0	33.0	955.0	1,139.0
816.0	246.0	1,180.0	2,321.0	910.0	458.0	1,288.0	823.0
3,917.0	1,544.0	799.0	5.18	7.0	1,417.0	1,348.0	873.0
3.60	7.72	4.92	7.25	7.93	2.2	8.48	9.12
Trace	0.0	0.0	0.0	0.0	Trace	Trace	0.0
0.18	0.09	0.16	1.30	1.79	1.1	1.06	0.51
1.57	0.97	0.16	0.26	0.35	1.1	.61	1.90
1.20	0.18	22.70	3.32	2.07	10.14	12.12	.38
2.07	1.05	1.85	1.97	3.24	1.67	3.03	10.89
0.65	0.18	0.69	0.67	0.34	0.28	0.76	0.76
120.0	44.0	175.0	---	62.1	125.0	182.0	304.0
6.04	2.02	0.0	0.0	0.0	0.43	0.45	9.75
1,576.0	2,631.0	0.0	0.0	262.0	166.0	152.0	3,291.0
27.60	52.46	17.78	0.47	3.03	29.03	28.64	21.14
12.44	23.33	39.79	0.26	5,80	33.89	33.48	28.23
1.34	2.37	25.56	2.28	6.69	21.53	25.61	9.12
143.0	263.0	0.0	0.0	0.0	28.0	15.0	3,468.0

TABLE 19.3. Essential Amino Acid Composition of Some Cereals and Legumes Compared to Reference Patterns[a]

Amino Acid	Reference Pattern: Ideal Protein (NRC, 1974)	Reference Pattern: 70 KG Adult Male[b]	Corn	Oats	Wheat	Rice	Rye	Peanut flour	Kidney bean	Soy Flour
Histidine	17	--	27	21	23	23	22	24	28	25
Isoleucine	42	12	(37)	(38)	(33)	42	(35)	(34)	42	45
Leucine	70	16	125	73	67	82	(62)	(64)	76	78
Lysine	51	12	(27)	(37)	(29)	(36)	(34)	(35)	72	64
Methionine	--	--	19	17	15	21	15	12	11	13
Cystine	--	--	16	27	25	15	19	12	8	13
TSAA	26	10	35	44	40	37	34	(24)	(19)	26
Phenylalanine	--	--	49	50	45	48	44	50	52	49
Tyrosine	--	--	38	33	30	32	19	39	25	31
TAAA	73	15	87	83	75	80	(63)	89	77	81
Threonine	35	8	36	(33)	29	(33)	(33)	(26)	40	39
Tryptophan	11	3	(7)	13	11	13	(7)	(10)	(10)	13
Valine	48	14	49	51	(44)	58	48	(42)	(46)	48
TEAA	373	89	430	393	(353)	403	(338)	(348)	410	418

[a]Mg amino acid per 100 g protein (nitrogen x 6.25). TSAA = total sulfur amino acids (methionine plus cystine); TAAA = total aromatic amino acids (phenylalanine plus tyrosine); TEAA = total essential amino acids (includes non-essential amino acids cystine and tryosine). Values in parentheses are below those of the reference pattern for an ideal protein.

[b]Calculated from estimated amino acid requirements for adults (12) and the Recommended Dietary Allowances of 0.8 g protein (N x 6.25) per kg body weight per day (21). Values for protein sources calculated from data given by FAO (22).

TABLE 19.4. Effects of Plant Protein Sources on Blood Lipid Status

References	Protein Source(s) Evaluated	Subjects	Results
Carroll et al. (28)	Soy protein	College-age females	Slight, but statistically significant decrease in serum cholesterol levels
Sirtori et al. (27)	Soy protein	Adults	Decrease of 20% in serum cholesterol levels after two weeks
van Raaj et al. (29)	Soy protein vs. casein diets	Adults of various ages	No effects observed
Shorey and Davis (30)	Soy protein vs. animal protein (low cholesterol diets)	Young adults	Decrease in serum cholesterol levels observed on both animal and soy protein diets
Bodwell et al. (31)	Soy protein vs. animal protein	Adult males	No differences between diets after 5 weeks
Walker et al. (32)	Wheat products, soy powder, rice, etc., vs. animal protein	College-age females	No differences between animal protein diet and vegegable protein diet after 6 weeks
Hodges et al. (33)	"Oriental" diet (soy protein and cereals) vs. "mixed" diet	Adult males	Significant decrease in serum cholesterol after 8 weeks of oriental diet
Anderson et al. (34)	Wheat gluten vs. egg white protein	College-age males	No differences in serum cholesterol after 4 weeks on each diet

TABLE 19.5. True Digestibility by Adults of Protein in Some Common Protein Sources (36)

Protein Source	Number of Reports	Digestibility (%) Mean	Range
Animal protein	41	96	90-106
Corn, whole	4	87	84-92
Corn, ready-to-eat cereal	5	70	62-78
Whole wheat	6	87	80-93
Wheat flour (white)	2	96	96-97
Wheat bread (white)	5	97	95-101
Wheat bread (coarse, brown or whole wheat)	2	92	91-92
Wheat gluten	4	99	96-104
Wheat, ready-to-eat cereal	9	77	53-88
Rice, polished	4	89	82-91
Rice, ready-to-eat cereal	3	75	77-85
Oatmeal	4	86	76-92
Oats, ready-to-eat cereal	4	72	63-89
Peas, Alaskan field	1	88	---
Peanut flour, butter	4	94	91-98
Soy flour	5	86	75-92
Soy isolate	3	95	93-97
Soy protein, spun	2	104	101-107

The true digestibility of protein of animal origin ranges from 90 to 100% (Table 19.5). Generally, digestibility is between 80 and 90% for cereals or cereal products and above 90% for most legumes. Digestibility can be lower (<70-80%) for highly processed cereals, such as some ready-to-eat breakfast foods, and for severely or improperly processed legumes.

Effects on Mineral Utilization

Cereals and legumes contain phytic acid and fiber that have been implicated as agents that depress the biological availability of minerals. The numerous studies of phytic acid and fiber have been reviewed (37-50).

The following discussion is thus limited to a consideration of the results from a few studies on the effects of phytate or fiber on zinc and iron utilization and to a consideration of the effects on iron utilization of soy protein when consumed as the primary protein source or when used to extend ground beef.

In two studies, the effects of phytate from wheat bran were separated from the effects of the bran fiber. Bran or dephytinized bran was provided in muffins in both mineral balance studies and iron absorption tests. In the iron absorption tests, muffins containing no bran were also used. The results of the balance studies are summarized in Table 19.6. The apparent absorptions of zinc, iron, manganese and copper were not improved by the removal of the phytate. In the studies of iron (Table 19.7), absorption was markedly inhibited (74%) when about 12 g of whole bran was included in the test meal (Study I). When ascorbic acid was added to the meal (Study II), inhibition

TABLE 19.6. Effects of Removal of Phytate from Bran on Apparent Absorption (mg/day) of Trace Minerals (51)[a]

Trace mineral	Whole bran muffins fed	Dephytinized bran muffins fed
Zinc	3.1 ± 0.4	2.4 ± 0.9
Iron	2.9 ± 0.7	2.0 ± 1.1
Manganese	1.4 ± 0.2	0.8 ± 0.7[b]
Copper	0.59 ± 0.09	0.56 ± 0.06

[a]Values (means ± S.D.) are for 10 subjects.
[b]Significantly lower (5% level) than value for period when bran muffins were fed.

TABLE 19.7. Effects of Whole Bran (Wheat) or Dephytinized Bran on Non-Heme Iron Absorption in Single-Meal Tests (52)

	Absorption	(mg/day; Geometric Mean)	Inhibition %
Whole Bran	With Bran	Without Bran	
Study I (n=10)	0.62	2.39	74
Study II (n=13)	1.69	3.46	51
Dephytinized Bran	Whole Bran	Dephytinized Bran	
Study IV (n=10)	1.29	1.37	6
Study V (n=18)	3.22	2.43	--

was still 51%. Removal of the phytate from the bran did not improve absorption (Studies IV and V). Those studies indicate that the fiber and not the phytic acid in wheat bran inhibits iron absorption.

In rats, a molar ratio of phytate to zinc of greater than 10-12 to 1 causes a marked decrease in zinc utilization (53). However, the data in Table 19.8, from a study at Beltsville, suggest that soy phytate does not greatly alter zinc utilization by humans. Cossack and Prasad (54), however, reported conflicting results. Five subjects were in negative zinc balance when consuming a high-soy-protein diet but were in positive balance when consuming animal protein as their primary

TABLE 19.8. Effects of Soy Protein on Zinc Balances in 16 or 17 Adult Men Consuming Approximately 1.6 g Protein/Kg Body Wt/Day (37)

Primary (>70%) Protein Source	Daily Intake (mg)	Molar Phytate to Zinc Ratio	Apparent Daily Balance (mg)	
			Days 15-21	Days 29-35
Textured soy	16.6	22.3	+0.8 ± 3.1	+0.3 ± 3.4
Soy isolate	11.8	14.2	+1.3 ± 2.8	+0.8 ± 2.2
Animal Protein	18.2	3.4	+3.6 ± 2.8	+2.7 ± 1.8[a]

[a]Significantly higher (5% level) than values for textured soy or soy isolate diets; values for days 15-21 not significantly different; balance values are $\bar{X}$ ± S.D.

source of protein. Apparently, the soy diets were "EDTA-washed" which might have affected zinc absorption if the EDTA were not completely removed.

Iron balances were also determined in the Beltsville study (37). Results are summarized in Table 19.9. Apparent iron balances were markedly lowered following ingestion of the diet containing high levels of soy isolate.

In a series of studies (Table 19.10), Cook and co-workers (56-58) observed marked inhibitions of non-heme iron absorption from test-meals containing various types of soy protein. Whether these observations and the results of the Beltsville study have practical implications is not certain. In the Beltsville study (Table 19.9), very high levels of protein were fed, and the findings on absorption (Table 19.10) are from single-meal tests only and might not predict long-term effects. However, the results of the four studies are disturbing and indicate that further research is needed to determine the effects of soy protein *per se* on iron utilization.

TABLE 19.9. Effects of Soy Protein on Iron Balances in 16 or 17 Adult Men Consuming Approximately 1.6 g Protein/kg Body wt/day (55)[a]

Primary (>70%) Protein Source	Daily Intake (mg)	Apparent Daily Balance (mg)	
		Days 15-21	Days 29-35
Textured soy	21.3 ± 3.3	-2.9 ± 6.7	-1.4 ± 5.8
Soy isolate	23.4 ± 3.8	-8.4 ± 5.9[a]	-7.9 ± 4.9[a]
Animal protein	17.5 ± 2.9	-0.3 ± 3.4	-0.8 ± 1.5

[a]($\overline{X}$ ± S.D.) for textured soy or animal diets: values for textured soy and animal diets not significantly different.

TABLE 19.10. Effects of Soy Products of Non-Heme Iron Absorption in Adult Men

Reference	Summary of Results
Morck et al. (57)	Marked reduction in absorption of non-heme iron from infant food supplements containing soy protein.
Cook et al. (56)	Inhibition (81-83%) of non-heme iron absorption from a test meal containing soy isolate, in comparison with a test meal containing casein or egg albumin.
Cook et al. (56)	Inhibition (82, 65, 92%, respectively) of non-heme iron absorption from test meals containing full-fat soy flour, textured soy flour, or soy isolate, in comparison with a test meal containing egg albumin.

Cook et al. (56) also reported that extending ground beef with soy protein markedly inhibited non-heme iron absorption (Table 19.11). However, the soy used was not re-hydrated in the usual manner and was included at a much higher level than is allowed by either the School Lunch Program (replacement of up to 30% of the ground beef with reconstituted soy products) or the Department of Defense (replacement of 20% of the ground beef with reconstituted soy concentrate).

Similar results to those of Cook et al. (56) were obtained by others when 50% replacement level was used (Table 19.11). However, in the most recent tests by Cook et al. (presented by Bothwell et al., 59; Table 19.11), with a 30% replacement level, reduction in total (both heme and non-heme iron) iron absorption was only 18%. If allowance is made for the dilution of heme iron in the soy extended beef, then the reduction due to the soy protein would be only about 10 to 12%.

Extended beef is widely used in the School Lunch Program and by the Department of Defense. When the results of Cook's earlier studies were made known, together with the results of the Beltsville balance study, considerable concern was expressed in relation to possible deleterious effects on the iron status of school-lunch participants and of military women of child-bearing age that might be caused by consumption of beef extended with soy protein. Consequently, a practical, large-scale, 6-month study was initiated at Beltsville. The general protocol of this study is given in Table 19.12. Clinical indices of iron status (total iron binding capacity, percent transferrin saturation, levels of hemoglobin, serum iron, and free erythrocyte protoporphyrin) were measured prior to and at the end of the study. Serum ferritin levels (a measure of body iron stores) were determined at about 45-day intervals in all participants (children, women and men). In the adult men, iron absorption tests were also conducted

TABLE 19.11. Effects of Extending Ground Beef with Soy Protein on Non-Heme Iron Absorption From Single Meals

Reference	Description	Results
Cook et al. (56)	All beef compared to extended with textured soy to provide (A) 25% and (B) 33% of protein from soy	(A) 61% and (B) 53% inhibition of <u>non-heme</u> iron absorption
Hallberg and Rossander (60)	50% replacement of beef protein with textured soy	30 to 40% reduction in <u>total</u> iron absorption
Stekel et al. (Bothwell et al., (59)	50% replacement of beef with reconstituted soy isolate	33 to 42% reduction in <u>total</u> iron absorption
Cook et al. (Bothwell et al., (59)	30% replacement of beef with reconstituted textured soy	18% reduction in <u>total</u> iron absorption

TABLE 19.12. Beltsville Study on Effects of Consuming Beef Extended with Soy Protein on Iron and Zinc Status

Duration: 180 days	
7 Products (meat patties):	
All beef	
Textured soy extended (FNS; "Fe and Zn fortified")	
Soy isolate extended (FNS; "Fe and Zn fortified")	
Soy concentrate extended (FNS; "Fe and Zn fortified")	
227 Participants Consuming "Meat Patty" Meals	
50 adult males	(9 meals/week)
41 menstruating adult females	(9 meals/week)
115 "School-lunch" participants	(7 meals/week)
21 other	(7 or 9 meals/week)
62 Controls (blood samples only)	

at the beginning and end of the study. Clinical parameters of iron status or utilization, ferritin levels, and iron absorption were not affected (or were improved) during the study (61-63). The data indicated that consumption of soy protein, at the levels studied, by school lunch participants or by military men or women would not impose a risk in those population groups.

Effects of Dietary Fructose

Fructose is one of the most lipogenic sources of carbohydrate. Compared to glucose, two rate-limiting steps mediated by enzymes (hexose kinase; phosphofructose kinase) are bypassed in its conversion to triglyceride. In rats, fructose has been implicated in the development of diabetes, elevated triglyceride levels, increased insulin response, and decreased insulin sensitivity (64-67). In copper-deficient rats, death rate from spontaneous heart rupture was high in rats fed fructose (48,68,69), but not in rats fed either starch or glucose. Although those results cannot be extrapolated directly to humans, fructose consumption by human subjects impaired their insulin binding and insulin sensitivity-effects that are similar to those observed in rats (70). Thus, some caution may be prudent in relation to making claims about the nutritional benefits of fructose in corn syrup and other foods.

Anti-Nutrients and Other Factors

In addition to phytic acid, most plant sources of food contain various other anti-nutrients or other undesirable factors. Results from studies of those factors and their possible significance have been reviewed (71-82).

Some of the more common anti-nutrients and toxic or undesirable substances present in legumes and cereals are listed in Table 19.13. The hemagglutinins are heat-labile. The heterosides (goitrogens) are usually heat-labile. Flatulence factors *per se* are not a nutritional problem but are of concern in relation to the extensive utilization of legume products.

TABLE 19.13. Anti-Nutrients and Toxic or Undesirable Substances in Cereals and Legumes

Substances	Occurrence Reported in	Heat-Labile
Hemagglutinins	Most legumes, wheat	Yes
Heterosides	Most legumes, wheat	Usually
Phytic Acid	Most legumes and cereals	No
Flatulence Factors	Most legumes	No
Protease Inhibitors	Most legumes and cereals	Variable

Various protease inhibitors are present in most cereals and legumes. Of particular current interest is the induction of hyperplastic nodules in the pancreases of rats fed soy protein, containing trypsin inhibitor, for 15 to 24 months (83). The problem was investigated in a series of studies by Unilever in Britain and by the USDA and Liener (University of Minnesota) in the United States. These unpublished studies, have confirmed the results of Morgan et al. (83). Furthermore, in short-term feeding trials with rats, trypsin inhibitors from other plant foods enlarged the pancreas and may also be demonstrated to induce hyperplastic nodules. Effects of trypsin inhibitors should be investigated in other animal species. The level of trypsin inhibitor is generally high in Western diets (84), but association with disease of the pancreas has not been reported in humans. Becker (85) reported that trypsin inhibitors, especially those from soybeans, inhibit cancerous tumor formation in tumor-prone mice, and Yavelow et al. (86,87) suggested that protease inhibitors from tofu, chick-peas and kidney beans may suppress tumor formation in humans. Only further study can determine the role of inhibitors in human health and disease.

Allergens, saponins, and low levels of estrogenic substances may also be present in plant proteins. If products are improperly processed or heat treated, toxic compounds such as lysinoalanine may be produced or the protein nutritive value may be lowered. Other reactions, such as the conversion of methionine to methionine sulfoxide or sulfone may also occur (88).

SOME IMPLICATIONS FOR THE FUTURE

As noted above, it is highly likely that the U.S. consumption of cereals and legumes and their products will increase during the next few years. This increase will be primarily due to the perception that reducing intakes of fat and increasing intakes of complex carbohydrates and fiber will result in significant long-term health benefits.

It is also likely that for both economic and health reasons, use of protein products, derived from soy or other legumes and from cereals, for extending meat or meat products will increase. During 1980 and 1981, between $10 and $11 million were saved by extending ground beef for use in the military and it can be calculated that, conservatively, more than $30 million is saved annually by extending ground beef for use in the School Lunch Program. Use of high quality plant protein products and limiting the level of replacement to about 20% would probably also greatly facilitate the acceptance and use by the general population of beef extended with plant protein.

Future research should investigate the significance and possible modulation of the effects of plant protein on mineral utilization and the effects of anti-nutrients and toxic or other undesirable substances on human nutrition. Research should be extended to legumes other than soybeans and to the cereal grains. The effects of high dietary levels of fructose on human nutrition also should be clearly defined.

REFERENCES

1. Fauconneau, G., in "Vegetable Proteins for Human Food" (C. E. Bodwell and L. Petit, eds.), p. 1. Martinous-Nijhoff/B. V. Junk, The Hague, The Netherlands (1983).
2. U.S. Senate "Dietary Goals for the United States," U.S. Government Printing Office, Washington, D.C. (1977a).
3. U.S. Senate "Dietary Goals for the United States." (2nd edition), U.S. Government Printing Office, Washington, D.C. (1977b).
4. USDA/HEW "Nutrition and Your Health, Dietary Guidelines for Americans," U.S. Dept. of Agri. and Dept. Health, Education and Welfare, U.S. Government Printing Office, Washington, D.C. (1980).
5. FNB (Food and Nutrition Board) "Toward Healthful Diets." National Academy of Sciences, Washington, D.C. (1980).
6. Bodwell, C. E., and Anderson, B., in "Muscle as Food" (P. J. Bechtel, ed.), p. 321. Academic Press, Inc., New York, New York (1983).
7. USDA Consumer Nutrition Division. "Composition of Foods: Pork Products; Raw, Processed, Prepared"; USDA Agri. Handbk. No. 8-10, Government Printing Office, Washington, D.C. (1983).
8. USDA Consumer and Food Economics Institute. "Composition of Foods: Poultry Products; Raw, Processed, Prepared," USDA Agri. Handbk. No. 8-5, Government Printing Office, Washington, D.C. (1979).
9. USDA Consumer and Food Economics Institute. "Composition of Foods: Dairy and Egg Products, Raw, Processed, Prepared." USDA, Agri. Handbk. No. 8-1, Government Printing Office, Washington, D.C. (1976).
10. USDA Consumer Nutrition Center. "Composition of Foods: Breakfast Cereals, Raw, Processed, Prepared." USDA Agri. Handbk. No. 8-8, Government Printing Office, Washington D.C. (1982).
11. McQuilkin, C. and Matthews, R. H., Provisional Table on the Nutrient Content of Bakery Foods and Related Items. USDA, HNIS. (1981).
12. NRC (National Research Council). "Improvement of Protein Nutriture," Food and Nutrition Board, NRC. Natl. Acad. of Sciences, Washington, D.C. (1974).
13. Bodwell, C. E., J. Am. Oil. Chem. Soc. 56, 165 (1979).
14. Bodwell, C. E., in "Protein Quality in Humans: Assessment and In Vitro Estimation" (C. E. Bodwell, J. S. Adkins, and D. T. Hopkins, eds.), p. 340. Avi Publishing Co., Inc., Westport, Connecticut (1981).
15. Bressani, R., Torun, B., Elias, L. G., Nauarrete, D. A. and Vargus, E., in "Protein Quality in Humans: Assessment and In Vitro Estimation" (C. E. Bodwell, J. S. Adkins, and D. T. Hopkins, eds.), p. 98. Avi Publishing Co., Inc., Westport, Connecticut (1979).

16. Scrimshaw, N. S. and Young, V. R., in "Soy Protein and Human Nutrition" (H. L. Wilcke, D. T. Hopkins and D. H. Waggle, eds.), p. 121. Academic Press, New York, New York (1979).
17. Torun, B., in "Soy Protein and Human Nutrition" (H. L. Wilcke, D. T. Hopkins and D. H. Waggle, eds.), p. 101. Academic Press, New York (1979).
18. Young V. R., J. Am. Oil Chem. Soc. 56, 110.
19. Torun, B., Pineda, O., Viteri, F. E. and Arroyone, G., in "Protein Quality in Humans: Assessment and In Vitro Estimation" (C. E. Bodwell, J. S. Adkins, D. T. Hopkins, eds.), p. 374. Avi Publishing Co., Inc., Westport, Connecticut (1981a).
20. Torun, B., Viteri, F. E. and Young, V. R., J. Am. Oil Chem. Soc. 58, 400 (1981b).
21. NRC (National Research Council) "Recommended Dietary Allowances," Ninth Edition. Natl. Acad. Sciences, Washington, D.C. (1980).
22. FAO (Food and Agriculture Organization). "Amino Acid Content of Foods and Biological Data on Proteins." FAO, Rome (1970).
23. Kimura, K. K., "The nutritional significance of dietary fiber." Life Sciences Research Office. Federation of American Societies for Experimental Biology, Rockville, Maryland (1977).
24. Talbot, J. M., "The Role of Dietary Fiber in Diverticular Disease and Colon Cancer." Life Sciences Research Office. Federation of American Societies for Experimental Biology, Rockville, Maryland (1980).
25. Vahouney, G. V. and Kritchevsky, D. (eds.), "Dietary Fiber in Health and Disease," Plenum Press, New York (1982).
26. NRC (National Research Council), Diet and Cancer. Natl. Acad. of Sciences, Washington, D.C. (1982).
27. Sirtori, C. R., Gatti, E., Mantero, O., Conti, F., Agradi, E., Tremoli, E., Sirtori, M., Fraterrigo, L., Tauazzi, L., and Kritchevsky, D., Am. J. Clin. Nutr. 32, 1645 (1979).
28. Carroll, K. K., Giovannetti, P. N., Huff, W. M., Moase, O., Roberts, D. C. K., and Wolfe, B. M., Am. J. Clin. Nutr. 31, 1312 (1978).
29. van Raaij, J. M. A., Kataan, M. B., West, C. E., and Hautrast, G. A. J., Am. J. Clin. Nutr. 35, 925 (1982).
30. Shorey, R. L. and Davis, J. L., Fed. Proc. 38, 551 (1979).
31. Bodwell, C. E., Schuster, E. M., Steele, P. D., Judd, J. T., and Smith, J. C., Fed. Proc. 39, 1113 (1980). (Abstract).
32. Walker, G. R., Morse, E. H., and Overly, V. A., J. Nutr. 72, 317 (1960).
33. Hodges, R. E., Krehl, W. A., Stone, D. B., and Lopez, A., Am. J. Clin. Nutr. 20, 198 (1967).
34. Anderson, J. T., Grande, F., and Keys, A., Am. J. Clin. Nutr. 24, 524 (1971).
35. Kritchevsky, D., J. Am. Oil. Chem. Soc. 56, 135 (1979).
36. Hopkins, D. T., in "Protein Quality in Humans: Assessment and Hopkins, eds.) P. 169. Avi Publishing Co., Inc., Westport, Connecticut (1981).
37. Bodwell, C. E., Cereal. Fds. Wld. 28, 342 (1983).
38. Erdman, J. W., Jr., J. Am. Oil Chem. Soc. 58, 489 (1981).
39. Hallberg, L., Ann. Rev. Nutr. 9, 123 (1981).
40. Jaffe, G., J. Am. Oil Chem. Soc. 58, 493 (1981).
41. Janghorbani, M., Istfan, N. W., Pagounes, J. O., Steinke, F. H., and Young, V. R., Am. J. Clin. Nutr. 36, 537 (1982).
42. Kelsay, J. L., Cereal Chem. 58, 2 (1981).

43. Monsen, E. R., Hallberg, L., Layrisse, M., Hegsted, D. M., Cook, J. D., Mertz, W., and Finch, C. A., Am. J. Clin. Nutri. 31, 134 (1978).
44. Morris, E. R., Fed. Proc. 42, 1716 (1983).
45. Morris, E. R., and Ellis, R., in "Nutritional Bioavailability of Zinc." (G. Inglett, ed.), Am. Chem. Soc. Symposium Series 210, p. 159. American Chemical Society, Washington, D.C. (1983).
46. Morck, T. A., and Cook J. D., Cereal. Fds. Wld. 26, 667 (1981).
47. O'Dell, B., in "Soy Protein and Human Nutrition" (H. L. Wilcke, D. T. Hopkins, and D. H. Waggle, eds), p. 187. Academic Press, New York (1979).
48. Smith, J. C., Fields, M., Ferretti, R. J. and Reiser, S., Fed. Proc. 42, 528 (1983). (Abstract).
49. Turnland, J. R., Cereal Fds. Wld. 27, 152 (1982).
50. Young, V. R., and Janghorbani, M., Cereal Chem. 58, 12 (1981).
51. Ellis, R. and Morris, E. R., Cereal Chem. 58, 367 (1981).
52. Simpson, K. M., Morris, E. R., and Cook J. D., Am. J. Clin. Nutri. 34, 1469 (1981).
53. Davies, N. T., in "Dietary Fiber in Health and Disease", (G. V. Vahouny and Kritchevsky, D., eds.), p. 105. Plenum Press, New York (1982).
54. Cossack, Z. T. and Prasad, A. S., Fed. Proc. 41, 383 (1982). (Abstract).
55. Bodwell, C. E., Smith, J. C., Judd, J., Steele, P. D., Cottrell, S. L., Schuster, E., and Staples, R., XII Internat. Cong. Nutri., p. 45, San Diego, California, Aug. 16-21 (1981). (Abstract).
56. Cook, J. D., Morck, T. A., and Lynch, S. R., Am. J. Clin. Nutri 34, 2622 (1981).
57. Morck, T. A., Lynch, S. R., Skikne, B. S., and Cook, J. D., Am. J. Clin. Nutr. 34, 2630 (1981).
58. Morck, T. A., Lynch, S. R., and Cook, J. D., Am. J. Clin. Nutr. 36, 219 (1981).
59. Bothwell, T. H., Clydesdale, F. M., Cook, J. D., Dallman, P. R., Hallberg, L., Van Campen, D., and Wolf, W. J., "The Effects of Cereals and Legumes on Iron Availability", The Nutrition Foundation, Washington, D.C. (1982).
60. Hallberg, L., and Rossander, L., Am. J. Cli. Nutri. 36, 514 (1982).
61. Bodwell, C. E., Miles, C. W., Morris, E. R., Mertz, W., Canary, J. J. and Prather, E. S., Fed. Proc. 42, 529 (1983). (Abstract).
62. Miles, C. W., Bodwell, C. E., Morris, E. R., Mertz, W. Canary, J. J. and Prather, E. S., Fed. Proc. 42, 529 (1983). (Abstract).
63. Morris, E. R., Bodwell, C. E., Miles, C. W., Mertz, W., Prather, E. S. and Canary, J. J., Fed. Proc. 42, 530 (1983).
64. Cohen, A. M., Teitelbaum, A. and Rosenman, E., Metabolism 26, 17 (1977).
65. Sleder, J., Chen, Yii-Der E., Cully, M. D. and Reaven, G. M., Metabolism 29, 303 (1980).
66. Zavaroni, I., Sander, S., Scott, S. and Raven, G. M., Metabolism 29, 970 (1980).
67. Blakely, S. R., Hallfrisch, J., Reiser, S. and Prather, E. S., J. Nutri. 111, 307 (1981).
68. Fields, M., Reiser, S. and Smith, J. C., Fed. Proc. 42, 528 (1983). (Abstract).
69. Reiser, S., Fields, M. and Smith, J. C., Fed. Proc. 42, 528 (1983). (Abstract).

70. Beck-Nielsen, H., Pedersen, O. and Lidskov, H. O., Amer. J. Clin., Nutri. 33, 273 (1980).
71. Anderson, R. L., Rackis, J. J., and Tallent, W. H., in "Soy Protein and Human Nutrition" (H. L. Wilcke, D. T. Hopkins, and D. H. Waggle, eds.), p. 209. Academic Press, New York (1979).
72. Ferrando, R., in "Vegetable Proteins for Human Food" (C. E. Bodwell and L. Petit, eds.), p. 251. Martinous-Nijhoff/B. V. Junk, The Hague, The Netherlands (1983).
73. Goulding, N. J., Gibney, M. J., Gallagher, P. J., Morgan, J. B., Jones, D. B. and Taylor, T. G., Qual. Plt.-Plt. Fds. Hum. Nutr. 32, 19 (1983).
74. de Groot, A. P., Slump, P., van Beek, L., and Feron, V. J., in "Evaluation of Proteins for Humans" (C. E. Bodwell, ed.), p. 270. Avi Publishing Co., Inc., Westport, Connecticut (1977).
75. Liener, I. E., in "Evaluation of Proteins for Humans" (C. E. Bodwell, ed.), p. 284. Avi Publishing Co., Inc., Westport, Connecticut (1977).
76. Liener, I. E., J. Am. Oil Chem. Soc. 58, 406 (1981).
77. Martinez, W. H., in "Evaluation of Proteins for Humans" (C. E. Bodwell, ed.), p. 304. Avi Publishing Co., Inc., Westport, Connecticut (1977).
78. Ory, R. L. (ed.), Food and Nutrition Press, Inc., Westport, Connecticut (1981).
79. Rackis, J. J., J. Am. Oil Chem. Soc. 58, 495 (1981).
80. Rackis, J. J. and Gumbmann, M. R., in "Antinutrients and Natural Toxicants in Foods," (R. L. Ory, ed.), p. 203. Food and Nutrition Press, Inc., Westport, Connecticut (1981).
81. Struthers, B. J., J. Amer. Oil Chem. Soc. 58, 501 (1981).
82. Struthers, B. J., Dahlgren, R. R., Hopkins, D. T. and Raymond, M. L., in "Soy Protein and Human Nutrition" (H. L. Wilcke, D. T. Hopkins, and D. H. Waggle, eds.), p. 235. Academic Press, New York (1979).
83. Morgan, R. G. H., Levinson, D. A., Hopwood, D., Saunders, J. H. B., and Wormsley, K. G., Cancer Letters 3, 87 (1977).
84. Doell, B. H., Ebden, C. J. and Smith, C. A., Qual. Plt.-Plt. Fds. Hum. Nut. 31, 139 (1981).
85. Becker, F. F., Carcinogenesis 2, (11), 1213 (1981).
86. Yavelow, J., Gidlund, M., and Troll, W., Carcinogenesis 3(2), 135 (1982).
87. Yavelow, J., Finlay, T. H., Kennedy, A. R. and Troll, W., Cancer Res. 43(5 Suppl.), 2454S (1983).
88. Marable, N. L., Todd, J. M., Korslund, M. K., and Kennedy, B. W., Qual. Plt.-Plt. Fds. Hum. Nutr. 30, 155 (1980).

20

Soy Products in Food Service

JOSEPH RAKOSKY

INTRODUCTION

Soy products have a long history of usage in foodservice. Most foodservice operators are familiar with these products, particularly soy proteins. Although this paper is confined to soy proteins and soy lecithins, the assumption is most foodservice technologists are familiar with the proteins but not the lecithins.

Soy products are utilized to an advantage in a number of food applications. The subject is too broad to cover fully, so certain areas will be highlighted where there is the most interest. Since soy products are additives in various food systems, it would be well to make a few brief comments about additives in general.

From the regulatory standpoint a food additive must be both safe and functional. Before additives are used in processed meat products, they must be approved by the U.S. Department of Agriculture. To obtain this approval the additive must be shown to be safe and to perform a useful function, and it must be shown that the amount permitted will produce the desired effect, and that it can, not only be detected in the meat product, but its amount can be determined. This is considered essential in products such as those that have standards of identity and which have water-added restrictions. In nonspecific products such as the beef pattie, regulations state that extenders and binders may be used with or without the addition of water in amounts so that the characteristics of the product are essentially those of a meat pattie.

Other than these regulatory requirements one might ask, "Why would we want to use additives, such as soy, in various foods?" The reasons can be grouped into three broad categories: functional--the product has a particular useful property; nutritional--it provides a nutritional benefit; and economical--there is a cost saving.

What is functionality? I propose the following definition: functionality is that property of a food additive that interacts with one or more components in a food system, thereby changing the system in some way. This change may be desirable or undesirable.

In our considerations we are looking for desirable changes. However, in some cases it might be necessary to put up with some minor undesirable effects to take advantage of the one we want. Some undesirable effects that one might encounter are color, texture and,

Cereals and Legumes in the Food Supply, edited by Jacqueline Dupont and Elizabeth M. Osman

to some degree, flavor. What are some examples of functional properties? The additive can be a stabilizer or an emulsifier; it can increase solids or viscosity. (In the case of lecithin products, the viscosity of oils may be lowered.) The additive might be an essential ingredient. It might also prolong the shelf life of a product; improve the product's eye appeal; improve texture and even flavor. The additive may also act as a binder of water, fat or even particles.

Nutritional benefits may be realized when fat is replaced, thereby lowering calories. The protein level can be increased and, in many instances, protein quality can be improved. An improvement in the nutrient profile of the food preparation can be obtained through the contribution of various nutrients such as vitamins and minerals.

In looking for economic benefits we are looking for ways to lower costs, hopefully without lowering product quality. Costs can be lessened by a partial or complete replacement of an expensive ingredient. In some cases lower costs can be achieved through processing improvements, as well as a reduction in the rework load.

Ideally, it would be desirable to use an additive that has all of the above-named attributes, at least the ones wanted. Unfortunately, this is not possible. Each additive has its own set of attributes that emerges under the processing conditions used. What is to be realized is that trade-offs must always be considered. With the trade-off, the questions is asked, "What is it worth to have one attribute at the expense of another?"

The food processor desires to make a product that the consumer will like and buy. The consumer needs to be satisfied with appearance, texture, taste and flavor. In the case of an extended meat product, most people feel that the all-meat product is best, particularly if it is composed of prime cuts. At the other extreme are cuts that may be tough, fatty, high in collagen or, for some other reason, would make formulated products highly undesirable. Such products cost less. If the economic pressures are on the producer, which they usually are, he/she will attempt to reformulate with different cuts to arrive at some quality level for a particular price. The producer can only go so far with this approach.

An alternative is to use additives such as cereal binders. Although they will reduce costs, will they add to or detract from product quality? Are there any other benefits, other than price, in using the additives? Through trial and error the foodservice operator learned that soy protein products not only lowered costs but contributed to product quality.

Compare the foodservice operator with the homemaker in looking at economic benefits of using protein additives. In the early days when the price differential between an all-meat product and an extended product was only a fraction of a cent per pound, the foodservice operator became interested, but the homemaker did not, because, in her eyes, the saving was insignificant. This was not the case for the foodservice operator who was dealing in large volume. Over a period of a year a fractional saving amounted to many thousands of dollars. This was the incentive for the foodservice operator to continue using the additives and to learn how to use them effectively. As the price differential between the meat and the additive got to be greater, so did the cost savings. At what point did the homemaker become interested? We found that the homemaker became interested when the price differential was about 10 cents per pound. When it got to 20 cents there was definite interest. We can be sure that such a saving for foodservice operations is even more attractive.

With this background, we turn to soy products, first to characterize them and then to see how they can be used in the various food applications in foodservice. I will discuss soybean processing only a little because some of this is covered in other chapters. I will cover only those aspects that I feel are necessary to help make decisions in their use.

SOY PRODUCTS

Soybeans are composed of three main fractions: 18% lipid material or oil, 38% protein and 30% carbohydrate. Of the three fractions, soy carbohydrates have little or no commercial value in human foods except for the fiber content. There is some value for the carbohydrate as an energy source found in the sucrose portion, which is about 5% of the whole soybean (1). As part of the oligosaccharide portion of the soybean, beside the sucrose are stachyose (3.8%) and raffinose (1.1%). These latter sugars are not digested but, are attacked in the lower digestive tract by bacteria with the productions of gas. These two sugars are considered to be the flatus factors of soy.

Lecithin

Although Dr. Erickson presented soybean oil and its composition in Chapter 4, I would like to point out that the lecithin content of soybeans is about 0.5%. On an oil basis this is about 2.8%. Lecithins are referred to as gums from the standpoint of oil refining. The gums are removed either in the early stages of oil refining by a hydration step or they are removed along with the soap stock in the alkali refining step. In the latter instance the lecithin is not isolated. When crude soybean oil is hydrated, lecithin comes down as a gum. It is then dried and often processed further into upgraded lecithin products. The composition of a crude or natural grade lecithin is a mixture of 15% phosphatidylcholine, 13% phosphatidylethanolamine, 9% phosphatidylinositol, 5% phosphatidic acid, and 2% phosphatidylserine (2). *Crude lecithin* can be refined into upgraded products having varying properties. Szuhaj (2) has classified these products into six types.

Clarified lecithins are essentially filtered products.

Fluidized lecithins are made more fluid by the addition of calcium chloride, fatty acid, oil or some other type of carrier.

Compounded lecithins are made by the incorporation of certain additives to give them special properties. These additives may be other surfactants.

Hydroxylated lecithins are modified products which are highly dispersible in water. They are made by a reaction of hydrogen peroxide and lecithin in the presence of a weak acid.

Deoiled lecithins are essentially freed of oil through the use of warm acetone. The products are almost pure phosphatides and are usually granular in appearance.

Fractionated lecithins are products obtained from a natural grade lecithin or a deoiled product using alcohol as a solvent. This results in two fractions alcohol insoluble and alcohol soluble products. As might be suspected, these lecithins have special functional properties.

Soy Protein Products

Soy protein products may be derived from two soybean fractions. One is the dehulled soybean cotyledon, often referred to as a fullfat soybean chip; the other is from defatted meal or flakes. In the former instance, the full-fat soybean chips are heat treated and ground into a flour resulting in a natural full-fat soy flour. This product has a protein content of about 42% and a fat content of about 22%. Little if any of this type of product is made in the U.S. except on special order. It has been used in special feeding programs in which both protein and calories were an important consideration.

Most of the soy proteins available today in the United States are derived from defatted flakes. Right after extraction, during and after the time the solvent is removed, the flake is subjected to varying amounts of moist heat to give the product certain properties. As might be suspected this treatment has an effect on the protein in that there are various degrees of heat denaturation taking place. If little or no moist heat is used the protein in the flake is highly dispersible in water, while at the other extreme strong or prolonged heat treatment will denature most of the protein so that little of it will disperse in water. This heat effect can be assessed in an indirect way by determining the percent of protein that will disperse. The results may be noted in terms of percent protein or percent nitrogen. In either case we talk about the protein dispersibility index (PDI) or the nitrogen dispersibility index (NDI). Type 1 is subjected to little or no moist heat, as a result it has a NDI of 70% to 90%. Type 2 is obtained by exposing the defatted flake to a light heat treatment. Its NDI ranges from 50% to 70%. Type 3 receives a mild treatment resulting in an NDI between 25% and 50%. Type 4 is the result of heavy moist heat treatment. This product has a NDI below 25%, more often than not, around 13%.

Moist heat treatment has more of an effect on the product than just dropping the NDI. Some fundamental knowledge is needed in considering these products for use as food additives. Moist heat will affect the product's functionality, its enzyme activity, its nutritional properties, its color, flavor, and even product yield in certain processes. This heat treatment actually predetermines the product's functional characteristics. These soy flours and grits are rarely referred to as types 1, 2, 3, and 4. The usual referrals are by their NDI or PDI, or by the names enzyme active, white, cooked, and toasted. Although we talk about a toasted product this ia a misnomer in that the product is subjected to moist heat, not dry heat, as is usually the case in toasting. It is called a toasted product because it has the appearance of a toasted product, i.e. brown in color. The importance of this will be brought out when soy product applications are discussed.

Three basic soy protein products are derived from the defatted soybean flake. They are soy flour and grits, soy protein concentrate, and soy protein isolate. The protein level for the soy flour and grits is about 50% to 55% on an as-is basis. To qualify as a soy protein concentrate, the protein level must be at least 70% on a moisture free basis. For the isolate it is 90% minimum protein on a moisture free basis.

The basic soy proteins may be described as follows: The first is soy flour and grits. These products are nothing more than defatted flakes that are ground to size. The finely ground product, 100 mesh or finer, is a soy flour. The coarse ground product, coarser than 100 mesh is a soy grit. As might be suspected the flours and grits retain the characteristics of the moist heat treated flake from which they are derived.

Recall that there is little if any natural full-fat soy flour produced in the U.S. There are other products being produced that are similar in many respects. These are called refatted or lecithinated soy flours. Such products have different functional characteristics from the regular soy flour. Also, they are considered as dry carriers of fat and/or lecithin. There are two types. One is blended with soybean oil containing about 2.5% lecithin; the other is a soy flour blended with a 50/50 mix of soybean oil and lecithin.

The soy protein concentrate is made from the defatted flake by an extraction process that is somewhat selective in that the protein is left behind together with other insolubles. There are a number of ways to make such products. One is to heat denature the protein and then extract with water. Another is to extract with an isoelectric wash. A third way is to wash with an aqueous alcohol solution. Soy protein is not soluble or even dispersible in the solvents used for the latter two processes just described, however, the sugars and other minor solubles are. There is a fourth method used that is similar to that in making the soy protein isolate. This method involves extracting the protein and the sugars together, separating the spent flake from the solution, precipitating the protein and removing the sugars, and then recombining the protein (either in the isoelectric state or as a proteinate) with the spent flake.

The third basic soy protein product is soy protein isolate. Since this product is discussed by Dr. Richert in Chapter 6, I will not further discuss its preparation.

There is another group of protein products that is derived from the three basic proteins that are textured. The products are referred to as textured soy flour, textured soy protein concentrate, and textured soy protein isolate. Without giving details, these products are produced by an extrusion cooking process. Another textured product is produced from the isolate by a spinning process. Spun fiber is produced and used in limited amounts.

The extrusion cooking process gives one a fair amount of latitude in producing a number of products. They may be of different sizes, shapes, colors, and textures. Some are expanded, compacted and even striated to look like meat fibers. Others are produced to look like nuts or dried vegetable products.

There are a number of soy protein products that are modified by a chemical or enzymatic hydrolysis that gives them special properties ranging from flavoring additives, e.g. hydrolyzed vegetable protein or soy sauce, to whipping agents and/or egg white replacers.

FOODSERVICE APPLICATIONS FOR SOY PRODUCTS

In considering foodservice applications for soy products, the discussion will be retricted to lecithin and soy protein.

It was stated earlier that there are three reasons for using additives in a particular food system: functional, nutritional and economic. It is conceivable that in a particular application, all three reasons will be realized as benefits. However, in many cases this will not be the case. It will be necessary to evaluate one benefit over another and, in some instances, consider certain disadvantages that will crop up. Hence, in using an additive a judgment must be made on whether or not there is enough merit for using it. Obviously, an overriding factor is economic; what is it worth to have certain benefits?

Lecithin

The composition of lecithin is quite complex in that it is composed of mixtures of the various phosphatides. A simple analysis does little to properly identify a particular lecithin product. To do a proper job in characterizing the product, it should be checked in a system similar to the one in which it will be used as well as subjected to certain laboratory analyses. One of these analytical methods involves the use of acetone. Since soybean oil or any triglyceride, fatty acid or sterol is soluble in acetone (whereas phosphatides are relatively insoluble) a method can be used to determine how much phosphatide is present in a lecithin product. (Recall that acetone was used to produce a deoiled or oil-free product.) A typical natural grade lecithin will have an A.I. (acetone insoluble) content of about 62%.

Lecithin has a number of interesting properties. It is a surfactant, an emulsifier, an emulsion stabilizer, and a pan release agent. It is an antispatter agent when it is used in margarine where it also functions as an emulsifier. The antispatter property is important in margarine because it allows one to fry in margarine similar to the way it is done with butter. Depending upon the grade of lecithin used and upon the degree of emulsion needed, the recommended level of usage ranges from 0.15% to 0.5%.

Lecithin also has the property of dropping the viscosity of an oil. This is an important functional property for chocolate coatings of candy centers and ice cream bars. In the case of enrobed candy centers, the chocolate coating needs to flow evenly and smoothly in a relatively short time. This requires low viscosity coating. Cocoa butter gives the coating the desired viscosity, but it is expensive. A less costly substitute is made from coconut oil or palm kernel oil, but their viscosities are too high. When lecithin is added to the substitute, however, the viscosity drops to a desired level. Lecithin is used at levels ranging from 0.3% to 0.5%. Higher levels are used in ice cream coating to improve moisture tolerance.

Lecithin is of value in many foods where emulsification, stabilization, wetting, instantizing, suspending, or surface active properties are important. In dairy-type applications lecithin helps to disperse malted milk solids and helps emulsify and stabilize fat solids in instant chocolate drinks. It also acts as stabilizer and emulsifier in ice cream.

In baking applications lecithin will improve dough tolerance and shelf life in yeast-raised goods, improve the eating qualities in prepared cake mixes and cake doughnuts, lessen fat absorption in doughnut frying, and will function as a pan release agent. I noted earlier that the natural grade lecithin has all the various phosphatide fractions and thus should have most of the properties one would expect of such products. So, how does one go about selecting the proper product? When little if any information is available as to which product to use, one might look first at a natural grade lecithin. Trial usage levels range from 0.25% to about 3.0% based on the oil in the formulation. In some applications the viscosity of the lecithin product is an important consideration, perhaps from a handling standpoint. Should it be plastic or fluid? This selection would depend on how it is used in the process. Should the A.I. be low or high? The selection of a natural grade lecithin often depends on how much soybean oil the finished product can tolerate. In some cases it might be better to use a product that has a carrier other than soybean oil.

Another important factor to be recognized is color. Will the dark color of the natural grade product cause a color problem? If so, then perhaps a single or even a double-bleached product might be used.

Although the natural grade of lecithin will function well in many applications, greater efficiency and better results may be achieved with one of the specialty products designed for specific applications. Some products are essentially hydrophilic, others lipophilic.

Obviously, the best product is the one that gives the desired results. It must be stressed that because good results are obtained from a particular lecithin product, we cannot assume that similar results will be obtained with a similar product from another company, even if the A.I. is the same. The best evaluation of various products between companies is to test the product in a particular application.

Soy Proteins

Of the three reasons for using an additive in a food system, the one reason that will trigger a search for an alternative is the rising cost of one of the ingredients in the present formulation or recipe. In most cases that ingredient is an animal product derived from egg, milk or meat. For this reason some of the earliest uses for soy proteins were: 1) in processed meat products--to extend or partially replace the lean meat; 2) in baking--to replace the nonfat dry milk and, in certain doughnut products, to partially replace the egg yolk solids requirement; and 3) in dairy-type products--to replace some of the milk products used. There are other situations in which these products are helpful when used to partially replace a scarce ingredient and/or an expensive one, or to help perform a particular functional job. Soy proteins can be used to partially replace the durum semolina in pasta or to help in the browning reaction. The major concern in this paper is processed meats with less attention to baking and dairy-like applications. The advantages in using soy proteins in these areas, should suggest some ideas that will apply in other food systems.

<u>Processed meats</u>. Various processed meats are used in foodservice such as hamburgers, a beef/lamb combination for the gyros, frankfurters, sausages, luncheon meats, pumped meats, etc. It is beyond the scope of this paper to get into each of these; instead I will describe only one application, the beef pattie, which probably is of greatest interest to foodservice technologists. What is presented can be used as a guideline in considering soy proteins in other processed meat products.

According to USDA regulations ground beef, chopped beef or hamburger cannot contain anything but meat and a little seasoning. No other additives are allowed, not even added fat, and the fat level cannot be any higher than 30%.

On the other hand, the beef pattie (or simply a pattie) can contain additives such as soy protein products and water of hydration, but only in amounts such that the product's characteristics are essentially that of a meat pattie. It is this type of product that I will describe in some detail. It is a popular item in the institutional trade.

Recognizing the fact that economic pressures cause the processor to use soy additives, the processor wants other benefits as well, but above all the product has to have good consumer acceptance. The additional benefits can be nutritional and/or functional in nature. When soy products such as soy grits, soy protein concentrate granules, and various textured soy proteins are used they do something more than simply extend meat. Soy protein has an affinity for water and/or meat juices, and because of this the patties do not shrink as much on grilling as the all-meat product. This attribute makes the pattie more juicy and flavorful. Considering the nutrients in the juices being lost on the grill, their retention makes the pattie more nutritious. It seems that this affinity for juices is somewhat selective in that the fat is not retained. Because of this affinity for juices or water, it is necessary for the protein product to be properly hydrated previous to its incorporation into the ground meat. If too little water is used, the pattie will be dry; if too much, it will not have the proper consistency and will appear watery. More will be said about proper hydration below.

Functions. In some people's eyes, because soy is giving such good results, "if little is good, more should be even better." This definitely is not the case. Soy products are essentially bland, or at least they do not taste like meat. Low level usage is acceptable because of the retained juices that contribute to some flavor. When one goes beyond a certain level this flavor dilution effect must be taken into consideration. Normally, a 10% extension will require nothing more than a little salt and pepper. With higher amounts, between 10% and 20%, I recommend in addition to the salt and pepper a little garlic and onion powder. At levels from 20% to 30% one should also include something like monosodium glutamate, protein hydrolysates, etc. With additive levels above 30% it is necessary to add meat-type flavor to help compensate for the flavor dilution effect.

The so-called secret in using soy products is this: first learn to use the products at low levels before going to higher levels. At low levels mistakes are minor with little, if any, adverse effects. At high level usage mistakes are greatly magnified.

Next, proper hydration is necessary, both as to the amount and the time to allow the moisture to be absorbed properly. It is well to hydrate the protein product so that the protein level approaches that of meat i.e., about 20%. This means for a 50% soy protein product such as soy grits and textured soy flour one part of soy should be hydrated with 1.5 parts of water for a total of 2.5 parts (50/2.5 = 20). For a concentrate of about 66% (as-is basis) it should be one part soy to 2.3 parts of water giving a total of 3.3 parts (66/3.3 = 20). In the case of an isolate having 90% soy it is one part soy to 3.5 parts of water for a total of 4.5 parts (90/4.5 = 20). (These soy-to-water relationships are discussed in economics.)

The next important thing to do in using soy in ground meats is to be sure to mix the hydrated product with a coarse ground meat and then grind the mixture together using a smaller plate opening. This assures one of a better blend, giving the best results in appearance and taste.

Lastly, one should not char cook the blended product because burnt soy protein will result in a burnt cereal flavor. Many people believe that the blended product should be grilled for a slightly shorter time than an all-meat item for best results.

Economics. So much for functional reasons, what about economic reasons? In working with these products one cannot make an initial judgment based on raw material costs alone. It is entirely possible that a high-cost additive may result in the lowest formulation cost. Hence, judgment should be withheld until all the facts are in and you have looked at the cost of the finished product.

To fully appreciate the economic advantages soy proteins can have when used in certain meat applications, take an example taken from a school lunch situation. A school lunch meat portion must contain 2 oz. of cooked lean meat. If the meat portion is a pattie, 17 pounds of ground meat are needed for 100 portions, taking into account a cooking shrinkage of 25%. If the meat costs $1.24 per pound, then the total cost for the 17 pounds is $21.08 or a little over 21 cents per portion. If there is an extension of 30% with a hydrated soy product, the meat cost would be $14.76 (11.9 pounds times $1.24). The cost of the textured soy flour (TSF) is about 35 cents per pound in quantity. A thirty percent extension, in this example, requires 5.1 pounds of hydrated TSF. With a hydration ratio of 1.5 to he cost per pound of the hydrated product is 14 cents or $0.714 total for the addition. The total cost for the blend is $15.47 which is 73.4% of the all-meat product. Each portion is now about 15.5 cents as opposed to 21 cents for the all-meat product.

Another interesting approach is to keep the 17 pounds of meat constant and then add the hydrated TSF so that the 70/30 ratio is still maintained. The required amount of hydrated product for this extension is 7.3 pounds, which adds $1.02 to the price of the meat. With this extension we now can feed 43 more children. This makes sense, but it only makes sense if the product is properly made and the children will eat it. This is why a proper job in making the extension must be done. If it is not and consumers develop an aversion to the extended product, we would have a difficult time winning them back, even if we eventually made an excellent product.

Next, consider the various soy proteins and make some economic comparisons. Soy grits can be used to extend ground meat as can textured soy flour. Both have the same protein content and about the same hydration requirement. So why not use soy grits, which are about one third less in cost? The answer is that soy grits can be used, but the use of grits is limited because of flavor and texture.

What about using a higher priced item such as a textured concentrate? Just on the basis of hydrated products alone, as to the amount of water the products take up, the cost relationship would be as follows: a textured concentrate could cost about 32% more than a TSF, and a textured isolate could cost 80% more. In comparing the concentrate with an isolate, the isolate may cost about 36% more. But these comparisons are not the whole answer, we must look at the finished product and then decide what the costs are compared to the various advantages that are observed. Another important consideration is that some types of products will hold more juices.

Sometimes frozen meat will lose much of its binding power, making it difficult for the pattie to hold together. The use of salt is a key factor in adjusting the degree of binding needed. Perhaps the addition of salt is not desired. In this case, we can take advantage of the binding properties of a functional concentrate or isolate which may be used in combination with the hydrated soy product.

A number of processed meats can utilize soy protein products. There is another area that also has application for textured products: topping-like products for salads, potatoes, pizzas, and the like. The foodservice operator should be aware that textured products come in various sizes, colors and even flavors. There are differences in the way they take up the water of hydration. There are some that one cannot, or would not want to eat as such because they are too hard and need to be hydrated. In the case of a bacon-like product for salad topping, it requires the opposite property--the textured product is eaten as is. It has that desirable crispy bite and taste one looks for in such products. Realizing these differences, it is relatively easy to find a product that will suit the food technologists' requirements.

Nutrition. What are the nutritional reasons for using soy proteins? As was pointed out earlier, the products can lower the calorie content if fat is replaced; protein quality can be improved, especially when used with cereal products; and its nutrient content can contribute to the overall nutrient profile. Probably, everyone agrees that soy protein is one of the best of all vegetable protein products, as far as protein quality is concerned. We know how valuable soy protein is in animal feed. Because Dr. Bodwell covered the nutrition of soy proteins in Chapter 19, I will forego further discussion.

Other applications for soy proteins in foodservice. I discussed the effect moist heat had on the defatted flake, particularly on enzyme activity, trypsin inhibitor and flavor. It also affects product functionality. Functionality can be altered somewhat in that the protein structure is altered in respect to its relationship between active sites, crossbonding, coiling, etc. For this reason, users of soy flour should know what type of soy flour they are using, because the type of flour used will have an influence on its performance. Whenever a recipe calls for soy flour it should also make some reference to the type of soy flour needed. If it does not, the question should be asked, "What should the proper PDI or NDI be for this particular use?" The importance of this was impressed upon the author a number of years ago when certain users of soy flour insisted on obtaining one that fell within a narrow PDI range. When it did not, poor results were obtained. This happened to be a baking application.

High PDI soy flours are not recommended in food systems that will receive very little cooking because of flavor, enzyme and trypsin inhibitor activities. In such systems it is better to use a product that has a lower PDI. In baking it is proper to use a higher PDI flour, because in baking the dough, the soy flour receives the proper amount of heat treatment to destroy the enzymes, and the trypsin inhibitor and to improve the flavor.

One of the largest uses of soy flour today is in baking where it is usually combined with cheese whey as a nonfat dry milk (NFDM) replacer. This use became popular when the cost of NFDM became prohibitive. The soy flour normally used in this application has a PDI ranging from about 60% to 70%. The combination, which is usually spray dried, has a protein level similar to NFDM i.e., about 35%. There are combinations available that have different protein levels and that may be composed of other types of soy protein. Bread standards permit the use of non-wheat flour products such as NFDM, soy flour, and the like in the bread formulation. The maximum is 3% based on the weight of the wheat flour. If the replacer has a 35% protein level, usually there need be no adjustment in the recipe. If the protein level is higher or if soy flour is used with the whey, it is necessary to adjust the formula by using more water. If water is added to dough based on feel, this should not be a problem because the proper amount will be used. If the feel method is not used, a good guideline to follow is to add about 1.5% to 1.75% water for every percent of soy flour used. It is suggested that the amount of oxidant be increased slightly.

The advantages in using soy flour in this manner are: it has a tenderizing effect, it will increase shelf life and improve the nutritional benefits, it has increased absorption which results in increased yield, and it will cost less.

Soy flour may be used in specialty breads at levels higher than 3%, but these need to be labeled as specialty breads. If a high protein bread is being made in which 12% soy flour is incorporated, it is necessary to use a special emulsifier, such as sodium steroyl-2-lactylate or ethoxylated monoglyceride, as well as an increased amount of oxidant.

Enzyme-active soy flour may also be employed in bread, but its addition is restricted to a maximum of 0.5%. The active enzyme in this soy flour is lipoxygenase, which, as the name suggests, oxidizes fats. In oxidizing fat, a flavor is imparted that many people like. More importantly, lipoxygenase bleaches carotenoid pigments in the wheat flour.

Lecithinated soy flour can be added to a pancake mix at a level of about 3%. When this is done three functional advantages are seen: the pancake has a light fluffy texture, it is golden brown in color, and because of the presence of lecithin the pancake does not stick to the griddle on frying. The golden brown color is due to the rich source of lysine in soy flour that takes part in the Maillard reaction.

A similar use is made of lecithinated soy flour in cake doughnuts in which the product has a sparing action for at least half of the egg yolk requirement. The advantages noted are: better browning, lighter texture, fewer cripples, better star formation, less fat absorption, prolonged shelf life, and a cost saving in the amount of egg yolk solids not used.

Mr. Leviton discusses tofu and soy milk preparations in Chapter 7. These products are the first known imitation dairy products. From the foodservice standpoint, it is to be recognized that many university cafeterias are now serving such products to students who are on vegetarian diets.

Soy protein products, especially functional concentrates and isolates, can be used effectively in producing a number of imitation dairy products such as coffee whiteners, milk-like beverages, diet drinks, frozen desserts, dips and spreads, whipped toppings, fermented drinks, and even imitation cheese products. In most of these applications milk products such as nonfat dried milk, cheese wheys, casein, etc. are used to make these products. Certain soy proteins can replace the milk proteins either completely or, at least, partially.

An important point to remember is that it is virtually impossible to produce a soy protein that will completely replace a milk product in all applications. If it did, its protein structure would need to be identical to the milk product. What we can do is use one type of soy protein for several applications, and another type in others. This also applies to similar products made by different companies. It is conceivable that if one company's product does not function properly in some application, another company's product might perform ideally.

SUMMARY

In this brief review I present some background information and cover a number of applications in which soy products are used to an advantage. These products offer three benefits that are broadly classified as functional, nutritional and economic. Guidelines are given in the selection and use of the various products as well as cautions. It is important to realize that there are subtleties in using the products that, if not taken into account, can lead to less than desirable results.

There are differences in the products that are available and there are sometimes subtle differences in products from different companies. A little background, such as is covered, should help the food technologist obtain optimal results in the use of these additives.

It is suggested that you seek the help of those who supply soy proteins and lecithins in utilizing their products. They have the experience and the knowhow to offer suggestions, and in many cases offer you technical help.

REFERENCES

1. Smith, A. K., and Circle, S. J., in "Soybeans: Chemistry and Techology. Volume 1 Proteins." The AVI Publishing Co., Inc., Westport, Connecticut (1972).
2. Szuhaj, B. F., JAOCS 60:306-309 (1983).

V

Consumer Acceptance and Regulations

21

Cereal and Legume Research: Economic Implications for Consumers

JEAN KINSEY

THE ROLE OF CEREALS AND LEGUMES IN FOOD CONSUMPTION

In 1981, U.S. consumers spent $329 billion on food; 11% of that expenditure was for grains and bakery products (1). The 1977-78 National Food Consumption Survey of U.S. households showed that in a given week, 41 cents out of each dollar spent for food at home was spent for cereals, mixes, bakery products, beans, lunchmeats, and meat extenders (2). Looking at the nutritional value of foods consumed by individuals in those households revealed that 22% of their protein and 28% of their calories came from legumes or grain products (3). The role of some cereals and legumes in the American diet is increasing while the role of others appears to be decreasing. Nourishment from these sources is not always readily identified because cereals and legumes enter the diet in multiple forms and at multiple stages of processing. Sometimes they enter under their own identity such as for rice and sometimes they are disguised, as in bologna, and oftentimes they are one of many ingredients in a complex product like bread. Cereals and legumes are treated to milling, bleaching, extruding, toasting, puffing, isolating, soaking, drying and refrying, among other things, before consumers encounter them as edible food. Consciously or not, however, they continue to provide a stable, relatively inexpensive source of calories and protein for consumers around the world with some promising new developments in process.

Technological innovation in the cereal and legume industry is a long and arduous process beginning in chemistry and biology laboratories where breeding technologies are developed and applied and more recently, genes are isolated, examined and spliced. Biotechnologies that lead to greater crop productivity are one phase of the total innovative effort in food and agriculture. Food processors are continuously trying to design new and better foodstuffs that will not only bring them profits but will fill the needs of consumers. Consumers' food needs consist not only of nutritional fuel but of satisfactory taste, variety, and convenience for their life style all at a price they are willing and able to pay. Since food habits are always culture bound and difficult to change, a critical extension of technological changes in food production and processing is an examination of what foods consumers want, how they can best be informed

Cereals and Legumes in the Food Supply, edited by Jacqueline Dupont and Elizabeth M. Osman © 1987 Iowa State University Press, Ames, Iowa 50010

about new choices, and how they can best be protected from current and yet unknown hazards to their health and safety.

Consumers' welfare is ultimately changed by how well their preferences are met by the products they consume. Potential changes in consumers' welfare resulting from activities in the cereal and legume industry will be the focus of this paper. First, evidence about consumer preferences for cereal and legume products will be briefly reviewed. The effect of innovations in production and processing on consumers' nutritional and economic well-being will be followed by a discussion of the economics of safety and information as it applies to product labeling. We will see that price effects are critical and that changes in consumers' welfare will depend upon research priorities and the dissemination of accurate information at all levels of the market.

Consumer's Preferences

Looking at trends in consumption of cereals and legumes in the United States between 1960 and 1981 (1) shows that per capita consumption has decreased for beans, peanuts and soy products, dry beans and dry peas with dry beans leading the decline (Table 21.1). Peanuts account for much of the decline in the first category. A study of the food service industry indicated a 20% drop in the quantity of peanuts served in bars and restaurants between 1969 and 1979, while there was a 530% increase in other snacks (4). In 1980 alone, expenditures on snack foods increased 12% to over $10 billion which is about 3% of all food expenditures (5). Crackers, a subgroup of snack foods, is reported to be the fastest growing segment of the cereal industry (6). With respect to major categories of cereals and legumes listed on Table 21.1, corn is the clear winner with a 154% increase in pounds per capita. This is attributed primarily to the recent innovation and quick industry adoption of high fructose corn sweeteners. Corn sweeteners have been relatively successful because they are competitive on price and have a reputation for having health advantages.

Recent studies of consumer food preferences in the United States (7) and in Europe (8) show that consumers prefer foods that are healthy, natural, and safe and they are willing to pay more for such foods.

TABLE 21.1. Percentage Change in Per Capita Consumption of Cereals and Legumes, 1960-1981, for United States

Food Type	Percent Change in Pounds per Captia	Pounds per Capita in 1981
Beans, Peanuts, Soy products	-9	15.0
Edible Dry Beans	-44	4.1
Dry Peas	-30	0.4
Cereal and Flour	+3	151.3
Fats and Oils--Vegetable	+67	48.4
Sugars and Sweeteners	+24	134.9
Corn Syrup	+439	55.0
Dry Corn Sweeteners	+284	44.6
Rice	+80	11.0
Corn	+154	120.5

How consumers define healthy, natural and safe is not always obvious. In Europe, "healthy" meant foods that were associated with high protein, high fiber and low calories. It also implied an absence of excessive amounts of refined white sugar and preservatives or artificial coloring. Europeans considered healthy foods to include yogurt, protein fortified snacks and pasta, low calorie salad dressings, bran and brown bread. The retail price differential between white and multigrained bread in England reflects a willingness to pay over 13 cents more for the same size loaf of a so-called healthy bread (8). In the United States, "natural" foods are identified as those which have minimum additives and are minimally processed. Confidence in processed foods has diminished partly, it is thought, because of extensive advertising promoting "natural" foods which leaves the impression that processed foods are unnatural and therefore inferior (7). On both continents consumers were vitally concerned with avoiding excess calories, harmful ingredients and hazardous substances. The latter are universal issues whereas only in highly developed countries are excess calories likely to be a problem. Useful technological innovation in food stuffs for post-industrial consumers might lie in finding ways to provide food with more bulk as well as more nutrients per calorie.

Studies of consumer food preferences in developing countries are less readily available but experience has shown that they, like us, prefer to eat their traditional foods. Educating individual households to use new foods is tedious but it may be the only way to influence the adoption of new food technology. The American Soybean Association is trying this approach in Mexico in an attempt to introduce high protein soy flour into home baked tortillas (9). Mexican families reported liking the soy flour tortillas. A study of Columbian consumers showed, however, that a favorable price ratio was not enough to induce large numbers of Columbians to purchase combination soy-meat products even though basic meat requirements were not being met by their normal diet (10). Traditional foods are terribly important to people's sense of security and identity. One of the major concerns with high technology food research is that it will not be applied to improving crop yields and nutritional content of foods that people actually want to eat but will be applied only to crops with high dollar value for commercial agriculture and international trade (11,12).

Innovation and Consumers' Welfare

Economic welfare. One goal of plant breeding and more recently of genetic manipulation is to engineer edible plants with more desirable characteristics. Biotechnology applied to food and agriculture holds promises of increased crop yields by making plants resistant to drought, herbicides, salt water, viruses, heat and cold. Plants may also be designed to have multiple flowers, stems and seeds and may be adapted to new environments such as tropics and seacoasts. Building in nitrogen fixing mechanisms can decrease fertilizer costs while insect resistance reduces costs of pesticides. Engineering for specific characteristics of the final food product is also under way. Tomatoes, for instance, appear to be the "white mice" of agricultural biotechnology as they are subjected to genes that make them pulpy for catsup, hard for shipping, resistant to cold temperatures for northern climates, and uniformly ripe for harvesting (13). Many consumers would argue that tomatoes seem to have genes that specialized in everything except flavor. Increasing the protein quality in rice and corn are examples of improving nutritional characteristics of food crops wich could be of great benefit to worldwide consumers. Other examples of things made possible through this research are fast rising yeasts, amino acid sweeteners, increased gluten strength, hard kerneled corn and edible protein from tobacco leaves.

Will consumers benefit from this type of research? Will they face new risks? The answer to both questions is yes. Menz and Neumeyer (14) argue that "the consequences of genetic engineering for commercial maize production are likely to be less in the direction of yield enhancement and more in the direction of cost reduction". To the extent that new plant characteristics decrease the costs of production, food prices should decrease over the long run. Declining food prices releases part of consumers' budgets for other goods and services and improves their well-being. The economics of this is illustrated in Figure 21.1 with the supply and demand curves for a genetically improved crop. Before improvement, consumers paid price P_0 for quantity Q_0. After improvement, decreased production costs shift the supply curve to S_2 (more can be produced at the same cost) and consumers now pay a lower price P_1 for a larger quantity Q_1. The area above the price line and below the demand curve is called consumer surplus; it approximates the value of the utility consumers receive from purchasing a given quantity in excess of what they actually spend. At the old price, P_0, consumer surplus was equal to area a and at the new lower price, P_1, consumer surplus is area a+b+c+d. When consumer surplus increases, consumers are better off (ceteris paribus).

Increased yields of agricultural products have served consumers well over time by making greater quantities of foods available at lower prices. This not only tends to improve health and nutrition, but increases consumer surplus. This can also be seen on Figure 21.2 where consumers originally pay price P_0 for quantity Q_0, the price and quantity where supply equals demand. Technological advances in agricultural crops have the effect of lowering the market clearing price to P_M for quantity Q_M and consumer surplus increases by the area designated b+b'+c+d+d'+e+f. Unfortunately, in many cases, price P_M is not high enough for farming to be a profitable business given the number of resources devoted to agriculture, in particular the number of farmers. For a variety of social and political reasons, the government often wishes to slow the rate of out-migration from rural and farm sectors. Subsidizing farm production is one way to slow that out-migration. Suppose the government raises the price farmers receive to P_T. This induces them to stay in agriculture and increases their output along supply curve S_2 up to quantity Q_T. At price P_T consumers will purchase quantity Q_D and relative to their position at price P_M they will have lost consumer surplus equal to area d+d'+e+f. They pay d+d'+e more for quantity Q_D; area f is an allocative loss because P_T is above the market clearing price, P_M. The difference between the quantity purchased by consumers, Q_D, and quantity supplied, Q_T, is subsidized by a government transfer of the difference between P_T and P_M and is equal to the darkened areas f+g+h. Assuming, first, that the surplus quantity sells in the export market, the net social cost of this subsidy is equal to area f. Area f is the result of subtracting the loss in consumer surplus plus the subsidy payment equal to areas d+d'+e+2f+g+h

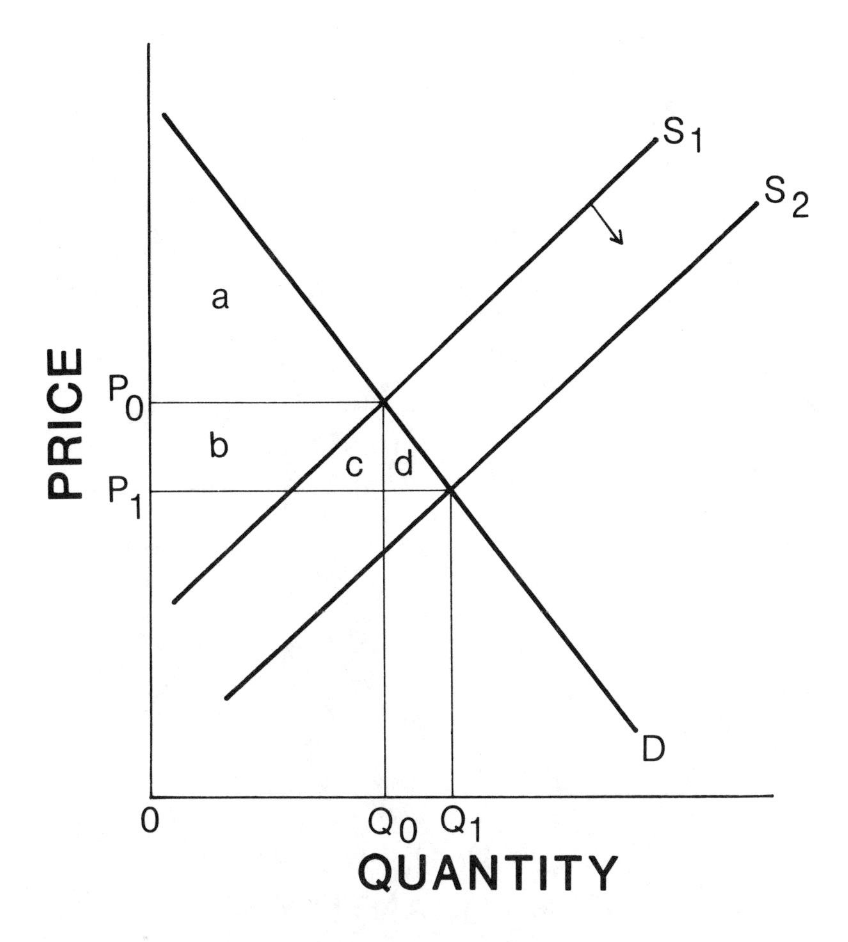

FIG. 21.1. Supply of genetically improved crop and changes in consumer surplus.

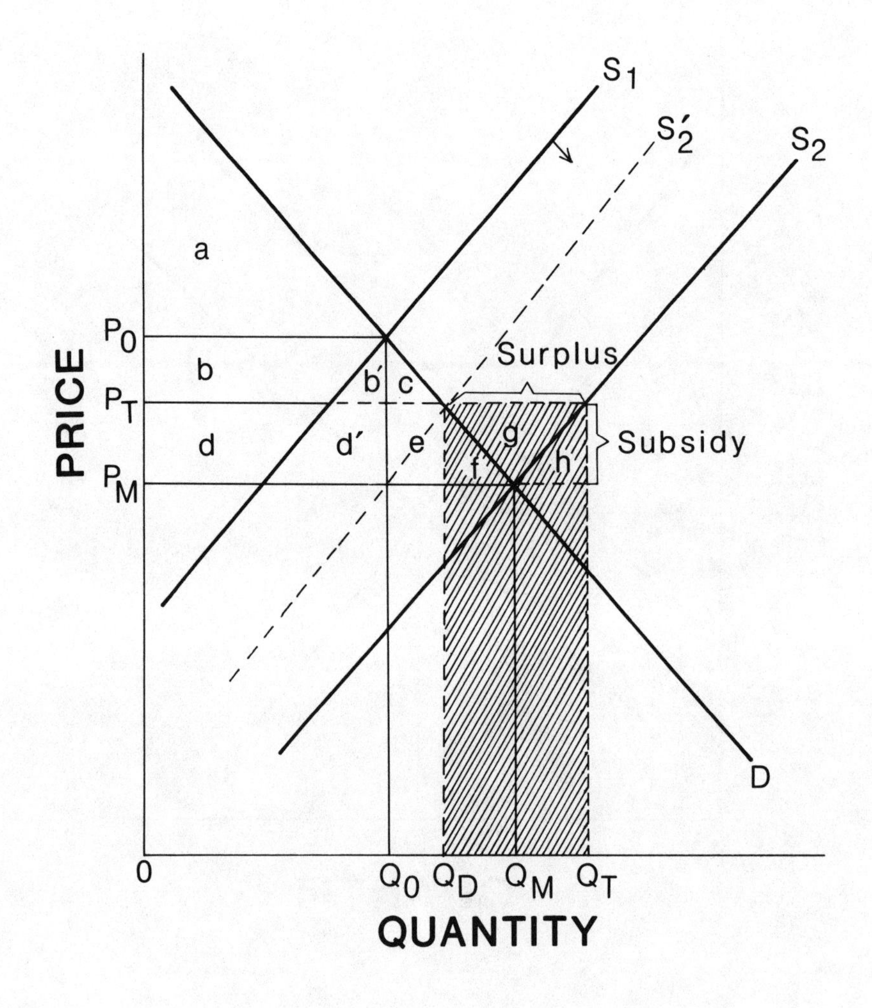

FIG. 21.2. Social costs of surplus agricultural products.

from the gain in producer surplus which is area d+d'+e+f+g plus a cash payment equal to area h.

Now, assume that the surplus does not sell on the export market, but is bought by the government. In that case the government transfer equals the entire darkened area plus whatever must be spent for administration, storage and distribution of excess food. All taxpayers share in these transfer payments and subsidy costs.

In the absence of the subsidy, the supply curve S_2 would drift back towards S_1 as resources, including farmers, migrated out of agriculture. Assume S_2 settled at S'_2, P_T would now be the price where supply equalled demand at Q_D. Consumers would still lose the same consumer surplus area (d+d'+e+f) and producer surplus would increase by d+d'+1/2e for a net social loss of f+1/2e. This presumes that the resources leaving agriculture found more profitable employment elsewhere.

The good news is that increasing crop yields increases the supply of food, which should improve the general level of nutrition, lower consumers' direct cost of food and increase their welfare. The bad news is that U.S. consumers will not purchase all the available supply at the government supported price. For U.S. agricultural commodities to be exported, they must be able to compete with the prevailing world market price. If exports are not sufficient, U.S. consumers may spend billions of tax dollars for agricultural subsidies and enormous surpluses of agricultural products may develop without other programs to restrict output. In the absence of greatly expanded exports, further improved crop yields in the United States implies that fewer resources, including farmers, will be used to produce our food or the social costs of food production will increase. On the other hand, if prime agricultural land is being destroyed by erosion or urban sprawl or too intensive use, high yielding crop seeds may become necessary just to meet the demand of domestic consumers and current levels of export trade. Under any circumstances the benefits that consumers might realize from most of the current genetic research are at least a decade ahead of us.

Technical and personal risks. If the quality of food is improved by biotechnology and by more conventional food research techniques, and prices do not increase disproportionately, consumers will be better off, assuming they adopt the food with the new characteristics. There is a tendency for consumers to think that any new food is not as good as old and familiar, natural foods. Furthermore, they are suspicious of unknown hazards lurking in laboratory concocted ingredients. This fear is not entirely irrational for one of the great unknowns about newly engineered foods is the human health hazards they may pose. Many of these hazards cannot be foreseen during food development stages and many will not be descovered until years after humans ingest them. We cannot be too careful about checking and double-checking for potentially disasterous ramifications. We cannot assume new discoveries do not bring new health hazards.

Other risks posed by genetic engineering, especially with respect to seeds, is that the world's suppply of germ plasm is being diminished and monopolized. Germ plasm is the material in every seed that determines the personality of the plant and by selecting personality traits that are commercially desirable, the existing variety of plants is diminishing. Unique varieties, perfect for isolated growing areas, may cease to exist, while fewer widely used seed varieties run the risk of exposure to an unknown blight or insect, which they have not been engineered to resist. Under these circumstances a major portion of the world supply of a new superior crop could be wiped out, causing financial disaster for farmers and drastically increasing food prices for consumers. In addition, lacking a wide variety of available germ plasm for further research, engineering new seed varieties to meet new requirements will become more difficult.

The total supply of germ plasm is also endangered by the monopolization of the seed industry (15). Allowing scientific knowledge to be patented diminishes its usefulness to the larger community and allows the patent owners to extract economic rent from all consumers of products eminating from that patent. Government funded and operated depositories for germ plasm have provided a partial answer to this long run problem (16), but divorcing incentives for basic research from commercial profits is a challenge that must be met if society as a whole is to realize long run payoffs from biotechnology in agriculture.

Food innovation: soy protein. Technological advances in food processing also expand food choice, improve foods' convenience and, in some cases, improve nutritional quality and reduce costs. The development of textured vegetable proteins (TVP), particularly soy protein, is a good example of innovation that may expand food choice. Soy can be separated from soybean oil in the form of flour, grits, textured protein, protein concentrates or isolates. Soy proteins are substituted for other ingredients in familiar foods because they impart superior characteristics such as improved protein quality, increased shelf life or better quantity and quality control. The wholesale price of soy protein varies with its form from about $.20 per lb. for flour to $.35 for textured soy protein, to $1.00 for soy isolates (17). When it replaces ingredients which cost more, food production costs decrease and consumers could benefit not only from high quality, low fat protein but from a lower price.

The 1977-78 National Food Consumption Survey found that individuals in the United States ate an average of 75.5 g of protein per day which is 165% of the recommended daily allowance (3). About 22% of their proteins came from vegetable sources and 47% from meat (3). Americans are not short on protein but they consume it in expensive forms. Prices per pound of protein in June 1980 were approximately $12.40 from beef (round), $3.25 from chicken and $1.03 from soy protein concentrate (17). If consumers obtained all their protein from just one of these sources each day it would cost them $2.11 for beef protein, $.55 for chicken protein and $.17 for soy protein. Except for tofu, American consumers rarely eat soy protein identified as itself. It is not a familiar food but its versatility has allowed it to simulate or imitate familiar foods, such as bacon bits, and it can be blended with processed meats on a one for one basis with virtually no obvious change in flavor or character. It has been used most extensively as a meat extender or substitute in institutional cooking. In 1971, the U.S. Department of Agriculture allowed the addition of 25% to 30% TVP to hamburger used in the school lunch program producing a cost savings of $36 to $39 per 100 pounds of meat used (17).

There are several reasons why Americans are not consuming more soy protein. For one thing, they are consuming more than adequate protein already and the form in which they consume it is familiar, tasty and affordable. If they are to consume more soy protein it will likely be as a substitute for meat or milk products. If soy protein was to be substituted for some portion of meat protein on a one for one weight basis, and consumers purchased the combined products, they could realize a savings in food expenditures with no loss in protein quality. Potential cost savings on meat expenditures by U.S. consumers resulting from a 10% and 25% soy protein substitution for the 1981 total meat consumption are reported on Table 21.2. Savings for the total population from a 25% substitution for the total amount of meat, fish, and poultry consumed would be $12,254.4 million or 18.8% less than expenditures for pure meat. This represents a savings of $53.32 per capita for one year. The ratio of the price of a pound of soy protein to a pound of meat varies with the type of meat and the form of soy protein but most calculations result in a ratio between $.13 and $.30. Prices used for these calculations on Table 21.2 have a ratio of $.25 for all meat but a ratio of .50 for edible offals. Edible offals were chosen because they are used in processed meat products like sausages and lunch meats--products to which soy protein can be easily added. The cost savings in these meat products is less because the soy protein-meat price differential is smaller and many fewer pounds of these types of meats are consumed. With a 25% substitution, however, a 12.5% savings could be realized but this amounts to only $.70 per person per year.

In order for soy protein to be consumed in quantities that will impart substantial savings, American consumers must first be convinced of their benefits and assured of their safety. Second, a substantial cost savings must appear. These two conditions are illustrated in Figure 21.3 which depicts, initially, two meat products, a and b. Vector oa represents a hypothetical, technical ratio of protein to calories in all meat and ob represents the same technical ratio for a meat-soy protein combination. Both meat products are the same price. Line ab represents the least cost frontier along which a rational consumer will choose product a or b or some combination depending on where the indifference curve U_1 is tangent to ab (18). Consumer's indifference curve U_1 is drawn tangent to b, implying the consumer will purchase product b. Product c represented by vector oc then enters the market at the same price (length of oc = length of ob and oa). Product c has more soy protein than product b but it will not be purchased by the consumer unless it is offered at a substantially lower price. Extending oc to oc' represents lowering the price of product c by about one-third. Line c'a becomes the new least cost frontier and consumers move to a higher indifference curve (U_2) at point d where (s)he still purchases some of product a but mostly product c. In this example, the price of a soy-meat mixture had to be considerably lower to attract consumption. Gallimore (19) found the sale of soy-beef combinations to be very sensitive to the price of close substitutes. A 10% increase in the price of hamburger led to an 11-16% increase in the sale of a soy-beef product. The soy-beef blend was also sensitive to its own price as a 10% decrease in the price of soy-beef blends led to a 16-18% increase in their sales.

One other way to entice consumers to purchase product c in Figure 21.3 is to change their relative preferences for protein versus calories by providing them with new information and experience. If, for instance,

TABLE 21.2. Potential Savings on Meat Expenditures by U.S. Consumers with Soy Protein Substitutes of 10 and 25 Percent ($ million)

	1981 Pounds per Capita[a]	Total Cost - No Soy Protein[b]	Cost with 10% Soy Protein Substitute[c]	Cost with 25% Soy Protein Substitute	Total Savings (Dollars per Captia)	Percent Saved
All Meat, Fish	237	$65,357.1	$60,372.6		$ 4,985.5 ($21.70)	7.6
Poultry	237	$65,357.1		$53,102.6	$12,254.4 ($53.32)	18.8
Edible Offals[d]	9.4	$ 1,296.1	$ 1,231.3		$ 64.8 ($.28)	4.9
	9.4	$ 1,296.1		$ 1,134.1	$ 162.0 ($.70)	12.5

[a] Reference 1, p. 2 and p. 33.
[b] Assumes meat costs average $1.20/lb. and edible offals cost average $.60/lb. (17).
[c] Assumes soy protein costs of $.30/lb. and soy protein substitutes one for one in pounds of meat (17).
[d] All meat includes edible offals.

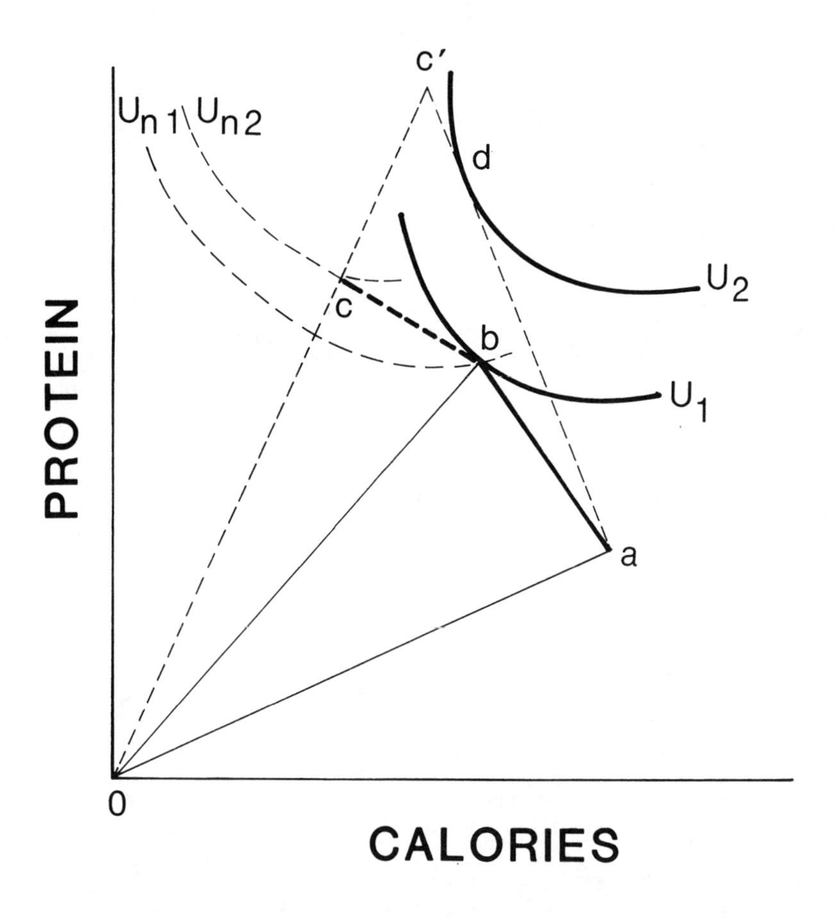

FIG. 21.3. Adoption of a new product by decreasing price or changing preferences.

consumers learned that calories were bad for their health and proteins prevented aging, a new indifference curve like U_n might appear. Before product c appeared on the market, consumers with preferences represented by U_n would purchase product b on U_{n1} and after c appeared on the market they would purchase product c on U_{n2}. In this case the price of c would not need to be lower than the price of a or b for consumers to purchase it. Changing consumer preferences for basic food attributes is not easy and generally not desirable unless some new health related information is discovered. But new information about the health benefits or hazards of foods and food additives has served to change consumer's preferences over the years.

Providing accurate information about the true attributes of different food products is always in the consumers' interest and improves their well-being. This is illustrated in Figure 21.4. Suppose oa again represents pure meat and ob represents consumer's beliefs about the ratio of flavor to "naturalness" in product b which shall be designated a soy-meat combination. The consumer believes that for f units of flavor, product b is half as "natural" as product a. Given the indifference curve U_1, all meat is purchased on the least cost line at point a. Suppose further that the true ratio of flavor to "naturalness" in product b is represented by ob'. With true information, the consumer would purchase b' on a higher indifference curve U_2 and be better off.

CONSUMER INFORMATION AND LABELING REGULATIONS

The question of consumer information and regulations regarding food labeling is a safety issue and an economic issue. The economics of product safety suggests that the party in the best position to know the characteristics of a product and to take action which will avoid accidents should be designated the "best cost avoider". U.S. manufacturers of food and drink in sealed containers were deemed the "best cost avoiders" and held strictly liable for the safety of their products under implied warranty law as early as 1913-1914 (20). An implied warranty on food and drink is that the product is safe for human consumption and will nourish not diminish one's health. To the extent that any food ingredient, additive, processing residue or container may be hazardous to someone's health or safety, they are at least entitled to know about its existence. The top line on product safety is to eliminate the hazard; the middle line is to inform when elimination is not feasible; and the bottom line is to hold the manufacturer liable for damages caused by foods consumed.

Determining the feasibility of eliminating hazards is mostly an economic issue, not a technical question. Product hazards can always be eliminated by removing the products from the market. We have seen products recalled when they were found to contain deadly poisons such as botulin, ptomaine, or cyanide. In addition, we've seen consumers who have purchased contaminated products be warned not to consume them. Consumers are able to react rationally to this type of information because: (1) they are informed specifically about its existence in a particular product, (2) it is easy to assess the consequences of consuming that product, and (3) they have alternative products to consume which are safer. These three conditions are necessary for consumers to make rational choices and act in the market to screen out those products they consider too risky.

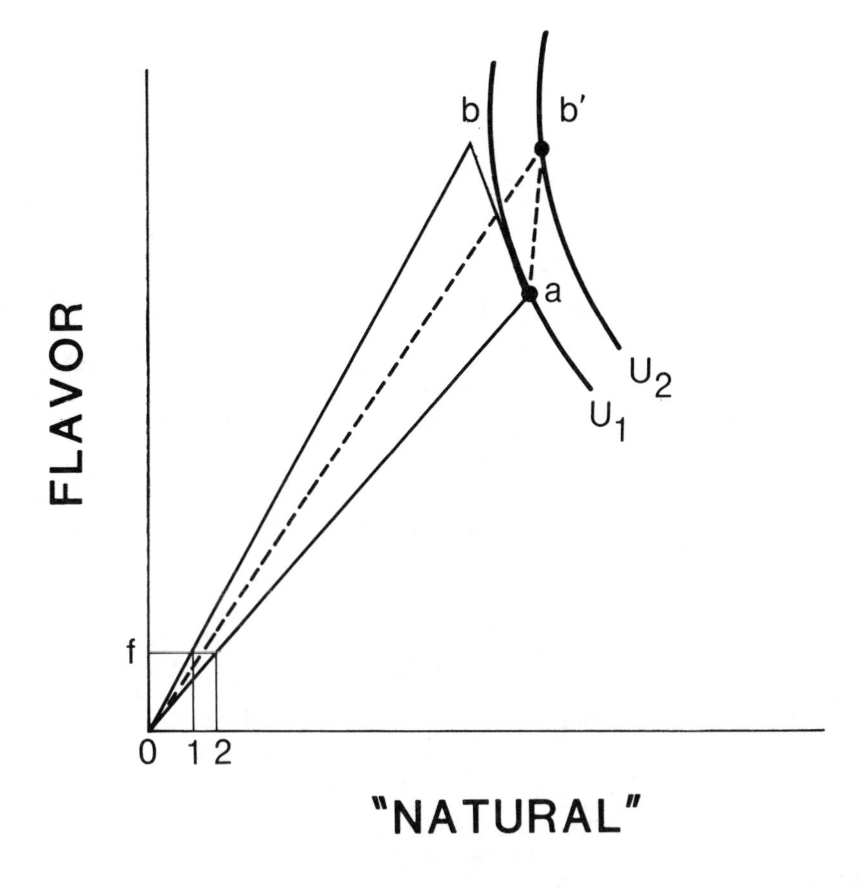

FIG. 21.4. Accurate information improves consumer welfare.

Some substances in food products are harmful to only a few individuals with specific health problems such as allergies or high blood pressure. They generally meet the second and third criteria for rational market selection; the first criterion can be met by informative labeling. One criterion for mandating that a particular type of information appear on food labels is to require the information when the present value of the cost of providing information is less than the present value of the probability of harm times the cost of injury or death summed over all individuals. The importance of this approach is that it accounts for the probability of harm and for the costs incurred by the injured. It also allows consumers to make informed choices; it does not automatically ban a substance because it was found harmful in isolated cases.

Although mandatory labeling seems to be an anathema to manufacturers, it can yield many social benefits. Usefulness of labels cannot be judged by the number of people who read them or by how often they are read in any one time period. Information is semi-durable and consumers do not need to read all labels on every shopping trip. Furthermore, the non-use benefits of mandatory information are distributed over all consumers (21). These benefits accrue in the form of generally safer and more predictable quality products because compliance with mandatory labeling imposes a discipline on the industry. Labeling the presence of sodium, for instance, requires testing for its quantity and a monitoring of that quantity over time. Even minimal enforcement of ingredient labeling would make most manufacturers hesitate to use TVP when their labels say "all beef." Besides industry discipline, mandatory labeling raises the conscientiousness level of both buyers and sellers and carries great potential for long-run improvement in nutrition and health.

Two areas where food labeling laws and demands for information are especially lax are for snack foods and food served away from home. As long as consumers associate food with recreation as much as nourishment, they seem to abandon their concerns for precise information. Ironically, snack foods and food away from home are two of the fastest growing segments of the food industry and will undoubtedly be asked for more precise information in the future.

Economic theory indicates that when consumers make purchases on the basis of incorrect information they cannot maximize their utility. An allocative efficiency loss occurs whether the product turns out to be better or worse than they expected (22,23). This is illustrated in Figure 21.5 for a product that turned out worse than expected. Originally consumers purchased quantity Q_0 at price P_0 along demand curve D_0. Experiencing inferior product performance, this demand curve shifts to D_1. Consumers would have been willing to buy only quantity Q_1 at price P_0 if they had known its true characteristics. They have already paid an amount equal to areas a+b+c for quantity Q_0 and expected consumer surplus equal to area d+e. Under their new demand curve, D_1, they received consumer surplus equal to d and are willing to pay areas a+b for Q_0. Area e represents consumer surplus not received. Area c represents dollars spent for which no utility was received. This is sometimes referred to as a deadweight loss since this expenditure cannot be transferred to other persons or products in the economy; it is simply lost.

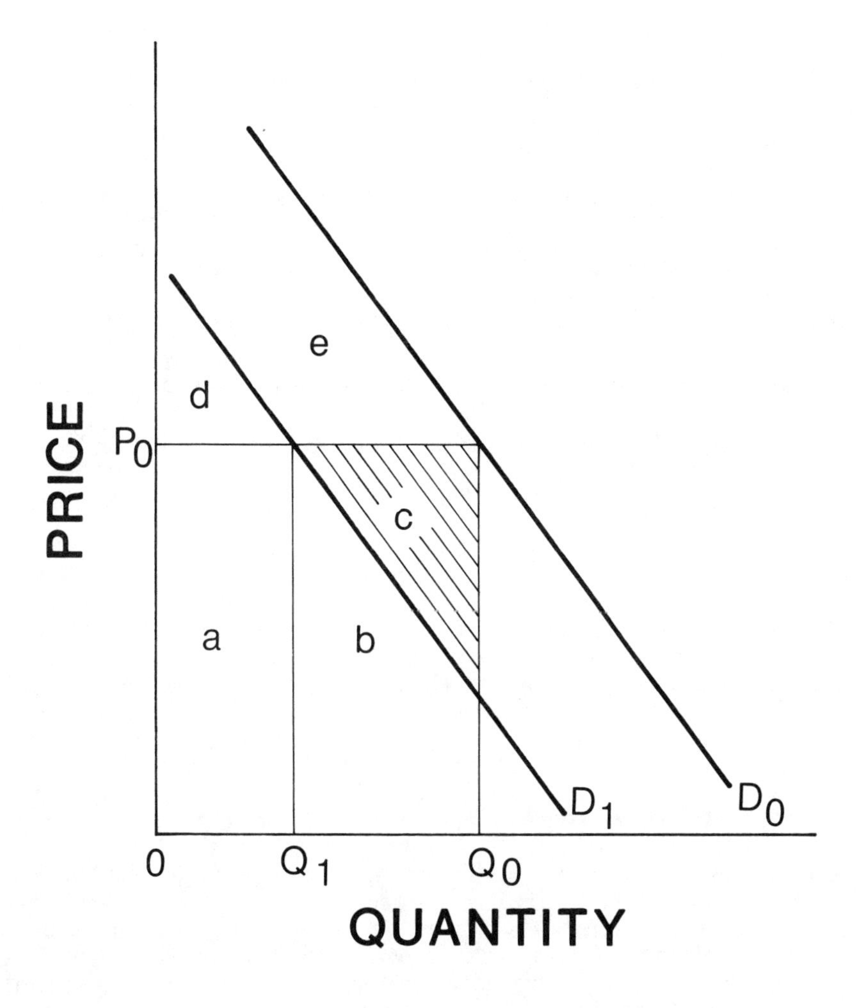

FIG. 21.5. Loss in consumption efficiency due to inaccurate information.

The question of how much consumers are willing to pay for information is often asked. Theoretically they would be willing to pay an amount equal to the deadweight loss (area c) to avoid losing utility on that expenditure. In fact, consumers pay for information in many ways: (1) increased cost of well labeled products, (2) purchase of newspapers or magazines or other informative literature, (3) use of their time (and direct costs of transportation) for comparison shopping and time to read the literature and the labels and talk to experienced users, and (4) experiment with the product and see how well it performs. When all of these search costs begin to outweigh the expected benefits in terms of decreased utility and disappointment, consumers stop gathering information. Minimizing consumers' search costs is one of the purposes of food labeling and standardization. Accurate food labels provide a relatively inexpensive way for consumers to be informed about their food choices. An even less expensive way for consumers to be informed about the characteristics of a food product is to create "standardized" foods, with standard identities and recipes or ingredient mixes. We have many standardized foods in the United States; peanut butter, mayonnaise, starch based salad dressing, and margarine are a few. Some have suggested that vegetable protein products should somehow be made a "standardized" product that could be marketed under its own identity rather than as "imitation" this or "simulated" that (24). The advantage of this approach lies not only in inexpensive consumer information but improved ability to promote consumption of an identifiable product. For example, both margarine and salad dressing now sell under their own names and not as imitation butter or imitation mayonnaise. Others have suggested that a standard be set for the percent of the original food, such as meat, that can be replaced by vegetable protein (12). This is analogous to the standard for peanut butter, 10% of which can be something other than ground peanuts.

Designing labels that serve the informational needs of consumers and also promote research and development of new food products is a tricky business. It has received international attention with respect to vegetable proteins. The Food and Agriculture Organization and the World Health Organization of the United Nations have a joint program on food standards which is operating chiefly through the Codex Alimentaris Commission. This commission is an international governmental body whose purpose is to protect consumers against health hazards in food and against fraud. The commission has 116 member governments and a special committee working on definitions and worldwide standards for vegetable protein products derived from soybeans, cottonseed, groundnuts, and cereals (25). With increased international trade and travel, international standards of safety and identity for food products is a concept whose time has come.

SUMMARY

Trends in consumption of cereals and legumes by U.S. consumers have been reviewed and their stated preferences for natural, healthier foods free from potential hazards have been examined. The economic benefits of technological change in basic agriculture and the food processing industry were examined in terms of how they might impact consumers' welfare. Benefits from new innovation can easily be exaggerated if not examined in the cold light of consumer demand. For example, promoters of soy protein argue that it can provide low income American households with less expensive and adequate protein. It turns out that individuals in low income households (<$6,000 per year) in the 1977-78 National Food Consumption Survey were already eating 154 percent of the recommended daily allowance of protein (3). Improving the quantity or quality of protein in American diets is not a large problem looking for a solution. Basic research need not and should not concern itself with commercial profitability, but scientists and marketers alike need to ask what foods consumers of the world need and want lest we find ourselves with a set of solutions looking for a set of problems.

New technology, especially in the consumer market, brings with it the inevitable problems of identity, adaptation and information. The economics of consumer information tells us that better informed consumers are better able to maximize their welfare. Since information is not free, consumers will bear the costs of search until the last unit of search costs more than the expected benefit from more information. Lowering search costs is one of the purposes of food labeling along with promoting health and safety and industry discipline. The benefits of accurate labeling are available to all food consumers. They are not diminished by someone else's use and they are paid for by both private and public dollars. They are both a private and a public good implying that accurate labeling is not only in the consumers' best interest but in the best interest of the public at large.

REFERENCES

1. U.S. Department of Agriculture, Economic Research Service, "Food Consumption Prices and Expenditures, 1960-1981." Statistical Bulletin No. 694 (1982a).
2. U.S. Department of Agriculture, Human Nutrition Information Service, "Food Consumption: Households in the United States, Spring 1977." Nationwide Food Consumption Survey 1977-78, Report No. H-1 (1982b).
3. U.S. Department of Agriculture, Science and Education Administration, "Food and Nutrient Intakes of Individuals in One Day in the United States, Spring 1977." Nationwide Food Consumption Survey 1977-78, Preliminary Report No. 2 (1980).
4. Van Dress, M. G., U.S. Department of Agriculture, Economic Research Service, Statistical Bulletin No. 690 (1982).
5. Scales, H., Cereal Foods World 27, 203 (1982).
6. Faubion, J. M., Haseney, R. C., and Seib, P. A., Cereal Foods World 27, 212 (1982).
7. Chou, M., Cereal Foods World 27, 243 (1982).
8. Healthy food market in Europe, Cereal Foods World 27, 507 (1982).
9. The Wall Street Journal, How a nutritionist teaches Mexicans to use soybeans to fortify tortillas. 5/20, p. 6 (1983).
10. de Buckle, T. S., Journal of American Oil Chemists Society 58, 433 (1981).
11. Hardin, C. M., Journal of American Oil Chemists Society 56, 173 (1979).

12. Ward, A. G., Journal of American Oil Chemists Society 56, 196 (1979).
13. Hughey, A., More firms pursue genetic engineering in quest for plants with desirable traits. Wall Street Journal 5/10, 56 (1983).
14. Menz, K., and Neumeyer, C., Evaluation of five emerging biotechnologies for maize. Staff Paper P81-28, Department of Agricultural and Applied Economics, University of Minnesota (1981).
15. Schapiro, M., Experts find seeds of agricultural crisis. Minneapolis Star and Tribune 3/30, 13A (1983).
16. Comacho, L. H., Journal of the American Oil Chemists Society 58, 125 (1981).
17. Langsdorf, A. J., Journal of American Oil Chemists Society 58, 338 (1981).
18. Lancaster, K. J., Journal of Political Economy 74, 132 (1966).
19. Gallimore, W. W., Journal of American Oil Chemists Society 56, 181 (1979).
20. Leigh-Jones, P. M., Cambridge Law Journal 27, 54 (1969).
21. Padberg, D. I., Journal of Consumer Policy I, 6 (1977).
22. Kinsey, J., Roe, T., and Sexauer, B., Imperfect information, consumer theory and allocative error in consumption. Staff Paper P80-8, Department of Agricultural and Applied Economics, University of Minnesota (1980).
23. Sexton, R., Journal of Consumer Affairs 15, 214 (1981).
24. Lambert, E. I., Journal of American Oil Chemists Society 56, 234 (1979).
25. Hutchinson, J., Journal of American Oil Chemists Society 56, 227 (1979).

22

Government Regulations Related to Cereal and Legume Products

F. EDWARD SCARBROUGH

Cereal and legume products are subject to a myriad of government regulations which range from controls on the kinds of pesticides that can be used on crops in the field, through current good manufacturing practice requirements in the processing plant, to an array of information either required or permitted on the label of the product as it appears on the grocery shelf. And to these regulations established under the Food, Drug, and Cosmetic Act must be added those established under other authority, such as regulations administered by the U.S. Department of Agriculture (USDA) or certain state requirements for the sale of enriched cereal-grain products. It would be impossible to discuss this wide variety of government regulations within the space allotted. However, from a Food and Drug Administration (FDA) perspective, when regulations related specifically to cereal and legume products are mentioned, the ones most often referred to are the food standards for these products. Therefore, I will confine my review primarily to the food standards, with a brief review of where FDA now stands relative to those standards and how the Agency arrived at its current status, followed by some discussion of the direction in which Agency policies appear to be evolving and of some of the potential problems FDA may be facing in the near future.

The concept of food standards in the United States is not new. In fact, by 1646 Massachusetts Bay Colony had passed a law to prevent the sale of underweight loaves of bread (1). However, the first federal general food law, the Pure Food and Drug Act, was not enacted until 1906 (2) and did not contain provisions for the establishment of food standards. This act established definitions of adulteration and misbranding and permitted the seizure of foods within these categories. The 1906 Act did not require foods to bear a statement of ingredients on the label. Thus the government was unable to deal with fabricated foods containing several ingredients and marketed under a distinctive name.

Between 1906 and 1938 far-reaching changes were occurring in American society. From a primarily agricultural base, society was evolving into an industrial, urban culture, increasingly separated from the production sources of foods and faced with a food supply system controlled by national corporations distributing coast to coast. It

Cereals and Legumes in the Food Supply, edited by Jacqueline Dupont and Elizabeth M. Osman 1987 Iowa State University Press, Ames, Iowa 50010

became apparent that some type of federal response was required to protect consumers against economically adulterated products which could not be detected by traditional sensory methods. Also it was important to protect responsible manufacturers against unscrupulous competitors who resorted to unethical practices to increase profits (3).

The enactment of the Federal Food, Drug, and Cosmetic Act (FD&C Act) in 1938 provided for the first time for the establishment of standards of identity for foods (4). This Act enabled FDA to prohibit the marketing of foods from which traditional constituents had been removed or in which cheaper ingredients had been substituted. The legislative history of the Act indicates that Congress was addressing primarily economic problems. Consumers were being defrauded by the sale of inferior products being offered as traditional foods. Standards were to provide a yardstick against which such economic adulteration could be measured.

Between 1938 and today, changes in American society have continued at an ever-increasing pace. For example, supermarkets have largely replaced the corner grocery store for the distribution and sale of food, a further depersonalization of the food supply system. Also, today Americans spend more than two thirds of their food dollars on processed foods and foods eaten away from home rather than on fresh agricultural products. Further, there is an increasing use of non-traditional food ingredients in or as substitutes for traditional foods (e.g., vegetable protein meat extenders and dairy product analogs). The makeup of the food supply has become more complex and despite the existence of food standards, consumer knowledge of food composition has become less thorough. It is not surprising, therefore, that the proper role and continued use of food standards are becoming subjects of greater current debate.

Not only has the American food supply changed, the objectives and uses of food standards have been broadened during the past forty-five years. The Congressionally intended function of food standards was to protect consumers against economic fraud and deception, as reflected in the language of the 1938 Act, which states that the purpose of food standards is to "promote honesty and fair dealing in the interest of consumers" (FD&C Act, section 401). However, as an example of another use of food standards, manufacturers have used them to limit substitutes for their products, because once a standard is adopted no product may "purport to be" the standardized food unless its composition conforms to the standard. Also, recipe food standards have permitted manufacturers of existing products to erect barriers of entry for new products by restricting competition largely to advertising to establish brand name recognition. Thus, existing producers may develop a vested interest in maintaining food standards because of the resultant restriction of competition. However, such an approach also results in discouraging the development of new food products that may benefit consumers in an economic sense.

FDA has clearly used food standards for other than economic purposes. The 1938 Act did not give FDA the authority to approve food additives before they were introduced into the food supply. Therefore, the agency used the food standard provisions to prohibit the use of untested additives, at least for standardized foods. The passage of the food additives amendment in 1958 eliminated the need for this use of food standards but the vestiges of this approach remain with us today (FD&C Act, section 409).

FDA has also used food standards as a vehicle for maintaining the nutritive value of our food supply. Fortification and enrichment programs have been used to combat malnutrition for some time. A few examples from our experience in this country will serve to illustrate this point.

The first fortification program in the United States was the addition of iodine to table salt to prevent goiter. A movement was launched in the 1920s to provide iodized salt to the public, and by the mid-1950s the incidence of goiter had dropped dramatically (5). Although FDA regulations do not require that salt be iodized, the presence or absence of iodine is required to be clearly labeled on salt packages (6).

For centuries, rickets, a bone disease of infants, was a major health problem. However, after the discovery of vitamin D and its use in various forms of dietary supplements, rickets, at least in exaggerated degrees, has largely disappeared in the United States (5). This decline has been attributed in part to the addition of vitamin D to the milk supply. FDA regulations provide for the addition of vitamin D to essentially all standardized forms of milk (7). Milk was considered the most appropriate vehicle for vitamin D fortification because of its high natural calcium and phosphorus content which, along with vitamin D, are all required to prevent rickets.

Perhaps the best known examples of the use of food standards to maintain or improve nutritional status or to provide the solution for a nutritional problem are the enrichment standards for milled grain products. It was recognized in the 1930s and 1940s that a substantial portion of Americans were suffering from a variety of nutrient deficiencies (8). The country was made acutely aware of this situation by the great number of young men who did not pass physical examinations for military service in the years just prior to World War II. The first FDA standard for enriched flour and farina was published in the *Federal Register* in May 1941 (9). It defined levels for thiamine, riboflavin, iron, and calcium. Later (1943) niacin was included (10) and enrichment was extended to cornmeal and rice in 1947 (11). Fortification with the B vitamins and iron was introduced to prevent beriberi, ariboflavinosis, pellagra, and iron-deficiency anemia. In 1935 there were more than 3500 deaths due to pellagra, whereas by 1959, fewer than 50 deaths were attributed to this disease and today there are essentially none (National Center for Health Statistics). Although the large number of pellagra deaths in the 1930s may be attributed, in part, to poor diets as a result of the Depression, the virtually complete elimination of this disease speaks for the success of cereal grain enrichment programs and of the therapeutic use of niacin by physicians.

In 1974, the Food and Nutrition Board of the National Academy of Sciences (NAS) conducted a review of cereal-grain enrichment programs and in its booklet "Proposed Fortification Policy for Cereal Grain Products" (12), the Academy suggested levels of specific nutrients with which all major cereal grain products, primarily wheat, corn, and rice, could be fortified. The nutrients recommended for inclusion in the fortification scheme were selected principally on the basis of potential dietary insufficiencies that were determined by food consumption surveys.

In spite of the inadequacies of certain information available to the NAS, FDA recognized the significance of their recommendations. However, the Agency was concerned about the technical problems involved in such fortification. Therefore, FDA contracted with the Food Protection Committee of NAS's Food and Nutrition Board to research the matter. The Committee submitted its report, "Technology of Fortification of Cereal-Grain Products", to FDA in March 1978 (13). The Committee concluded that many of the recommended nutrient additions are technologically feasible but that some significant difficulties remain. For example, calcium and magnesium need additional evaluation before general addition to cereal grains is feasible, zinc fortification presents some taste problems, and vitamin A stability requires further study.

In October 1973, with knowledge of what NAS intended to propose in its fortification policy, FDA published an order revising the standards of identity for enriched flour, enriched self-rising flour, and enriched breads to adopt the NAS recommended levels for thiamine, riboflavin, niacin, and iron (14). The Agency met with no opposition to the specified levels for the vitamins, and these are provided for in the current standards of identity. However, FDA was thwarted in its efforts to increase iron enrichment levels because of a series of objections and requests for a hearing in response to the 1973 order. The principal objections raised were: possible iron toxicity, unsubstantiated clinical benefit, and unsubstantiated efficacy in ameliorating iron deficits.

After holding public hearings, the Agency issued a tentative final order in November 1977 calling for iron levels in the enriched grain products that were substantially lower than those proposed by NAS (15). This decision was based primarily on the lack of adequate studies to refute the questions raised. On February 2, 1982, FDA issued a final rule amending the standard of identities for enriched bread and flour (16). This changed the requirements for iron enrichment from a range with maximum and minimum levels, as established in the 1977 tentative final rule, to a single-level requirement. The single level was set at the previously allowed maximum for bread with a corresponding figure for flour. FDA also undertook a major clinical nutrition research program to address the issues raised relative to iron enrichment of bread and flour. Depending on the results of this study, it may be possible for the Agency to repropose increasing the levels of iron used to enrich basic staples.

In the recent past, increasing the fortification levels of cereal-grain products has been the subject of several requests or petitions to FDA and of a Congressional hearing (17). FDA has taken the position that there are insufficient data for major modifications to the standards for cereal-grain products. FDA's policy, discussed below, calls for fortification on the basis of demonstrated nutritional deficiency, while the recommended increased levels are based on food consumption data. Currently it is not possible to correlate food consumption data with measures of health or nutritional status except in extreme circumstances not encountered in the United States. Dietary methodologies generally reflect short-term consumption, whereas clinical and biochemical measurements reflect longer term aspects of health and nutrient intake. Attempts to correlate nutrient intake with biochemical or clinical parameters on an individual basis have not been successful. Until dietary collection methodology is improved and developed to a point where it relates to measures of health and nutrition status, it is not valid to base fortification policy on dietary data alone. For example, food consumption data indicate low average caloric intake for age groups 55 and older. However, a study of the caloric intakes of the elderly segment of the U.S. population indicates that information from national surveys may be underestimating caloric intakes (18). This may be indicative that intakes of other nutrients calculated from food consumption information are also underestimated.

Another example of a potentially complicating factor is the estimate that over 40% of the population 16 years old or older now regularly take some form of vitamin and/or mineral dietary supplement (19). The contribution of these supplements to nutritional status is not reflected in current food consumption data.

Clinical and biochemical measurements to assess nutritional status are severely limited. From a public health point of view, the most significant studies are the National Health and Nutrition Examination Surveys (NHANES) conducted by the National Center for Health Statistics--NHANES-I covering the period 1971 to 1975 and NHANES-II for 1976-1980. The text of the 1974 NAS publication acknowledged that their suggested fortification program was not based on "well documented clinical evaluations" because such data were not available. The nutrients selected for inclusion by NAS were "based on the best information available, with a view toward adequately meeting the nutritional needs of those groups of the U.S. population that appear to be at the greatest nutritional risk."

Since the NAS-recommended levels for some nutrients have been implemented and because of the controversy surrounding the iron increase, Dr. Robert O. Nesheim, chairman of the NAS panel that drafted the 1974 proposal, recently concluded that future modifications to fortification regulations can be accomplished only where adequate data exist to satisfy the scientific community. Dr. Nesheim cautioned that "the cereal industry must move carefully to develop its nutrition case prior to actual changes in fortification levels" (20).

The recent "General Policies for Nutrient Addition" by the American Medical Association, NAS, and the Institute of Food Technologists states that sound fortification policies require knowledge of national dietary patterns and of the nutritional status of the population (21).

It has been suggested that maintaining the nutritive value of the food supply should not be the major purpose of food standards (22). And in recent years, the trend within FDA is to rely less on food standards to safeguard the nutritional profile of the diet. For example, in January 1980, FDA published in the *Federal Register* the Agency's general policy for the addition of nutrients to food (23). Nothing has happened since publication to change this policy; if anything, it has been strengthened because it is fundamentally a voluntary program designed to serve as a basic guide to industry in the practice of modern, rational nutrition science. These guidelines state that FDA does not encourage indiscriminate addition of nutrients to food, nor does it consider appropriate the fortification of fresh produce; meat, poultry, or fish products; candies; or carbonated beverages. The four principles to be used when fortifying foods are as follows:

1) Nutrients may be added to a food to correct a dietary insufficiency recognized by the scientific community to exist and known to result in nutritional deficiency disease, if sufficient information is available to identify the nutritional problem and the affected population groups and if the food is suitable to act as a vehicle for the added nutrients;
2) Nutrients are added as permitted or required by existing regulations;
3) Nutrients may be added to restore such nutrients to levels representative of the food prior to storage, handling, or processing; and
4) Nutrients may be added to a food in such a way so as to balance the vitamin, mineral, protein, and caloric content.

The second part of the policy, addressing technical aspects, states that a nutrient added to food must be 1) stable under conditions of storage, distribution, and use; 2) physiologically available; 3) present at a level which will not result in excessive intakes; and 4) suitable for its intended purpose.

We at FDA believe that such statements of general policy, rather than detailed case-by-case regulations, may well be used as a model for addressing the subject of the nutritional quality of the food supply.

Although not a new initiative, another policy of major importance in maintaining the nutritional quality of foods is FDA's imitation/substitute policy (24). This pertains to any food that resembles and substitutes for a traditional food. The distinction between a substitute food and an imitation food is whether the food is nutritionally inferior to the traditional food. Nutritional inferiority is any reduction in the content of an essential nutrient, provided the nutrient is present in the traditional food at a level of at least 2% of the U.S. Recommended Daily Allowance. A reduction in calories, fat, or sodium is not considered when determining nutritional inferiority. If a food is nutritionally inferior the policy requires that it be labeled "imitation", thus encouraging manufacturers to produce substitute foods that are not nutritionally inferior to avoid having to use the "imitation" label. Market research indicates that the common perception of "imitation" is as a pejorative term.

Another regulatory initiative with direct impact on the nutritional quality of the food supply is nutrition labeling, with which we now have had over a decade of experience (25). As part of a broad educational program, nutrition labeling provides the consumer with information in a consistent format to allow nutritional comparisons in order to facilitate the appropriate selection of foods, thus contributing to good health. FDA surveys reveal that, on a dollar volume basis, approximately 55% of packaged processed food sales carry nutrition labeling (26).

FDA is concerned about the maintenance of the nutritional quality of food in light of the changing characteristics of the food supply. For example, nutrient interactions and nutrient bioavailability are areas of continuing Agency concern. Recently FDA has focused attention on the possible interference with mineral absorption by cereal and legume protein sources, sparked particularly by the recent clinical research observations of major decreases in nonheme iron absorption in the presence of soy protein (27). Similarly, FDA is preparing to publish regulations for vegetable protein products which will require elevated levels of zinc fortification because it is known that phytate affects the bioavailability of zinc (28).

A special Subcommittee of the International Nutritional Anemia Consultative Group was funded to assess the effect of cereals and legumes in the diet on the bioavailability of iron and to develop a background paper based on this assessment. This was part of a joint effort by several Federal Government agencies: the Department of Defense, the Agency for International Development, USDA, and the FDA. The subcommittee was charged with reviewing the state-of-the-art of the measurement of iron bioavailability, reviewing data on the modification of mineral bioavailability by cereals and legumes, reviewing experimental data as it may affect human feeding programs, developing preliminary guidelines for use by policy makers, and making recommendations for additional research. A final report was submitted to the government agencies in June 1982 (29).

Earlier chapters dealt with affecting the nutritional quality of cereal and legumes through breeding and genetic engineering. The prospects of significantly changing nutritional characteristics of staples in the food supply is, of course, of major interest to FDA. However, I think it is fair to say that the Agency is now just beginning to come to grips with the scientific questions raised by such processes and, characteristically, is approaching the issues cautiously.

An example in this area, although not related to cereal products, is FDA's actions relative to rapeseed. Rapeseed normally contains erucic acid, a potentially toxic substance, and for this reason rapeseed oil has not been permitted as a food product in the United States. Recently low erucic acid-bearing varieties of rapeseed have been developed by plant breeding, and the Agency has published a notice that we now consider rapeseed oil from these varieties to be generally recognized as safe (GRAS) (30). This is an example of plant breeding being used to reduce a toxic component rather than to improve the nutritional profile.

An Agency task force has recently considered the role of FDA in the regulation of products of recombinant DNA technology. The task force concluded that, where consistent with Agency policy, new applications will be required for products obtained via recombinant DNA technology. This will be true even if identity is demonstrated with the natural substance or with a previously approved substance produced in a conventional way. However, each case will be handled on an ad hoc basis. The amount of data needed to support such applications will vary depending on a number of factors, including whether the product is identical to a previously approved product; the amount of previous experience with the product produced via conventional technology, and the amount of previous experience with recombinant DNA-derived substances (31).

These guidelines were developed considering the full range of FDA-regulated products, but discussion centered primarily on drug-type applications. Some modification of this approach would appear to be appropriate when basic food staples are considered.

In summary, the issue of the proper role of food standards has not received the public attention that many other issues before the Agency have. Large segments of industry are not calling for changes in food standard policy because of the perceived benefit as a means for reducing competition. Consumer and Congressional attention is focused on other issues, such as food safety legislation. However, within the context of a retrospective review of all regulations, FDA has begun a process of reconsidering the proper role of food standards, including their costs and effectiveness (32). Also, as announced by the Center Director, the Agency has contracted for the development of an economic model which can be used to predict the costs and benefits of food standards or common or usual name regulation (33). It is hoped that this model will provide a means of estimating all costs associated with the economic regulation of food, as well as the impact of regulation on nutrition.

The reevaluation of the role of food standards will be a difficult, complex undertaking but it is one that can no longer be avoided or postponed, especially in light of the increased emphasis placed on economic analysis of regulation.

REFERENCES

1. Schultz, H. W., Food Law Handbook p. 1, Avi Publishing Company, Inc. Westport, CN (1981).
2. Act of June 30, 1906, ch. 3915, 34 Stat. 768, repealed 52 Stat. 1059 (1938).
3. Merrill, R. A. and Collier, E. M. Jr., Columbia Law Review 74, 561 (1974).
4. Federal Food, Drug, and Cosmetic Act, 21 U.S.C. U.S. Government Printing Office, Washington, DC.
5. Aykroyd, W. R. Conquest of Deficiency Disease, World Health Organization, Geneva, Switzerland (1970).
6. Code of Federal Regulations, "Salt and Iodized Salt", 21 CFR 100.155, U.S. Government Printing Office, Washington, DC.
7. Code of Federal Regulations, "Milk and Milk Products", 21 CFR Part 131, U.S. Government Printing Office, Washington, DC.
8. National Academy of Sciences, Inadequate Diets and Nutritional Deficiencies in the United States, National Research Council Bulletin 109, Washington, DC (1943).
9. Federal Register, "Wheat Flour and Related Products; Definitions and Standards of Identity", 6 FR 2574, May 27, 1941.
10. Federal Register, Enriched Flour and Farina, Amendments to Definitions and Standards of Identity, 8 FR 7511, June 5, 1943.
11. Federal Register, Corn Flour and Related Products, 12 FR 3107, May 13, 1947.
12. National Academy of Sciences, Proposed Fortification Policy for Cereal-Grain, National Academy of Sciences, Washington, DC (1974).
13. National Academy of Sciences, "Technology of Fortification of Cereal-Grain Products", National Academy of Sciences, Washington, DC (1978).
14. Federal Register, "Improvement of Nutrient Levels of Enriched Flour, Enriched Self-rising Flour, and Enriched Bread, Rolls, or Buns", 38 FR 28558, October 15, 1973.
15. Federal Register, "Iron Fortification of Flour and Bread", 42 FR 59513, November 18, 1977.

16. Federal Register, "Iron Fortification of Enriched Bread and Flour; Standard of Identity; Confirmation of Effective Date", 47 FR 6425, February 12, 1982.
17. Miller, S. A., Statement Before the Subcommittee on Natural Resources, Agriculture Research and Environment, Committee on Science and Technology, House of Representatives, April 29, 1982.
18. Pennington, J. A. T., personal communication.
19. Stewart, M. L., McDonald, J. T., Levy, A. S., Schucker, R. E. and Henderson, D. P., J. Am. Diet. Assn. 85, 1585 (1985).
20. Nesheim, R.O., At the American Association of Cereal Chemists Fortification Workshop "Adding Nutrients to Food -- Where Do We Go From Here" Arlington, VA, January 19-20, 1982.
21. American Medical Association. Nutrition Reviews 40, 93 (1982).
22. Merrill, R. A., Food Drug Cosmetic Law Journal 39, 113 (1984).
23. Federal Register, "Nutrient Quality of Foods; Addition of Nutrients", 45 FR 6313, January 25, 1980.
24. Federal Register, Imitation Foods; Application of the Term "Imitation", 38 FR 20702, August 2, 1973.
25. Code of Federal Regulations, "Nutrition Labeling", 21 CFR 101.9, U.S. Government Printing Office, Washington, DC.
26. Food and Drug Administration, Nutrition Labeling in the Retail Processed Food Supply, Division of Consumer Studies (1984).
27. Cook, J. D., T. A. Morck, and S. R. Lynch, Am. J. Clin. Nutr. 34, 2622, (1981).
28. Oberleas, D. and Harland, B. F., J. Am. Dietetic Assn. 79, 433 (1981).
29. International Nutritional Anemia Consultative Group, "The Effects of Cereals and Legumes on Iron Availability", The Nutrition Foundation, Washington, DC, (1982).
30. Federal Register, "Direct Food Substances Affirmed as Generally Recognized as Safe; Low Erucic Acid Rapeseed Oil", 50 FR 3745, January 28, 1985.
31. Miller, H. I., "Role of the Food and Drug Administration in the Regulation of the Products of Recombination DNA Technology: Update 1983", Bandbury Report 14, p. 335, Cold Spring Harbor Laboratory, NY (1983).
32. Federal Register, "Review of Agency Rules", 46 FR 36333, July 14, 1981.
33. Miller, S. A. and Skinner, K. J., Food Drug Cosmetic Law Journal 39, 99 (1984).

23

Monitoring Cereals and Legumes for Chemical Safety: An Overview of the Food and Drug Administration's Chemical Contaminants Programs

ELLIS L. GUNDERSON

INTRODUCTION

The Food and Drug Administration (FDA) conducts several monitoring programs designed to ensure that the consuming public is not exposed to excessive residues of hazardous chemicals. These activities form an important part of the overall Federal responsibility for all aspects of food safety. FDA's chemical contaminant monitoring activities include coverage of pesticides, industrial chemicals, natural toxins, and toxic elements in or on a wide variety of food commodities, including cereals and legumes. The Agency's overall activities are described, with the greatest emphasis given to pesticides, since the majority of the effort is devoted to them.

REGULATORY RESPONSIBILITIES

The Environmental Protection Agency (EPA) is responsible for the registration of pesticides under the Federal Insecticide, Fungicide, and Rodenticide Act (FIFRA). Registration is prerequisite to commercial use of pesticides, and involves determinations relating to product efficacy, environmental and human safety, and residue tolerance level establishment. A tolerance is a legal limit on the maximum amount of a pesticide residue permitted in or on a food or feed. EPA has the authority to amend a product's registration, thus possibly altering use conditions and also tolerances in light of new information.

Pesticide residue monitoring activities for products in intrastate shipment are carried out by the states; FDA and the United States Department of Agriculture (USDA) are responsible for monitoring and enforcing tolerances for products shipped in interstate commerce. FDA's authority under the Federal Food, Drug, and Cosmetic Act extends to food and feed products, with certain exceptions pertaining to meat, poultry, eggs. Examination and enforcement with regard to these items are USDA's responsibility at "official establishments" under the Meat and Poultry Inspection Acts. FDA's authority generally extends to these products at other points in commerce.

FDA also monitors foods for industrial chemials such as polychlorinated biphenyls (PCBs), toxic elements such as lead, and natural toxicants such as aflatoxins. This paper stresses pesticide

Cereals and Legumes in the Food Supply, edited by Jacqueline Dupont and Elizabeth M. Osman 1987 Iowa State University Press, Ames, Iowa 50010

coverage, since the major share of resources is devoted to this area. Some smaller program efforts are not discussed.

FDA's CHEMICAL CONTAMINANTS PROGRAMS

FDA's chemical contaminants programs are designed to protect the consuming public from exposure to foods contaminated with excessive residues of hazardous chemicals. This goal is accomplished through conducting research, exploratory surveys of selected foods for certain contaminants, toxicological and epidemiological studies, standards setting, and monitoring and regulatory programs. We shall address selected programs in some detail; they direct the agency's monitoring activities in several areas of contaminant type or subject.

The initial overall program planning effort, coordination, continuing assessment and redirection, and evaluation necessary to provide effective national coverage is conducted by FDA's Center for Food Safety and Applied Nutrition for all areas except animal feeds, which are dealt with by the Center for Veterinary Medicine. Samples are collected by all 21 FDA Districts; sixteen of these are equipped to conduct analyses for pesticides and industrial chemicals.

Major Sampling Types

FDA's monitoring efforts may be divided into two major categories: (1) surveillance sampling programs covering domestic and imported foods and feeds and (2) Total Diet Studies (TDS) designed to determine dietary intakes of pesticides, toxic elements, some industrial chemicals, and selected nutrient minerals by specific population groups. Both efforts may induce additional regulatory activity. That is, previous routine monitoring sampling or TDS analyses may indicate a problem area requiring specifically directed sampling efforts. Results of the analyses may indicate that regulatory action (e.g., seizure; import detention) is warranted to remove the product from consumer channels.

Overall Objectives

The general objective of the chemical contaminants program is to assure the consumer's protection from undue risk because of excessive residues of chemical contaminants in the food supply, thus promoting consumer confidence. FDA's continuous national program of sampling and analysis of domestic and imported foods is a key Federal program designed to accomplish these goals.

Specific Program Objectives

Specific program objectives include: determination of the incidence and levels of an ever-increasing number of pesticides, elements, natural toxins, and industrial chemicals in or on a wide variety of products; enforcement of tolerances and elimination of food sources with unsafe residues; development of short and long term information on levels and intakes via the TDS, and sharing of this and other data with EPA, Congress and the public.

Means of Accomplishing Objectives (Pesticides)

Several major program features facilitate the attainment of these objectives. Sampling emphasis is placed on foods of major dietary importance, while analytical emphasis is given to pesticides with the greatest potential health hazard. FDA's evaluation of each pesticide to determine its relative potential hazard as a food contaminant (the Surveillance Index) will be discussed in detail later. For the sake of efficiency and effectiveness the agency must rely heavily on multiresidue analytical methods. These are methods capable of determining a group of chemicals simultaneously. Such methods currently used by FDA determine a total of more than 200 pesticides and industrial chemicals, and are detailed in the Pesticide Analytical Manual, Vol. I (1). Flexibility in sampling and/or analytical redirection is exercised. Data monitoring and assessment permit identification of information gaps in pesticide/commodity information. Finally, the Center for Veterinary Medicine's animal feeds program and the Center for Food Safety and Applied Nutrition's programs covering human foods are coordinated so that significant residue findings in animal feeds result in corresponding sampling of appropriate foods of animal origin, or vice versa.

MAJOR SAMPLING AREAS

The following discussion details the general areas in which FDA's efforts are directed; they are discussed in ascending order in terms of person years expended.

Mexican Produce

This program involves sampling and analysis of predominantly fresh produce imported from Mexico for determination of the incidence and levels of pesticide residues. Its principal objective is to prevent importation of shipments of commodities containing pesticide residues not covered by a tolerance, or levels exceeding established regulatory limits. FDA's Dallas and Los Angeles Districts are responsible for all sample collection and analyses. About 9 field person years were directed to this effort in fiscal year 1984. This allowed for collection and analysis of more than 2,000 samples. The legumes most frequently sampled are peas and green beans.

A separate effort distinguishing Mexican produce from other imported foods has been maintained since fiscal 1979. It was initiated because the problem of pesticide residues in imported Mexican produce presented a situation sufficiently unusual to warrant a separate program. Special coverage is desirable, since Mexican produce represents a substantial and increasing percentage of the winter produce consumed in the U.S. In addition, previous data indicated that a significant percentage of shipments sampled were in violation of the Federal Food, Drug and Cosmetic Act, particularly because of pesticide residues not covered by U.S. tolerances or other regulatory limits.

Commodity selection and analytical emphases are based on information from a number of sources. These include the Mexican government, U.S. Customs Agents, USDA commodity forecasts, brokers, shippers, importers, FDA's Mexican Liaison Staff, and FDA's Surveillance Index.

All samples are initially analyzed for pesticides by the rapid multiresidue procedure of Luke et al. (2). Additional analyses by other methods may be conducted, depending on the findings and other information.

As with all imported foods, violative shipments are detained according to Agency compliance guidelines. FDA's certification requirement--although applicable to all imported foods--has been particularly useful with Mexican produce. This certification requirement is invoked when at least two shipments from a single grower or shipper are found to contain violative levels of the same pesticide residue. Subsequently, all further shipments of the commodity from that grower or shipper must be accompanied by a certificate of analysis. This requirement is eliminated when, in FDA's judgment, that grower or shipper no longer has a residue problem.

This program has improved liaison with the Mexican government. FDA's Mexican Liaison Staff notifies appropriate Mexican officials at the first appearance of a pesticide residue problem. This allows for rapid corrective action by the Mexican government. Such action may involve sampling produce prior to harvest, condemning contaminated fields, and holding educational meetings with growers.

Other Imported Foods

The Agency's efforts in this area provide broad coverage of all imported foods (except Mexican produce) for pesticides, industrial chemicals, mercury, and aflatoxins.

Initial sampling criteria for pesticides include volume of product imported, dietary significance of product, previously identified residue problems, and pesticide/commodity coverage directed by FDA's Surveillance Index. Raw agricultural products and freshwater fish of dietary significance are given priority coverage. These processed foods also receive sampling emphasis: dairy products, particularly cheese; frozen, dried or otherwise preserved fruits and vegetables; processed grain products; eggs (fresh, frozen, or otherwise preserved); edible fats (margarine, shortening or other prepared fats).

The selection of products sampled is left primarily to the discretion of individual FDA Districts. Specific coverage of certain pesticide/commodity combinations may be initiated when residue problems of national concern emerge or when coverage of selected foods needs strengthening.

The program does not specify aflatoxin coverage for particular imported products. Sampling emphasis is based on known problem areas and/or current district information regarding potential contamination. Generally, past problems have involved some spices, seeds and nuts (including peanuts and peanut products from certain countries of origin).

In fiscal 1984 about 29 field person years (including aflatoxin work) were invested in this effort. This allowed for collection and analysis of about 2,800 samples. In addition, a small number of products used as animal feeds or feed ingredients (e.g., cottonseed, cottonseed meal) were sampled under the Mycotoxins in Animal Feeds Program directed by the Center for Veterinary Medicine.

Samples are usually analyzed for residues of organochlorine and organophosphorus pesticides and polychlorinated biphenyls (PCBs) by applicable FDA multiresidue methodology. The appropriate analytical method is generally selected by the districts, and is dependent on the product, suspected residues, and other available information. Aflatoxins are determined by use of various AOAC methods (3).

Products found to violate the Federal Food, Drug, and Cosmetic Act are detained in accordance with the Agency's compliance guidelines.

Domestic Foods and Feeds

FDA's coverage in this area consumes over two-thirds (about 100) of the field person years committed to the chemical contaminants programs. In a typical year, 7,000-8,000 samples are analyzed. Approximately 88% of the effort is directed towards food; the remainder is devoted to feeds and feed ingredients. Sampling is diverse, and includes selected domestic raw agricultural products, milk, cheese, eggs, grain, fish, and feeds and feed ingredients. Special provisions include: rapid identification and continuing assessment of emerging residue problems; redirection of resources and initiation of action to address specific problem areas; continuation of surveillance, with additional regulatory activity when adverse findings are encountered; and information exchange and coordination of activities within FDA and other federal and state agencies.

Generally, the products sampled, the number of samples of each, and the residues determined are left to the discretion of the District. Sampling efforts are concentrated on commodities of dietary significance, except in cases identified by Districts involving special contaminant/commodity problems.

Districts are provided time to perform special investigations, particularly as related to use of new pesticide chemicals or use of existing materials with revised application criteria, etc. Samples reflecting these conditions are collected as required.

Sampling occurs most often at the grower or wholesale level, with the exception of the Total Diet Studies. Sampling at these points in the distribution system allows more time to remove violative shipments from trade channels before they reach the consumer.

A wide variety of analyses are conducted. Pesticide and selected industrial chemical residues in foods are sought by multiresidue methods, as discussed previously. Single analysis methods are used as necessary. Again, the appropriate analytical method is selected from the FDA Pesticide Analytical Manual by the analyzing laboratory.

Aflatoxins may occur in susceptible products, such as tree nuts, shelled and milled corn products, and peanuts. The USDA is responsible for raw, shelled peanuts; FDA monitors processed peanuts. Grains other than corn are less susceptible to aflatoxin contamination. Domestic food products are analyzed by official AOAC procedures (3). The percentage of samples with detectable aflatoxin levels or levels above the 20 ppb guideline varies greatly from commodity to commodity and from year to year. A survey for vomitoxin (deoxynivalenol) in bread is currently underway. In fiscal year 1984, a survey to determine the incidence, levels, and factors affecting levels of vomitoxin in corn and wheat was conducted.

The information on animal feed contamination impacts on the human foods program and vice versa, since pesticide residue findings, in particular, in feeds and foods derived from animals are often interrelated. The Centers for Food Safety and Applied Nutrition and Veterinary Medicine cooperate fully in planning program approaches and in exchanging data on toxicants detected in sampled human and animal foods. The USDA is also involved in the latter activity.

In examining foods for toxic elements, FDA has placed the greatest emphasis on lead, cadmium, and mercury; the TDS includes analyses for these, plus arsenic, selenium and zinc. In recent years, additional work was performed under a Memorandum of Understanding (4) by FDA, USDA and EPA. Under this agreement, FDA was responsible for analyzing several thousand crop samples for various elements. Nearly 2,000 samples of lettuce, peanuts, potatoes, soybeans, sweet corn, and wheat from major production areas have been analyzed to determine background levels of cadmium, lead and several other elements. The sampled crops either represent major sources of cadmium in the human diet and/or are known cadmium accumulators. Generally, levels of cadmium and lead have been much lower than those previously reported. There are many possible contributing factors, including careful sampling and sample handling, and use of laboratory practices and equipment designed to minimize contamination of samples in the analyses (5).

FDA's Total Diet Studies, which are described more fully below, form the final domestic program element discussed. Approximately 18 field person years are spent in determining dietary intakes of many contaminants by various population groups. Analyses of table-ready foods are conducted. This program is invaluable in that it provides a final check on the effectiveness of the U.S. pesticide regulatory system.

In summary, diverse sampling areas are addressed by individual FDA programs. Since the largest effort is devoted to the domestic pesticide program, particularly as applied to human foods, I will present it in greater detail.

Domestic pesticide program elements and efforts. This program has the specific objectives of determining levels of pesticide and selected industrial chemicals in domestic raw agricultural products, milk, cheese, eggs, grains, and fish, and removing food with residues of regulatory significance from commerce. Since the general sampling and analytical options were previously detailed, this discussion is limited to specific efforts.

The "core" sample element is comprised of a minimum number of selected foods of animal origin (milk and eggs). A specified minimum sampling effort by each district is required, since these foods contain the majority of persistent environmental contaminants in the diet and since milk is so important in the diet of young children. Approximately 15% of this program's field effort is devoted to core samples.

The majority of FDA's domestic effort (more than 50 field person years) may be termed "district option surveillance". This involves objective sampling in which the districts, at their discretion, systematically survey commodities of dietary significance and local agronomic importance. Information on local pesticide usage is important in selecting pesticide/commodity combinations requiring emphasis.

An important element common to all FDA programs is compliance sampling and follow-up. Compliance samples are subjective in nature. They are collected with foreknowledge that there is likelihood of regulatory action because of pesticide misuse, etc. This information is derived from previous surveillance results or other sources. Sample findings and the case history may result in regulatory action (seizure, etc.).

The last major sampling and analytical area relates to special surveys or "directed surveillance". Several sampling and analytical areas are highlighted in recent years including: selected fish samples to be analyzed for tetrachlorodioxins; analysis of grains for the fumigants ethylene dibromide, carbon tetrachloride, and chloroform; examination of peanuts, grains and soybeans for malathion residues; determination of N-methyl carbamates in various commodities; analysis for paraquat in selected samples of wheat, dried beans, grapes and potatoes, and sampling of leafy vegetables (cabbage, spinach, lettuce) from fields known to have been treated with ethylenebisdithiocarbamate (EBDC) fungicides.

In addition to the specific coverage provided by the above sampling/analytical emphases, provision has been made for each district to utilize up to 20% of its domestic pesticide program resources in conducting special assignments involving specific pesticide/commodity combinations not otherwise covered. The most important concept which aids in directing these pesticide activities is the Surveillance Index.

Several years ago, FDA initiated a program designed to evaluate each of the more than 300 pesticides registered by EPA for agricultural use. The purpose of these evaluations is to determine the relative hazard of these pesticide residues in foods. The compilation of hazard evaluations is termed the Surveillance Index (SI) (6). About 180 assessments have been completed. Factors such as production volume, crop usage, environmental stability, significant impurities, alteration products, toxicity, and potential dietary exposure are assessed. Each pesticide is categorized in one of five hazard classifications, with Class I representing the greatest potential hazard. The classification is then used in determining priorities for selection of commodities and analytical methods in both this and other programs. These priorities may culminate in the form of specific sample collection and analytical efforts. For example, a survey involving phosphorodithioate pesticides was done in 1983. Three such compounds (phorate, disulfoton and terbufos), together with several toxic metabolites, were determined in either beans, lettuce, cole crops, tomatoes, or sweet corn known to have been treated with one of these pesticides. Because of the sample treatment knowledge, the survey data are particularly useful in assessing residue levels of these pesticides (and metabolites) identified by the SI as being worthy of special monitoring effort.

Districts also use the SI in planning and executing their local monitoring plans. The SI, together with local investigations, provides information necessary to formulate an effective and efficient plan of sample collection and analysis that focuses on those pesticide/commodity combinations posing the greatest potential health risk. This approach will receive increased emphasis.

<u>Total diet studies</u>. This program has as its major objective the determination of the levels of pesticide residues, PCBs, other industrial chemicals, and seleted toxic elements in the diets of various population groups (e.g., adult females and males, young children, infants, etc.). The daily dietary intakes of these contaminants are computed and compared with "acceptable" intake levels, e.g., FAO/WHO (Food and Agricultural Organization/World Health Organization) Acceptable Daily Intakes (ADIs). The intake data for the determined contaminants are then useful in determining trends.

The Adult Total Diet Studies from 1961 to March 1982 were based on information from USDA's 1955 or 1965 Household Food Consumption Surveys. The Infant and Toddler studies, initiated in 1975, were based on food consumption data from the latter survey.

These studies used a food composite approach; approximately 120 foods were represented. Foods representative of the particular diet were purchased at the retail level nationwide and prepared in a manner representative of that used in the home (washed, trimmed, boiled, baked, etc.). Groups of similar foods were composited (e.g., dairy products, leafy vegetables), using consumption figures to determine the weight of each component in the composite. The adult diet was represented by 12 composites, while the infant/toddler diet had 11. The foods and/or the composites from one collection site are often referred to as a "market basket". An exaggerated adult (teenaged male) diet (3900 kcal per day) maximized contaminant intake; however, the infant (6 month) and toddler (2 year) diets reflected average intakes. Generally, 20 adult and 10 infant/toddler "market baskets" were analyzed per year by FDA's Kansas City District Total Diet Laboratory.

The Total Diet Study suffered in recent years from being based on outdated food consumption data. A revised Total Diet Study has been conducted since April, 1982. Individual foods and ingredients necessary for preparation of recipe items are collected at retail markets 3-4 times each year. Each collection occurs simultaneously in 3 cities. Two hundred thirty-four ready-for-consumption items are analyzed. They are representative of the diets of eight age/sex groups: infants, young children, teenage females and males, adult females and males, and older females and males (7).

The 234 food items are analyzed for selected pesticides, PCBs, lead, arsenic, selenium, cadmium, zinc and mercury. Selected foods are analyzed for some of the more volatile chemicals. One market basket per year is monitored for radionuclides. Selected food items are separately analyzed by Dallas District for higher chlorinated (Cl_6--Cl_8.) dioxins. Additional analyses are made for selected minerals; this effort is covered under a separate assignment directed by the Bureau of Foods' Division of Nutrition.

Two national dietary surveys were used in formulating the updated diets, representing the foods consumed by the eight age/sex groups specified. They were the USDA's Nationwide Food Consumption Survey (NFCS) of 1977-1978 and the National Health and Nutrition Examination Survey (NHANES) II of 1976-1980 from the National Center for Health Statistics (NCHS). The average number of grams of each food consumed daily by each age-sex group was determined by averaging the consumption data from the two surveys.

Analysis of individual foods provides for future flexibility in assessing dietary intakes for other population groups. Also, contaminant intakes calculated on the basis of diets formerly used may be computed for comparative purposes. Elimination of the dilution effect resulting from compositing several food items is another distinct advantage of the revised study.

The increased analytical burden of the revised program necessitated a reduction from 30 market baskets to the 4 presently planned for analysis annually. (In effect, field resources available to conduct the revised study remained constant.) The reduction in market baskets is more than compensated for by the increased number of foods and age-sex groups represented and elimination of the dilution effect by the analysis of individual foods.

Analyses of the first four market baskets from the revised program have been completed. Because the number of samples to date is so limited, a detailed assessment of the findings would be premature. However, a greater frequency of low level findings has been noted, along with an increase in the number of different chemicals detected.

This is not unexpected, since the dilution effect of the composite approach is no longer a factor.

As previously stated, FDA's Total Diet Studies provide a final check on the effectiveness of the U.S. pesticide regulatory system. This is illustrated by the fact that only in the mid-1960s did the intake of a pesticide approximate the ADI established by the FAO/WHO. This occurred with dieldrin in 1966 (8). In recent years, there have been sharp reductions in the intakes for some chlorinated insecticides, such as dieldrin and DDT. The substantial decrease in intake levels of these persistent compounds, whose food uses in this country were eliminated about a decade ago, clearly demonstrates that reduction in dietary intakes can occur in a relatively short time.

Reports summarizing data obtained via this and selected other monitoring programs are available through the National Technical Information Service, U.S. Department of Commerce, Springfield, Virginia 22161.

SUMMARY

FDA's monitoring efforts in the area of chemical contaminants are diverse. Coverage of pesticides, industrial chemicals, toxic elements and natural toxins in a wide range of foods and feeds is provided. The scope of potential contamination is always much broader than that which can be monitored with available resources. Chemicals will undoubtedly play an increasingly important role in agriculture. FDA's chemical contaminants programs will continue, with emphasis directed towards the most potentially hazardous residues. The agency's overall efforts are directed to assuring food safety and in promoting consumer confidence in the food supply.

REFERENCES

1. McMahon, B. M., and Sawyer, L. D., eds., "Pesticide Analytical Manual," Vol. I. U.S. Dept. of Health and Human Services, Food and Drug Administration, Washington, D.C. (1983a).
2. McMahon, B. M., Sawyer, L. D., eds., Chapter 2, Extraction and cleanup, Sections 212.2 and 232.4. "Pesticide Analytical Manual," Vol. I. U.S. Dept. of Health and Human Services, Food and Drug Administration, Washington, D.C. (1983b).
3. Horwitz, W., ed., Chapter 26, Natural poisons. "Official Methods of Analysis," 13th ed. Association of Official Analytical Chemists, Arlington, Virginia (1980).
4. Anonymous. Fed. Reg. 44, p. 44940 (1979).
5. Wolnik, K. A., Fricke, F. L., Capar, S. G., Braude, G. L., Meyer, M. W., Satzger, R. D., and Bonnin, E., Agric. Food Chem. 31, 1240 (1983).
6. Reed, D. V., ed., "The FDA Surveillance Index", U.S. Dept. of Commerce, National Technical Information Service, Springfield, VA, item no. PB82-913299.
7. Pennington, J. A., J. Am. Diet Assoc. 82, 166 (1983).
8. Duggan, R. E. and Lipscomb, G. Q., Pest. Monit. J. 2, 153 (1969).

24

Industry Views of Regulations Concerning Cereal and Legume Products

ADOLPH S. CLAUSI

I am one representative of the food industry. My views perhaps reflect the perspectives of my company, but I certainly cannot presume to represent the industry as a whole for we are as diverse in our opinions as any population at large.

Because Post cereals are such an important part of General Foods' Business, I'm very much aware of the regulations and standards which apply to this part of the food business. And some of the discussions I remember from 20 years ago are still going on and in general it doesn't appear to me that we have moved the balls forward very much:

...For example, how will natural toxicants be handled--those that occur naturally or those that inevitably find their way into nature?

...What about indirect additives--the pesticides and agricultural chemicals that the grain farmer may use?

...Do standards of identity really benefit and protect the consumer?

...Is there need to further reconcile FDA and USDA responsibilities?

...And there are still lingering questions about antioxidants and ready-to-eat cereals--the use of BHA and BHT--as well as other concerns that some people have about additives. This applies particularly to ready-to-eat cereals and other grain or legume based processed packaged foods.

These questions obviously relate largely to safety, and the consumers' perception of safety, and I will address that issue. First, I would like to take a giant step backward and look at government's role with respect to the food industry, and to explore the nature of that relationship and how well it is working.

As I see it, the government should be the consumers' surrogate in assuring that a safe, wholesome and abundant food supply is available to the population at large without unduly restricting the free enterprise system. There are many dimensions to that statement, but let me highlight a few, which are governmental obligations:

1) The need for assurance that the food supply is "safe enough," recognizing that science has clearly demonstrated that "zero risk" is unachievable.
2) The need for assurance that economic fraud will not be perpetuated on the public, but recognizing that new technology will of necessity alter to some degree the nature of traditional foods.

3) The need for establishing an environment conducive to innovation and new technology so that the food supply can become more abundant, more economical and more coincident with emerging consumer needs.
4) The need to allow the free enterprise system to operate efficiently and cost effectively consistent with the above.

One area that stands out in this discussion is the role of the government in establishing standards. The intent was clear. If standards are established for various food categories, then unscrupulous merchants will be deterred from adulterating their products with inferior ingredients.

But what does this do to the need for flexibility, especially in the light of new technology?

As one member of the food industry, I was delighted to hear the Food and Drug Administration's Dr. Sanford A. Miller express his personal opinion for "flexibility" in food standards at the Food and Drug Law Institute's Food Update meeting in 1983, because as Dr. Miller said, the nature of our food supply has changed and it's now difficult to compare new products with traditional foods. So he called for innovation to make certain that food standards did not remain "counterproductive" or discourage competition and the development of new technology.

Food standards are not the only hindrance to our ability to use emerging food technologies that can provide new markets for cereal and legume products. Over the last 20 years or so, food technologists have been developing totally new fabricated food products which open new markets for these and other commodities. Unfortunately, however, many of these new foods compete directly with other established commodities. And when this competition occurs, special interests often seek legislative protection against market erosion. A clear example of this with which my company is familiar is the filled-milk statutes in several states.

Finally, I need only remind you of the long and often irrational debates over the nomenclature for poultry analogs of red meat products, and the suitability of fish protein for human consumption.

I've dwelled on the issues that standards raise in part to reinforce some of the statements made by Tom Scarbrough (Chapter 22), but also to emphasize the need for change. This need is accentuated by both economic and nutritional arguments for the increased utilization of cereal grains and legumes in a host of new products.

I would like to address the need for legislative reform in food safety. In all of my more than three-and-a-half decades in the food industry's technical research function, there is no issue that has caused so much comment--and often unjustified consumer concern--than that of food safety. Even for such relatively simple products as some of those processed from cereals and legumes. And I would like to explain some of my own views of food safety and the public interest.

My views have been tempered by the opinions of others who served with me on the Food Safety Council, a group which brought together representatives from both the public and private sectors and spent a total of six years developing what we believe is a better way to make food safety decisions.

From my work on the Food Safety Council as well as from my technical research experience, it's obvious that one concept that is widely held is making it difficult for regulators, processors and the public to come together on the resolution of food safety issues and concerns. And that is the concept of "zero risk" which some infer from current law, but which undoubtedly was never intended by the Congress back in 1958.

The well-known Delaney Clause provides the best example of this concept, of course. And most thoughtful people within the food industry and regulatory agencies will agree, I believe, that this "all or nothing" statute--which became law nearly 30 years ago--should be revisited in the light of today's technological expertise. Beyond that, other core food safety legislation and regulations also need updating. For it becomes clearer every year that the zero-risk concept is incompatible with modern science, and is not needed to achieve adequate protection of the public health. And the Food & Drug Administration has said time and again that it does not need this clause to bring action against substances it believes are unsafe.

The assumption that zero risk is an achievable state underlies the prevailing interpretation not only of Delaney but also of most of our complex maze of food safety legislation and regulations. In other words, "safe" is considered synonymous with zero risk, despite the tremendous advances that have been made in analytical methods since most of the food safety laws and regulations were enacted.

We now have the ability to determine parts per billion and per trillion with accuracy. And soon, perhaps, parts per quadrillion. Therefore, can we ever attain a state of no detectable presence of a substance which at an appropriately higher dose, can be shown to be harmful? Undoubtedly not, because our science will some day find "something" in everything!

By continuing to pursue the zero risk concept, I feel we are refusing to recognize the evolving concept of "insignificant risk," which has been reflected in some recent decisions by the Food and Drug Administration. Moreover, we are overlooking the promising role which risk assessment can play in assuring the safety of food additives-and its relevance to overall food safety as well. I believe we should stop kidding ourselves that an ideal state of "no risk" can be achieved, and should start deciding how we can effectively manage insignificant levels of risk.

One may feel that cereals and legumes have less to do with the zero risk concept than the highly processed foods we consume. That may be true by degree, but don't rest on it--everything gets involved, eventually, if "zero risk" mentality prevails.

The big question facing the Congress, regulatory agencies, industry, and the consuming public is not whether food safety legislative reform is necessary. It seems to me that this is a foregone conclusion, given the current concern and skepticism on this public health issue. Whatever our individual perspective, there's no denying that it is time for us to examine our U.S. food safety laws.

The real question that needs answering is how to effect legislative reform that still provides adequate protection to the consuming public. I believe that reform will be achieved in this technically complex, politically sensitive area only through a step-by-step process that is broad-based, bipartisan, and consensus building--one step at a time. One small step for mankind, but one big step for the food supply.

And I also believe that once we start down the road to reforms, we will allay many consumer fears and much confusion about what is safe to eat, and we will dispel food industry uncertainty about the ground rules on which the safety of products will be judged.

I readily admit to prejudice by saying that the years of comprehensive work of the Food Safety Council can provide some cogent answers to how we can deal with the challenge before us. Not only on the process that might be used to work out the needed reforms, but also on the reforms themselves.

The Food Safety Council's governing Board of Trustees was made up of an equal number of representatives from the public--academia, government, professional societies, and public and consumer interests--and from private organizations--primarily, those related to the food industry. It was a diverse group, with almost as many viewpoints as there were members. But it worked!

We came to respect each other's viewpoints as we continued to strive for interaction, for dialogue, and for a synthesis of perspectives. I think we demonstrated that by employing an open and interactive process, people of good will--knowledgeable people from different sectors of society--can work together to help resolve an important and complex problem. We were able to make what I have termed "an experiment in consensus-building" come alive.

I would suggest that any proposed legislative or regulatory reform could benefit from using this technique--for it allows consideration of more than just one point of view. And if those moving forward in this area can focus on broad-based opinions, I believe that a reasonable consensus will be the result. But the Congress must take the lead and begin moving the process forward. Similarly, moderates of both industry and consumer interest organizations must lay on the table the small steps they are willing to take.

Inherent in the many suggested modifications already existing is a useful starting point for a distillation of viewpoints. The bills in Congress, comments from various areas of the public sector, industry recommendations and, last but by no means least, the proposed food safety evaluation process developed by the Food Safety Council, all represent steps in the right direction.

The Council's final report identified four broad issue areas in which the current food safety system could be improved. It lays down how to ...

...Scientifically assess risks in the food supply,
...Determine socially acceptable levels of risk,
...Manage food risks, and
...Involve affected interests in the decision-making process.

These are the four major points which must be carefully considered as reform of food safety legislation is undertaken in the United States:

First, we must make sure that the best science is brought to bear on the problem.

By that, I mean that our current regulatory system, which operates under flexible, unwritten guidelines for testing a substance's potential risk, should be formalized to the extent that we end up with a systematic, organized testing pattern. I am not advocating a rigid testing regimen, but one that is more formal, more systematized, more predictable, than what we have now.

The decision-tree approach recommended by the Food Safety Council organizes a sequence of tests designed to generate the information that regulators need to decide whether a substance can be accepted, rejected, or if more tests are required. Because of the constant advance of scientific methods, the Council does not suggest that precise tests be specified in legislation or regulations. Rather, the Council's decision-tree approach provides the template for evaluating food safety in an orderly fashion--one that avoids bad decisions being made too quickly without adequate data while at the same time providing ground rules for thoroughness.

The Food and Drug Administration has since gone on record with a decision-tree approach of their own, which has been acclaimed. So several sources have suggested means of acknowledging in our laws and in our behavior not to put too much weight on any one test but rather on a systematic, organized testing pattern before decisions are made on substances in our food supply.

A second point is that the food safety legislative reform must provide for some type of risk/benefit assessment that is based on its relevance to humans.

Regulatory agencies currently use three approaches for determining the safe use of a substance in foods: the oft-employed safety factor, that is extrapolating from high dose animal studies; automatic prohibition; and occasionally specific restrictions on use in foods. Because of our increasing ability to identify and measure minute quantities of substances--and even understand some of the dimensions of the risks that substances may create--we need to develop a more innovative risk management strategy that employs other approaches in handling the potential risks to which consumers may be exposed.

We need to make broader use of quantitative risk assessment, especially for those difficult-to-evaluate potential chronic effects where we must go from small populations of animals to large populations of humans. We also need to formalize the use of limited risk/benefit analysis, which identifies and balances the risks versus all the benefits of having a substance in our food supply--especially where they are critical to the food supply, such as cereals and legumes.

A risk management strategy, properly developed, can effectively protect consumers while still allowing for a greater degree of regulatory flexibility. We in the United States have traditionally considered risk black or white--either allowing or banning/restricting a substance's use in food. Now, with the greater use of quantitative risk assessment, risk/benefit or risk/risk analysis, there is an opportunity to make the regulatory process more explicit, systematic, and realistic in applying science.

In addition to employing all of the advances that have ever been made in identifying and evaluating risks, the Council proposed that food safety decision-making can benefit from broader and more effective participation of all affected interests.

By added participation, we mean offering industry, the public and other interests the opportunity to participate in a timely way, without causing inordinate delays. Qualified individuals from all sectors of society who have relevant information to contribute can add to both the substance and the quality of food-safety deliberations.

Naturally, government is the final arbiter of what is best for society when it comes to food safety. But greater participation can ensure that the regulatory agencies' judgments reflect the concerns of the various interests in our society. And greater participation will not only improve decisions by making them more informed but will also enhance their perception and acceptance by society.

A third point is that peer review is an essential component when significant risk is involved.

Even though we believe that the Council's decision-tree approach will reduce uncertainties associated with toxicological testing, it cannot eliminate them completely. Test results can still lead to different and seemingly equally sound scientific interpretations on the type and magnitude of risks posed. So regulatory agencies, which will have to make decisions regardless of uncertainties, may find a second scientific opinion useful before proposing to regulate a substance.

Peer review can help resolve scientific questions concerning test results and assist an agency in understanding the scientific uncertainties associated with them. Because peer review is scientific in nature, it should be used primarily on technical issues, such as determining whether a test was well designed and properly conducted, as well as to assist the agency in interpreting the validity of the test results and not to render social opinions.

The final point is that we need a wider array of regulatory options than are now available.

Regulatory agencies have traditionally used two principal approaches to manage risks: banning a substance from the food supply, or establishing a safe level, depending on the food system in which it is used. In recent years, the FDA and the Congress have also employed labeling to provide consumers with information on a substance's potential risk and/or its presence in foods, such as what is required for saccharin-containing products today.

The Food Safety Council favors the employment of a wider array of approaches to a risk management strategy for food safety, which will not only protect consumers from existing potential food risks but also provide for an abundant and nutritious food supply within a flexible, innovative regulatory structure. For example, phase-in of new substances and phase-out of substances to be removed.

We also favor more open and balanced information from government-information that could help shift some of the responsibility for making food safety decisions from the government to consumers. Providing more or fuller information regarding possible risks of a substance, when discussion is appropriate, does not negate the government's responsibility to establish safe use levels. Rather it gives consumers the opportunity to better understand risk and have more say in what they eat.

As a member of a food company that has served consumers for decades, I feel that people should be allowed to make well-informed judgments on what to eat. Giving consumers more information--where practical and necessary--will certainly help make them aware that nothing in life is risk-free.

I admit to bias, but I believe that the Food Safety Council presents a useful framework of issues to deal with and suggests approaches to their answers. The Council never expected that all its positions would be immediately adopted--in fact each one of us, myself included, would probably want to make some changes--but it does represent a potential level of broad base consensus that we should all look at seriously.

This leads to a last point. As an educated society and led by government, we must recognize--and accept the fact--that nothing in this world is "absolutely safe". We need to make full use of all the scientific tools available--to be practical and realistic if we are to maintain the quality of our food supply and ensure its safety, wholesomeness and abundance at an affordable cost. Therefore, the policy we should pursue is one of making sure our food supply is not "risk-free" but "safe enough".

We must create an environment in our laws and regulations that will call for examining and reexamining our standards, our administrative procedures and our governmental responsibilities. What we need is an evolutionary, not a revolutionary, process. We need to continue to move the ball forward--at the same time we must avoid chiseling regulations, standards and agency roles into cement.

What it comes down to is an attitude of "let's talk"--a use of democracy and openness, not backroom politics and backbiting. Because no one will win unless we all give a little for the sake of advancing our ability to cope with the future.

VI

Discussion

25

Implications of Food Changes for Higher Education Programs

KENNETH A. GILLES

INTRODUCTION

Today, the United States seems to be blessed with an abundant food supply. This abundance has become a major problem; we are holding about 60% of the entire world's reserve of cereals. On one hand, we can say how fortunate we are to be able to do this; on the other, we must recognize that it is a severe economic depressant to the agricultural community. However, we in America were not always confronted with that type of problem. In general, we should recognize that our food production and processing concepts were introduced into the United States largely by immigrants who came from Europe. The techniques of cultivation, harvesting, storage, and, ultimately, processing--generally in small quantities--all were practices indigenous to the social culture of the settlers. In many respects, the development of grain production and utilization had changed little for many centuries. The general state of agriculture in pilgrim times was such that about 95% of the people were involved in other business, social, and political pursuits. As we all know, today the situation has changed, largely through innovation and application of the free enterprise system which permits approximately 3% of the people to be engaged in agricultural production; some 25% in the total agriculture infrastructure; and about 75% of our current population is involved in other business--artistic, social, and political pursuits.

It is my opinion that agriculture, food science, and technology, in their broadest aspects, are one of the United States' most important contributions to the improvement of lifestyle in the 20th century. Worldwide, the United States is noted for its progress in manufacturing autos, steel, aircraft, and more recently, electronics and microprocessors. These certainly have improved our lifestyle. However, I feel these technical advances were made possible because the innovators and creators of these advances were able to pursue their creative efforts because of freedom from concern about a wholesome, reliable, and economic food supply in the United States.

Within the last decade, remarkable changes in attitudes concerning food availability, production, and education have occurred. In 1975, the University of Minnesota hosted a "Conference on World Food Needs." This event occurred just a few years after the remarkable increase

Cereals and Legumes in the Food Supply, edited by Jacqueline Dupont and Elizabeth M. Osman 1987 Iowa State University Press, Ames, Iowa 50010

in grain exports to the Soviet Union sent shock waves throughout the grain industry. The fear, at that time, was based on the fact that the world's population would grow so rapidly that doubling of food production within the next 20 years probably would be a momentous task. Ambassador Edwin Martin remarked that the World Food Conference of 1974 concluded there was enough grain to feed the expanding population through 1985, providing economics, weather, and politics were favorable (1). He concluded that unless a sense of global community developed, food would be the most critical need in an overpopulated world by the year 2000. These concerns were very frequently articulated by people during the 1970s.

While we haven't reached 2000, as yet, the need to take land out of food production to the extent that occurred in the 1983 crop certainly causes one to reflect and reassess some of the concerns that were articulated a decade ago. Food production capability of the world can continue to be expanded if adequate incentives, technology, political stability, and economic assessments are attractive.

I believe we can readily agree that in today's agriculture, horizons are more expansive, the opportunities are more abundant, the problems more complex, and the challenges greater than ever before. These comments were addressed by Robert Parrott during the discussion on Changing Horizons of U.S. Agriculture (2). He noted that America is endowed with many natural advantages and resources for food production. Other countries, similarly endowed, exist on marginal diets. He concluded that the major difference was related to the American infrastructure, which includes the entire chain of the free market system, coupled with the profit incentive at each critical link, a competitive system of local markets, a transportation network of highway, railways, and waterways, interconnected by subterminal and terminal markets, a closely integrated marketing organization, high-capacity export facilities, and financial institutions to provide the credit that enables the product to flow smoothly through the system.

The remarkable expansion of food exports of the 1970s were dealt a serious blow by President Carter's imposition of the grain embargo on Russia in 1980. While directed at one country, this action renewed concern that the United States was not a reliable supplier of food products to the world; indeed, the value of food products exported reached its highest peak in 1981 and has been diminishing since. This slippage hurts the whole economy.

Agriculture is the nation's biggest business, and farm products are the nation's biggest export category, accounting for 18% of all U.S. exports. The Wall Street Journal (3) noted in 1982 that $39 billion in farm products were exported, and each $1 billion provided 30,000 jobs off the farm. The value of exports has declined. It has caused people in the grain industry to suggest that because of the loss of U.S. market shares in world trade, the grain industry's surplus shipping and handling capacity will probably persist until the late 1980s. Thus, the entire agribusiness system that a few years ago geared up to feed the world now is in serious straits.

You may wonder whether I am attempting to act as a Washington bureaucrat or as a person directing comments to the dynamic changing situation contronting higher education in America. May I make these types of comments: Those people who know me well will testify I am not a typical bureaucrat. Secondly, my viewpoints are tempered, just as yours may be, by the totality of past experiences we have individually and collectively enjoyed. My experiences and challenges as a student, instructor, food scientist, college administrator, and, now, a government administrator have enabled me to gain some insights into the complexities of the world food supply and the roles that cereal and legumes play in this food supply. I have attempted to enumerate in my opening comments that one of the greatest contributions of the United States has been in the area of agriculture production, food science, and technology. I believe this happened primarily because of the wisdom displayed in the post-Civil War expansion era. When the United States created the land-grant college system, the role of education, research and extension took on a unique importance in the United States, and, combined, these roles have provided stimulus for our abundant food supply.

HIGHER EDUCATION PROGRAMS

When one addresses the subject of higher education programs, various facets could be emphasized. Today, I would like to consider, primarily, the quality of the product and current concerns that are articulated by the press and public. I believe we could readily agree that the quality of a product generally reflects the quality of resources that are brought to bear in creating the product. However, the concept of quality changes with time and style. In the educational area, progressive education, open classrooms, and rote memorization, are terms that may reflect buzzwords. Fortunately, buzzwords frequently come and go, depending on people's attitudes and style. More specifically, as we speak of resources in higher education, I believe we are thinking of three major components: (A) people, (B) facilities, and (C) finances.

People

"Agricultural expertise is vital to the security and well being of this country. The United States cannot continue as the lead nation in agriculture without new efforts for the development of its human capital, the ultimate resource" (4).

Recently, the Resident Instruction Committee on Organization and Policy, Division of Agriculture, National Association of State Universities and Land Grant Colleges, published a brochure entitled "Human Capital Shortages: A Threat to American Agriculture." Some of the major problems relate to the fact that scientific literacy exhibited by students entering colleges and universities has declined dramatically during the past decade. Insufficient preparation of entering college students is a serious problem in developing agricultural expertise. Data published by the National Academy of Sciences in 1982 showed that about one-third of the high school graduates completed 3 years of mathematics; only one-fifth completed 3 years of science; only 10% had taken a course in physics; and fewer than 8% had completed a unit of calculus. Moreover, there is a tendency for high school students not to go to college. Minnesota and North Dakota are states that lead the nation with more than 85% of their high school graduates going on to institutions of higher education.

The concern for quality of secondary education has been expressed by a number of states which now require proficiency examinations before a high school diploma is awarded. The States of Florida and Virginia are among these. Recently, certificates of attendance were awarded rather than diplomas for those who failed to pass the proficiency tests. This is causing considerable trauma. Perhaps it is a trend that will be useful in the future. In 1983, the Washington Post carried an article (5) that indicated three out of four 9th graders failed to pass math proficiency. A series of questions were included, which suggested that the requirements for this proficiency were very modest, indeed.

One welcome trend, expressed in Time Magazine, suggested that many leading colleges are pursuing the star seniors as never before, offering scholarships based on achievement, talent, and promise rather than on financial need. The new celebrities are students with high SAT scores and good report cards. Competing colleges are offering scholarships rather than grants in aid. This departure bothers many educators. Yet, one wonders whether this change in college recruitment is not overdue. Why shouldn't academic achievement be rewarded? Perhaps this question was appropriately answered by Eric Engles, who, following a perfect 800 SAT score, accepted a scholarship to the University of Virginia, and said, "It's about time they gave the same attention and money to scholars that they do to athletes" (6).

Perhaps the United States has entered a new phase of educational style in which achievement and scholarship will be come meritorious objectives from the secondary and post-secondary educational system. As one who has observed changes in educational styles for several decades, this current trend is most welcomed.

Interesting data, compiled about graduates in higher education in food and agricultural sciences, has been published by the U.S. Department of Agriculture (7,8,9,10).

The second facet of the people resource relates to teachers. To those teachers who are enthusiastic, dedicated, knowledgeable, patient, and empathetic for the student, we should devise a means of recognition and acclaim. For the terms praise, recognition, and reward are, indeed, as meaningful to people in the academic as the business and political fields. Unfortunately, during my tenure as Chairman of the Legislative Committee for the Division of Agriculture, NALGCSU, it became painfully evident that professors of agricultural sciences were not among the most highly paid people at the university, and, in fact, many who had 12-month appointments found their average monthly paycheck to be less rewarding than those people in the science, literature and arts areas who were on 9-month appointments. Nevertheless, I am firmly convinced that the quality of people in agriculture and food science areas is certainly comparable to the quality of people in any of the other areas of the academic institution. Unfortunately, agriculture and food science are not glamorous fields; while recognized as needed, they lack the charisma of popular topics. However, viewed objectively over several decades, the professors of agriculture have enjoyed a far greater and more durable area of respect than those charismatic professions that ride popularity waves to pinnacles of momentary success and subsequent oblivion.

One of my concerns, long-standing, is related to the fact that many professors ascend to their current position by being good students, attending a good graduate school, becoming a teacher, and then ascending the academic ladder. This in-house type of training has some very strong merits, particularly if the student has had exposure to several different institutions. Unforunately, in the training of a professor, contact with the free enterprise system frequently has been minimal. It would seem to be that to strengthen the role and value of the professor, one should encourage greater interchange of ideas and contacts with the business and government community. Food science is not operated in a vacuum--it is interdependent--and this should be recognized, appreciated, and experienced by a professor. Perhaps the opportunity for on-the-job-training, for cooperative education, and for developmental leave should be encouraged with greater participation than appears to be currently practiced. Moreover, we need to protect a cadre of qualified instructors, particularly as there are deepening deficits of qualified university faculty members and industrial scientists, caused by economic conditions and retirement.

Within the next 5 to 10 years, a high proportion of current agricultural scientists and educators, who began their careers shortly after World War II, will retire. Some segments of the U.S. agricultural industry forecast a similar pattern. Replacements simply will not be available in many specialities (4). The number of doctorates graduated in science and engineering has dropped about 9% since the peak year of 1973. Currently, more than one in five doctoral degrees in science and engineering now goes to foreign citizens. In some fields, the figures are much higher. In 1981, foreigners got 52% of all American doctorates in engineering, 38% in agriculture, and 31% in computer science (11). Consequently, at a time when we are approaching high retirement and potential turnover, we are training fewer potential replacements.

One of the major challenges for agricultural science and education is attracting highly capable students. We need the brightest and best undergraduate and graduate students. It is suggested that colleges and universities should better communicate areas of critical shortages, and should redirect funds so that highly qualified applicants can be attracted. Incentives must be competitive with competing disciplines. Agricultural business and industry must provide added emphasis and visibility to emerging high technology careers through advertising and promotion campaigns. Professional organizations must increasingly articulate human capital needs to targeted groups outside of their membership. The Council of Agricultural Science and Technology (CAST) should be commended for the publication of Science of Food and Agriculture, a magazine which seeks to advance the public's understanding of food and agricultural science and technology. This publication represents the joint effort of some 25 scientific societies (12). Charles Black, Editor-in-Chief, as a member of Iowa State University faculty, was a driving force in this organization for more than a decade. The organization has focused its energy primarily on supplying national decisionmakers with scientific information. Indeed, this organization has contributed significantly during the past decade to these objectives.

Facilities

The second major resource, facilities, may be viewed by some as most controversial. In some instances, laboratory equipment at some universities is considered to be a decade or more behind that available in industry, according to Frank Press, President of the National Academy of Sciences, but he has indicated that future NAS budgets will contain some funds for new equipment. In the area of agriculture, Secretary Block consistently supported advances in agricultural research, and provided an increase in the agricultural budget. Generally, during a time of budget constraint, travel and equipment are two items considered to be discretionary expenses and are most liable to be reduced, in an effort to save necessary funds to provide for people and for supplies. We in the United States have gone through considerable belt-tightening in budgets in recent years; concomitantly, we have experienced during the last decade a tremendous increase in the cost of equipment. With the change from the 20% inflationary spiral of the 1980s to the less than 5% inflationary increases now, these problems will be reduced to a more reasonable perspective.

The other aspect of facilities pertains to the physical structures, which generally I consider to be very good, when viewed from the perspective of the quality in physical structures used for higher education throughout the world. Undoubtedly, changes will continue to be made, and continuous building improvements will be possible. However, perhaps the immediate emphasis should be on concentrating our efforts to improve the human resources.

Finances

The third resource is that of finances. I have already alluded to the fact that finances are extremely important when one considers facilities and personal compensations. The current tendency has been to have the states provide an ever-increasing share of the cost of higher education and the cost of agricultural research. If one compares the federal investment in agricultural research today with that of 40 years ago, one would note a very sharp reduction in percent of the total budget, but would notice a substantial increase in the total number of dollars that are devoted for this purpose. Consequently, from the national viewpoint, it is thought that increases in financial commitment for higher education and research are more appropriately directed by the states, with support by the Federal Government.

TRENDS IN THE FOOD INDUSTRY

As one views the trends in the food industry, one might simplistically cite several major areas: unit processing; integration; international activities; and, acquisition.

Unit processing primarily was directed to meet local needs. As the food industry became more affluent, it expanded into regional and national activities. Subsequently, the emphasis on food processing changed to that of integration, where the food processor was more concerned with production, processing, and marketing. In some cases, contract growing and marketing techniques provided better quality and more uniform products for specific markets. But, as the industry has grown further, emphasis on international activities has occurred. During the last 20 years, for example, the value of agricultural commodities exported increased from about 2 billion to 42 billion dollars, and in more recent times it has begun to decline. Because of the growth of business, a typical food company became attractive for mergers and acquisitions by many businesses not fundamentally associated with the food industry. Consequently, the trends that we have experienced in the last two decades have altered significantly the manner in which the food industry currently is structured. With the degree of complexity and challenge for the future, one can only anticipate that the trends where the big get bigger and the small experience acquisition or go out of business will continue.

TRAINING NEEDS

I would not be presumptuous and suggest a specific syllabus for training for food needs of the future. However, I would like to share a few points to ponder. I think as one looks at higher education, the training should emphasize primarily basic food science. This would include an emphasis in the hard sciences of biology, mathematics, chemistry, and physics. Secondarily, the emphasis should occur in communication skills. The need for English, speech, foreign language and culture, as well as interpersonal skills are areas that we need to explore for a strengthened educational training system. In a tertiary area, I feel strongly that exposure to accounting, economics, management theory, civics, or government--call it what you may--basics of how a government is organized and functions, what its regulatory agencies and responsibilities are, some exposure on foreign affairs and world history, on-the-job training, and lifelong recreation, are things that should be built into a curriculum. And, lastly, I would hope that educators would stop looking at other institution's curricula with the intent of comparing requirements so that we can give more relevant, easier courses.

OTHER POINTS TO PONDER

A very interesting booklet, "Four Hard Truths for Higher Education", should be read by those interested in the challenges for higher education (13). The four challenges include: (A) disinflation, (B) revitalizing the American economy, (C) back to basics, and (D) the resurgence of traditional values.

Disinflation

For years, higher education was swept along by the tide of inflation that was prevalent in the American economy. Colleges and universities added programs, provided new services, reduced workloads, raised salaries, participated in countless committees, freely indulged in discussions of governance, academic freedom, and tenure. The cost of these activities were defrayed by raising tuition and fees and prevailing upon state and federal governments to provide additional appropriations. As elsewhere in the American economy, there was little resistance to the price hikes. Now, however, inflation has slowed. The easy money is becoming tight. Enrollments are leveling off or decreasing. Two basic questions must be raised: Should you offer your service and product at a lower relative price? And can you increase productivity? Obviously, there are no pat answers to the questions, but in many instances the affirmative answer is possible in both questions. I recall an example where student-faculty ratios in a given department frequently were made up by one or two professors teaching large classes so that other professors might enjoy teaching one course per quarter. Is this fair to the students? Is this really imparting teaching quality and philosophy that is needed to maintain a high-quality scientific education?

Revitalize Economy

The second change that was noted is the need to revitilize American economy. From the standpoint of higher education, there is need to unleash individual creativity by encouraging and rewarding success and by recognizing and punishing failure. We do this in the marketplace; why not in the educational environment as well? It has been alleged that America has been graduating economic illiterates who can neither comprehend nor appreciate how the free enterprise system functions. Perhaps we should add to the definition of what comprises the educated man or woman to include comprehension of our economic system--of free enterprise, of capitalism, of opportunity, of risk-taking, and risk-rewarding.

However, risk-taking should not include a compromise of intellectual integrity. Recently, there have been stories of young scientists who have chosen to ignore contractual arrangements with universities or have compromised on basic principles and fabricated data in the search for increasing opportunities in publications. Public disclosure of these events does irreparable harm to the scientific community. Recently, John Darcy was forced to leave Harvard Medical School after a long and prolific career in data fabrication (14,15). The Darcy case raised serious questions about the way in which young investigators are supervised during their training. It appears desirable that all principals involved in research take full intellectual responsibility for the research, including examination of data, statistical analysis, and assessment of expert consultation. Without this simple series of activities, it is quite easy for a person inclined to dishonesty to achieve bare objectives. When this occurs, neither science nor the public perception of science is enhanced.

Back to Basics

The third change is the back-to-basics movement in elementary and secondary schools. As I mentioned earlier, recognition has grown, but the nation's secondary schools have been graduating illiterates and many more with barely competent basic skills. At the start of the 1980s, surveys indicated that approximately 30% of entering college freshmen were having serious difficulty with reading, writing, and mathematics. Minimum competency laws have been introduced in 35 states with the intent of enforcing better teaching of the basics. This action is desperately needed so that young students once again will develop a study discipline as an essential part of their education. As teachers, you must set high standards and display your enthusiasm for them. The need for remedial training is not only discouraging at higher education levels, but is particularly discouraging when young college graduates are found incapable of adding columns of figures and doing simple algebra. The agency that I head went through a rapid expansion period in 1978 through 1980. One of the major problems was finding qualified college graduates who could make simple mathematic calculations and accurately record data on ship's logs. The archaic employment system used by the Federal Government requires no proficiency test at initial employment. To compensate for this deficiency, the agency provides training for many newly-hired college graduates in remedial mathematics and remedial writing to provide basic skills.

Traditional Values

The fourth change is a resurgence of traditional or conservative values. These are seen reemerging in many key sectors of American life. They include renewed respect for economics, religion, and criminal justice. In addressing these values, Burton Pines (13) notes that at universities, particularly in the area of liberal arts, there seems to be an opposition, not merely neutral or critical, but hostile, on social issues, religion, partiotism, and relativism. He comments without recommendation, just the observation, that by remaining in opposition to the vast majority of your countrymen, academia risks losing credibility and relevance. Indeed, I believe these are some interesting points to ponder.

SOCIAL ISSUES

I should like to close on some thoughts that the agricultural and food scientists should become involved, not only in technological issues of importance to society, but in discussion of societal issues in the national political arena as well (16). Scientists have had the privilege of receiving a high level of education and of acquiring special technological skills. These privileges and skills carry with them a corresponding level of responsibility (17). Indeed, the subject of this book, "Cereals and Legumes in the Food Supply", conveys to me broad concerns not only of science, but of politics and the general welfare of the population. Food scientists' activities ultimately are directed to the benefit of society. Food is a basic societal need. If food scientists do not speak out, the opinions of others--the professional lobbyist, the persons in the news media, the consumer advocates, the charlatans, the anti-technology zealots, etc.--will prevail. Many of these individuals are less qualified to speak on food-related issues than the food scientist.

A former president of the Institute of Food Technologists, Owen Fennema (17), admonished food scientists to become involved in technological affairs beyond the point of just generating scientific knowledge and guarding the ethics of the scientific community. They need to assist in determining how scientific knowledge is interpreted and utilized by society. They need to influence the policies and practices of the Federal Government, food companies, and universities. They need to participate in professional group activities by publishing worthwhile technical papers, engaging actively in dissemination of educational material to the public, and discussion of conscientious societal issues.

Higher education should train young people potentially entering the food industry that their role extends far beyond that of a pure scientist, but as a qualified citizen with a major role in facilitating professional decisions relating to food science and food production. Certainly, as the trend of the food industry becomes more oriented toward an international emphasis, the role of the food scientist to assist in providing better food for a larger world population will prove to be a gratifying challenge and opportunity for professional and personal growth.

I am more than cautiously optimistic as I have been for more than a decade that, indeed, we will provide an adequate nutritious food supply for the peoples of the world well into the 21st century.

REFERENCES CITED

1. Trevis, James E., "Conference on World Food Needs", Cereal Science Today 20, 6:272 (1975).
2. Parrott, Robert B., "Changing Horizons of U.S. Agriculture", Cereal Foods World, 24, 5:176 (1979).
3. Shellenbarger, Sue and Birnbaum, Jeffrey H., "Long U.S. Dominance in World Grain Trade is Slowly Diminishing", Wall Street Journal, May 19, 1983.
4. Annon. Human Capital Shortages: A Threat to American Agriculture, Pub: Nat. Aaan. of State Universities and Land Grant Colleges, 1983.
5. Annon. "Three Out of Four Fail Math Test", Washington Post, May 1983.
6. McGrath, Ellie, "Top Dollar for Top Students", Time, May 2, 1983.
7. Coulter, Kyle Jane and Stanton, Marge, Graduates of Higher Education in Food and Agricultural Sciences: Vol. 1. Agriculture, Natural Resources and Veterinary Medicine. USDA, Washington, D.C. 1980.
8. Coulter, Kyle Jane and Stanton, Marge, Graduates of Higher Education in Food and Agricultural Sciences: Vol. 2. Home Economics. 1981.
9. Coulter, Kyle Jane and Stanton, Marge, Sex, Race and Ethnicity Characteristics of Students and Graduates and of Food and Agricultural Professionals: Vol. 3. 1983.
   Schmitt, Roland W., "Building R & D Policy from Strength", Science 220:1013, 1983.
10. Kaestle, Carl F., "Higher Education: The Past Reappraised", Science 220:814 (1983).
11. Hilts, Phillip J., "U.S. Science Losing Its Magic as Rivals Excel", Washington Post, May 21, 1983.
12. Black, Charles A., "A New Resource for Science Teachers", Science of Food and Agriculture 1:1 (1983).
13. Pines, Burton Yale, Four Hard Truths for Higher Education. The Heritage Lectures, 20, Heritage Foundation, Washington, D.C. (1983).
14. Culliton, Barbara J., "Emory Reports on Darsee's Fraud", Science 220:936 (1983).
15. Culliton, Barbara J., "Emory Reports on Darsee's Fraud", Science 22:1029 (1983).
16. Gilles, Kenneth A., Social Responsibilities of Scientists. A.A.C.C. President's Lecture (1972).
17. Fennema, Owen, Food Tech. 37, 5:8 (1983).

Index